Calculus&*Mathematica*

DERIVATIVES: Measuring Growth

Bill Davis
Ohio State University

Horacio Porta
University of Illinois, Urbana-Champaign

Jerry Uhl
University of Illinois, Urbana-Champaign

Addison-Wesley Publishing Company

Reading, Massachusetts • Menlo Park, California • New York
Don Mills, Ontario • Wokingham, England • Amsterdam • Bonn
Sydney • Singapore • Tokyo • Madrid • San Juan • Milan • Paris

ISBN 0-201-58466-2

1 2 3 4 5 6 7 8 9-CRS-98 97 96 95 94 93

Preface

Growth and change. These are the themes of life and measuring them is the theme of calculus. Learning how to measure rate of growth and learning how to use this measurement is what this part of Calculus&*Mathematica* is all about.

In this unit of Calculus&*Mathematica*, you'll work with the mathematics underlying measurements of growth rates and you'll use these measurements of growth rates to model processes naturally occurring in life. You'll have the chance to model the United States national debt, the United States population, growth of an animal, the blood alcohol level resulting from a given drinking schedule, the relationship between the growth of the height of an animal and the growth of the weight of the same animal, failures of O-rings on the space shuttle, credit card interest, personal finance, competing species, competing armies, spread of infection, and a lot more. Along the way, you'll get really good at *Mathematica* calculations and plotting. By the end of this part of Calculus&*Mathematica*, you'll be well on your way to mastering *Mathematica* as you begin to master calculus.

How to Use This Book

In Calculus&*Mathematica*, great care has been taken to put you in a position to learn visually. Instead of forcing you to attempt to learn by memorizing impenetrable jargon, you will be put in the position in which you will experience mathematics by seeing it happen and by making it happen. And you'll often be asked to describe what you see. When you do this, you'll be engaging in active mathematics as opposed to the passive mathematics you were probably asked to do in most other math courses. In order to take full advantage of this crucial aspect of Calculus&*Mathematica*, your first exposure to a new idea should be on the live computer screen where you can interact with the electronic text to your own satisfaction. This means that you should avoide "introductory lectures" and you should avoid reading this book at first. After you have some familiarity with new ideas as found on the computer screen, you should seek out others for discussion and you can refer to this book to brush up on a point or two after you leave the computer. In the final analysis, this book is nothing more than a partial record of what happens on the screen.

Once you have participated in the mathematics and science of each lesson, you can sharpen your hand skills and check up on your calculus literacy by trying the questions in the Literacy Sheet associated with each lesson. The Literacy Sheets appear at the end of each book.

Significant Changes from the Traditional Course

Your writing, plotting, and experimentation is the stock in trade of the course.

Experienced as a course in measurements heavily intertwined with other parts of science and the world.

Your emphasis is on linear and exponential growth from the beginning, before calculus begins. Linear functions are those with constant growth rates; exponentials are those with constant percentage growth rates.

You learn at the very beginning that exponential growth dominates power growth without appeal to the mysticism of L'Hopital's rule or any other calculus ideas.

You study functions not studied for their own sake, but rather for the measurements they make.

You learn about the derivative as a measurement of the instantaneous growth rate. As a result, the idea that functions with positive derivatives are increasing functions is available to you immediately without waiting for the Mean Value Theorem. The interpretation of the derivative as the slope of the tangent line is delayed.

You work with and analyze real world data on applications important to you.

Financial calculations recur on a regular basis.

You learn the meaning of the derivative as a measurement at the same time you're learning to calculate derivatives. This idea is reinforced by many plots that you produce and analyses of the graphs of $f[x]$ and $f'[x]$ on the same axes.

Although there is no formal "epsilon-delta" presentation of limits, you experience the limiting process visually by plotting the average growth rates $(f[x+h]-f[x])/h$ and the instantaneous growth rate $f'[x]$ as functions of x and watch what happens to the plots as they make h close in on 0.

The active form of the Mean Value Theorem, called the Race Track Principle, is introduced. Euler's method is explained in terms of the Race Track Principle.

Following Poincare, you do differentiation of functions of two variables with respect to each variable is done with no particular fanfare.

You do serious work with mathematical models involving derivatives. The benefits are twofold: Working with the models reinforces the idea of what the derivative is, and you can experience the tenacles of calculus outside the traditional calculus classroom.

Biological models are favored over physical models at the beginning because the derivative measures growth and growth is a natural biological process.

Linear dimension makes a decisive entrance into calculus.

Logistic growth is studied in some detail.

Qualitative analysis of the solutions of simple differential equations and the solutions of simple systems of differential equations enters a calculus course for the first time. Reasons: Studying them reinforces the meaning of the derivatives and they beautifully show the scope of calculus in science. You experiment with predator-prey, spread of infection, and Lanchester war models and try to explain the results in terms of derivatives.

Parametric plots in two and three dimensions are studied in the first course because they provide you with needed plotting freedom for what's to come.

Contents

DERIVATIVES: Measuring Growth

In normal use, the student engages in all the mathematics and the student engages in selected experiences in math and science as assigned by the individual instructor.

LESSON 1.01

Growth

Basics

■ B.1) Growth of line functions

A line function $f[x]$ is any function whose formula is $f[x] = a\,x + b$ where a and b are constants. Here's a plot of a line function:

In[1]:=

```
Clear[f,x]
a = 0.5; b = 4; f[x_] = a x + b;
Plot[f[x],{x,-2,8},
PlotStyle->{{Red,Thickness[0.015]}},
AxesLabel->{"x","f[x]"},
PlotLabel->"A line function"];
```

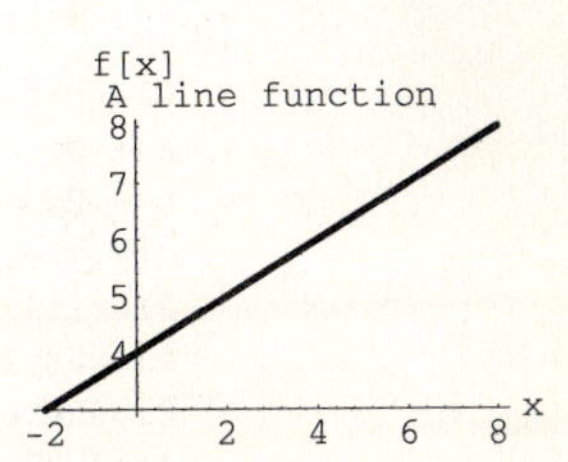

There's steady growth as x advances from left to right. Here's another:

In[2]:=

```
Clear[f,x]
a = -0.3; b = 2; f[x_] = a x + b;
Plot[f[x],{x,-2,8},
PlotStyle->{{Red,Thickness[0.015]}},
AxesLabel->{"x","f[x]"},
PlotLabel->"A line function"];
```

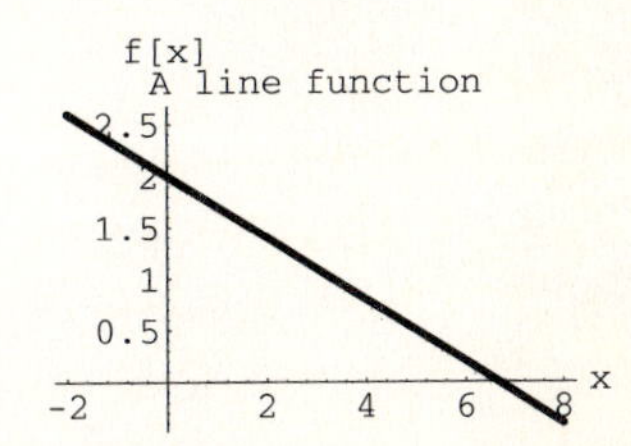

Steady (negative) growth as x advances from left to right. Play with other choices of a and b until you get the feel of a line function.

B.1.a.i) The most important feature of a line function $f[x] = a\,x + b$ is revealed by the following calculation.

In[3]:=
```
Clear[f,a,b,x,h]; f[x_] = a x + b; Expand[f[x + h] - f[x]]
```
Out[3]=
```
a h
```

> What feature of line functions is revealed by this calculation?

Answer: The calculation reveals that when you take a line function $f[x] = a\,x+b$, then you find that

$$f[x+h] - f[x] = a\,h.$$

This tells you that when x advances by h units, then $f[x]$ grows by $a\,h$ units. Consequently a line function $f[x] = a\,x + b$ has constant growth rate of a units on the $f[x]$-axis for each unit on the x-axis.

B.1.a.ii) As you saw above, the growth rate of a line function $f[x] = a\,x + b$ measures out to a units on the $f[x]$ axis per unit on the x-axis.

> What is the significance of a?

Answer: Big positive a's force big-time fast growth as the following true scale plot shows:

In[4]:=
```
Clear[f,x]
a = 8;
b = 2;
f[x_] = a x + b;
Plot[f[x],{x,-1,5},
PlotStyle->{{Red,Thickness[0.015]}},
PlotLabel->"Big a > 0",
AspectRatio->Automatic];
```

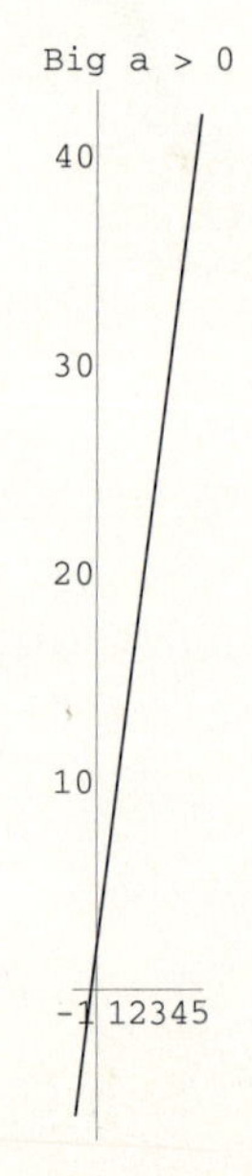

Small positive a's force slow growth as the following true scale plot shows:

In[5]:=

```
Clear[f,x]; a = 0.2; b = 2; f[x_] = a x + b;
Plot[f[x],{x,-1,5},
PlotStyle->{{Red,Thickness[0.015]}},
PlotLabel->"Small a > 0", AspectRatio->Automatic];
```

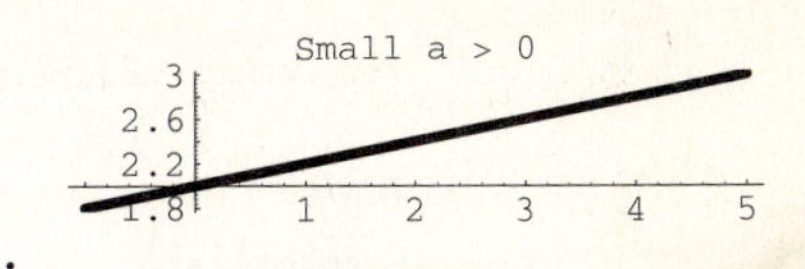

$a = 0$ forces no growth at all as the following true scale plot shows:

In[6]:=

```
Clear[f,x]; a = 0; b = 2; f[x_] = a x + b;
Plot[f[x],{x,-1,5},
PlotStyle->{{Red,Thickness[0.015]}},
AxesLabel->{"x","f[x]"},
PlotLabel->"a = 0", AspectRatio->Automatic];
```

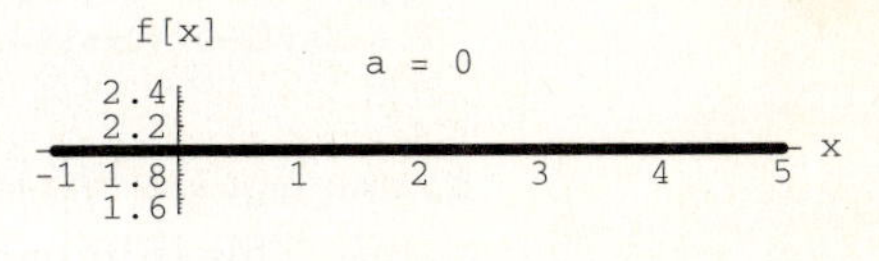

Small negative a's force slow (negative) growth as the following true scale plot shows:

In[7]:=

```
Clear[f,x]
a = -0.2; b = 2; f[x_] = a x + b;
Plot[f[x],{x,-1,5},
PlotStyle->{{Red,Thickness[0.015]}},
AxesLabel->{"x","f[x]"},
PlotLabel->"Small a < 0",
AspectRatio->Automatic];
```

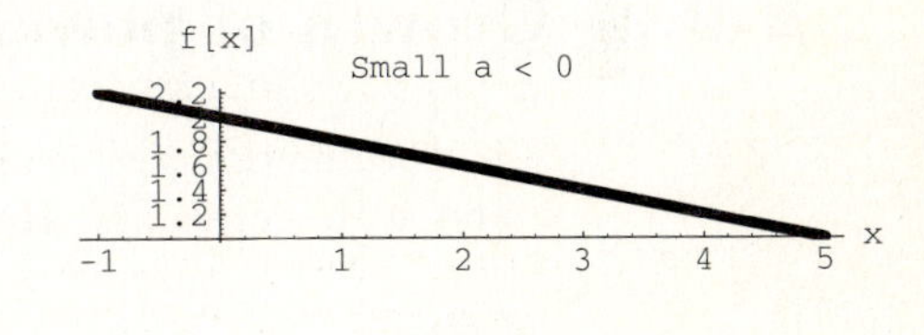

Big negative a's force fast (negative) growth as the following true scale plot shows:

In[8]:=

```
Clear[f,x]
a = -4;
b = 2;
f[x_] = a x + b;
Plot[f[x],{x,-1,5},
PlotStyle->{{Red,Thickness[0.015]}},
AxesLabel->{"x","f[x]"},
AspectRatio->Automatic];
```

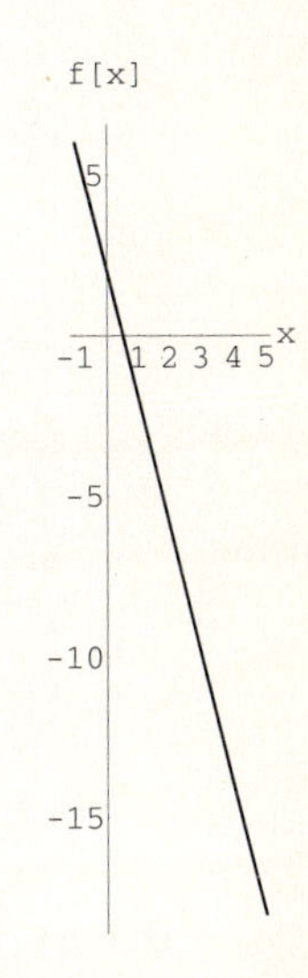

Not a handsome plot, but the message gets through.

B.1.b) Plot $f[x] = 0.5\,x + 3$ for $-1 \leq x \leq 6$ in true scale. $f[x]$ goes up how many times faster than x?

Answer: Here is a true scale plot:

In[9]:=

```
Clear[f,x]
f[x_] = 0.5 x + 3;
Plot[f[x],{x,-1,6},AxesLabel->{"x","f[x]"},
AspectRatio->Automatic,
PlotStyle->{{GrayLevel[0.4],Thickness[0.01]}}];
```

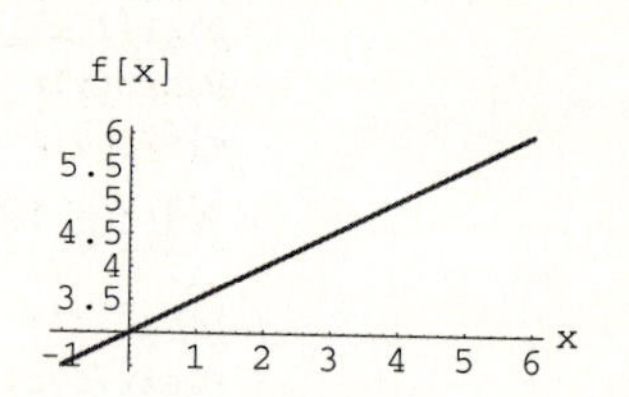

As you can see, $f[x] = 0.5\,x + 3$ goes up 0.5 units on the $f[x]$-axis per unit on the x-axis. This is in harmony with the fact that $f[x] = 0.5\,x + 3$ has growth rate 0.5. As a result, $f[x]$ goes up 0.5 times as fast as x goes up.

■ B.2) Growth of power functions $f[x] = a\,x^k$

A power function $f[x]$ is any function whose formula has the form $f[x] = a\,x^k$ where a and k are constants. Here's a plot of a power function:

In[10]:=

```
Clear[f,x]
a = 3; k = 2; f[x_] = a x^k;
Plot[f[x],{x,1,5},
PlotStyle->{{Red,Thickness[0.015]}},
AxesLabel->{"x","f[x]"},
PlotLabel->"A power function"];
```

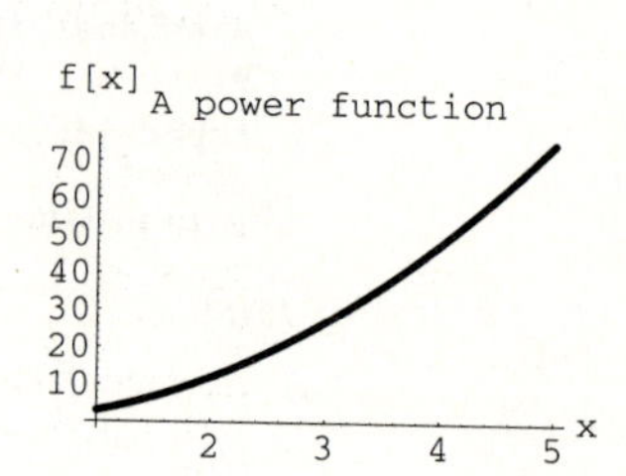

The growth increases as x advances from left to right. Here's another:

In[11]:=

```
Clear[f,x]
a = 3; k = -2; f[x_] = a x^k;
Plot[f[x],{x,1,5},
PlotStyle->{{Red,Thickness[0.015]}},
AxesLabel->{"x","f[x]"},
PlotLabel->"A power function"];
```

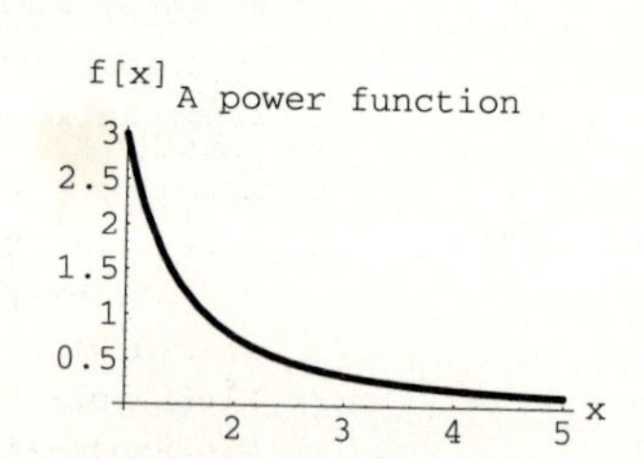

The (negative) growth decreases as x advances from left to right.

B.2.a) When you take a power function $f[x] = a\,x^k$, what is the significance of k?

Answer: Here is what a large positive k forces:

In[12]:=
```
Clear[f,x]
a = 3; k = 5; f[x_] = a x^k;
Plot[f[x],{x,1,4},
PlotStyle->{{Red,Thickness[0.015]}},
AxesLabel->{"x","f[x]"},
PlotLabel->"Big k > 0",
PlotRange->All];
```

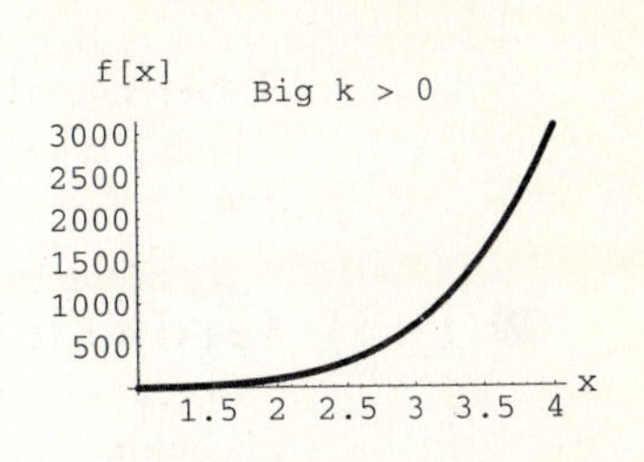

Look at those numbers on the vertical axis! Big positive k's force big-time growth. Here is what a small positive k forces:

In[13]:=
```
Clear[f,x]
a = 3; k = 0.2; f[x_] = a x^k;
Plot[f[x],{x,1,4},
PlotStyle->{{Red,Thickness[0.015]}},
AxesLabel->{"x","f[x]"},
PlotLabel->"Small k > 0",
PlotRange->All];
```

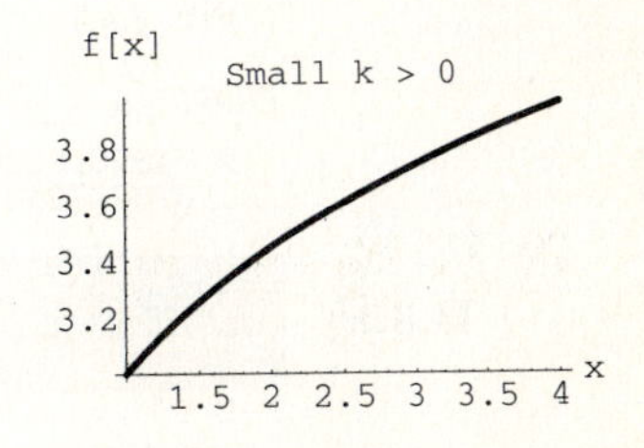

A small positive k forces small-time growth. Here is what $k = 0$ forces:

In[14]:=
```
Clear[f,x]
a = 3; k = 0; f[x_] = a x^k;
Plot[f[x],{x,1,4},
PlotStyle->{{Red,Thickness[0.015]}},
AxesLabel->{"x","f[x]"},
PlotLabel->"k = 0",
PlotRange->All];
```

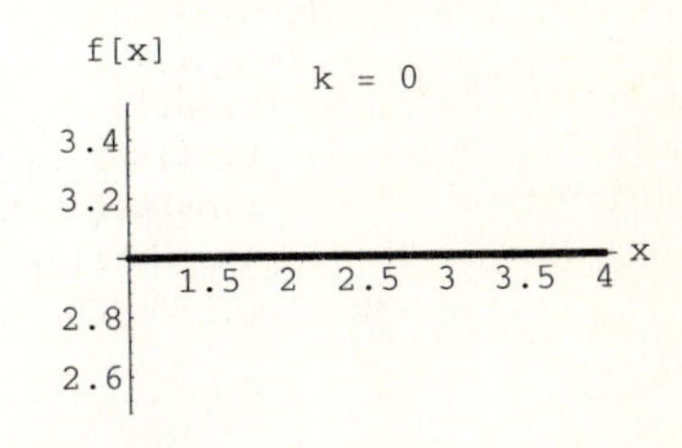

$k = 0$ forces no growth. Here is what a small negative k forces:

In[15]:=
```
Clear[f,x]
a = 3; k = -0.3; f[x_] = a x^k;
Plot[f[x],{x,1,4},
PlotStyle->{{Red,Thickness[0.015]}},
AxesLabel->{"x","f[x]"},
PlotLabel->"Small k < 0",
PlotRange->All];
```

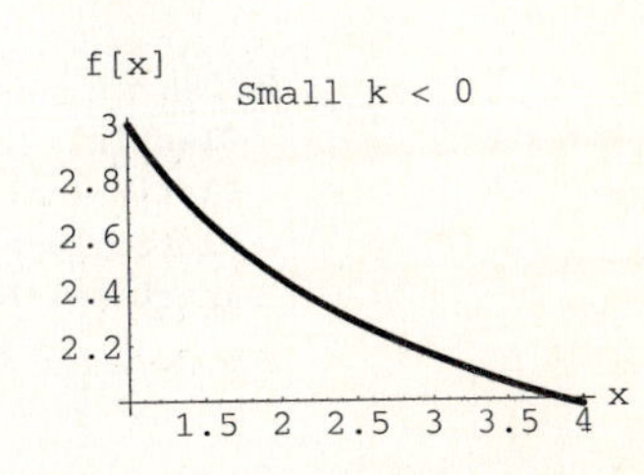

A small negative k forces small-time negative growth. Here is what a big negative k forces:

In[16]:=
```
Clear[f,x]
a = 3; k = -8; f[x_] = a x^k;
Plot[f[x],{x,1,4},
PlotStyle->{{Red,Thickness[0.015]}},
AxesLabel->{"x","f[x]"},
PlotLabel->"Big k < 0", PlotRange->All];
```

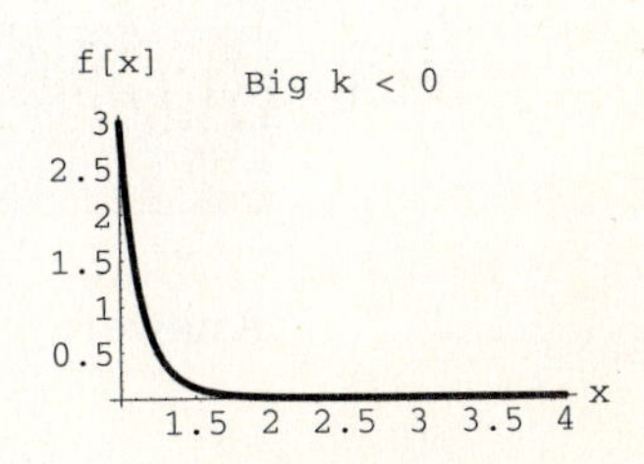

A big negative k forces big-time (negative) growth until x reaches the point at which $f[x] = a\,x^k$ peters out.

■ B.3) Growth of exponential functions $f[x] = a\,e^{rx}$

One difference between serious science and old-time classroom algebra is the significance of a certain squirrelly number called e.

In[17]:=

```
N[E,100]
```

Out[17]=

```
2.7182818284590452353602874713526624977572470936999595749669\
  6762772407663035354759457138217852516642 7
```

B.3.a) Plot e^x and then plot e^{-x} and describe what you see.

Answer: Here is a plot of e^x:

In[18]:=

```
Clear[x]
Plot[E^x,{x,0,7},
PlotStyle->{{Blue,Thickness[0.015]}},
AxesLabel->{"x","E^x"}];
```

Pristine exponential growth. Here is a plot of e^{-x}:

In[19]:=

```
Clear[x]
Plot[E^(-x),{x,0,7},
PlotStyle->{{Blue,Thickness[0.015]}},
AxesLabel->{"x","E^(-x)"}];
```

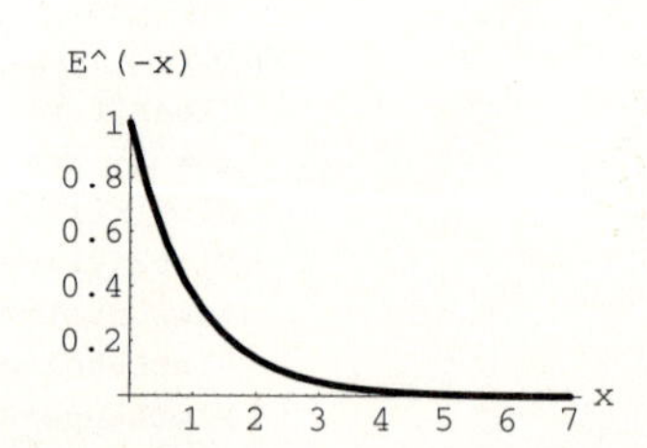

Pristine exponential decay.

B.3.b) When you take an exponential function $f[x] = a\,e^{rx}$, what is the significance of r?

Answer: Here is what a large positive r forces:

In[20]:=

```
Clear[f,x]
a = 3; r = 6; f[x_] = a E^(r x);
Plot[f[x],{x,0,4},
PlotStyle->{{Red,Thickness[0.015]}},
AxesLabel->{"x","a E^(r x)"},
PlotLabel->"Big r > 0",
PlotRange->All];
```

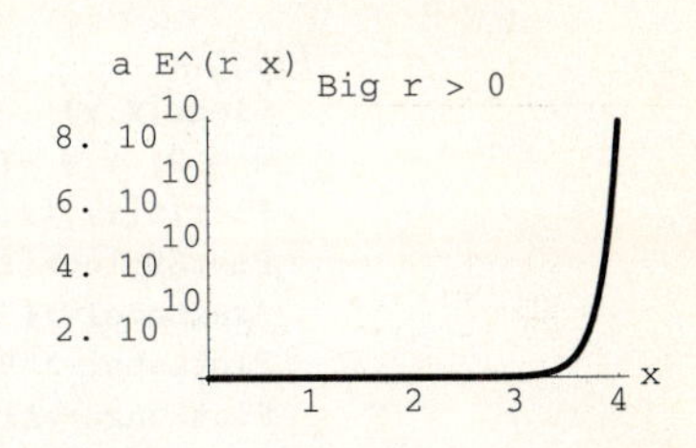

Look at those numbers on the vertical axis! Big positive r's force astoundingly big-time growth. Folks call this by the name "exponential growth."

Here is what a small positive r forces:

In[21]:=

```
Clear[f,x]
a = 3; r = 0.5; f[x_] = a E^(r x);
Plot[f[x],{x,0,4},
PlotStyle->{{Red,Thickness[0.015]}},
AxesLabel->{"x","a E^(r x)"},
PlotLabel->"Small r > 0",
PlotRange->All];
```

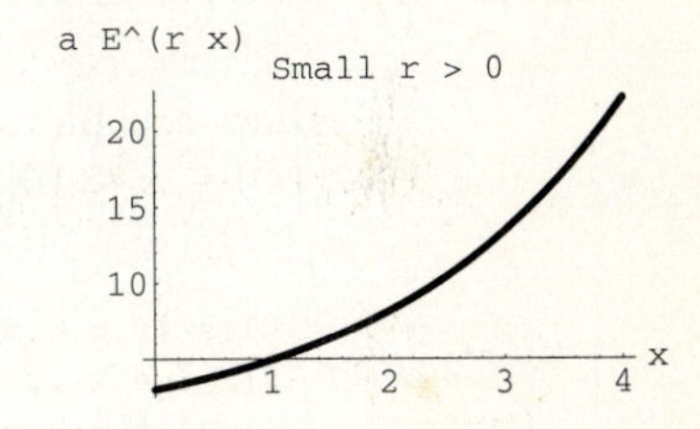

A small positive r forces small-time growth at first.

Here is what $r = 0$ forces:

In[22]:=

```
Clear[f,x]
a = 3; r = 0; f[x_] = a E^(r x);
Plot[f[x],{x,0,4},
PlotStyle->{{Red,Thickness[0.015]}},
AxesLabel->{"x","a E^(r x)"},
PlotLabel->"r = 0",
PlotRange->All];
```

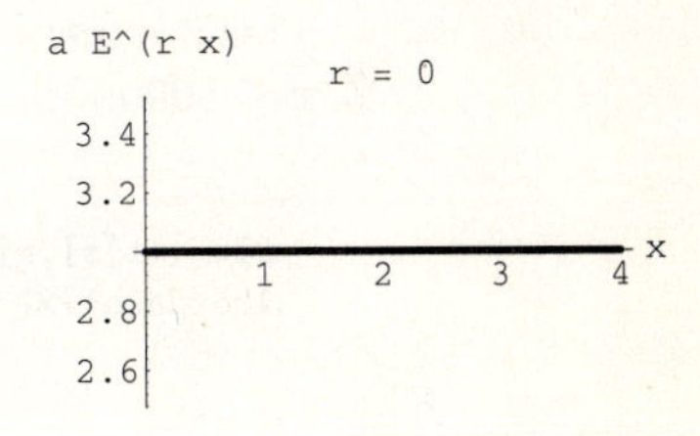

$r = 0$ forces no growth.

Here is what a small negative r forces:

In[23]:=

```
Clear[f,x]
a = 3; r = -0.2; f[x_] = a E^(r x);
Plot[f[x],{x,0,4},
PlotStyle->{{Red,Thickness[0.015]}},
AxesLabel->{"x","a E^(r x)"},
PlotLabel->"Small r < 0",
PlotRange->All];
```

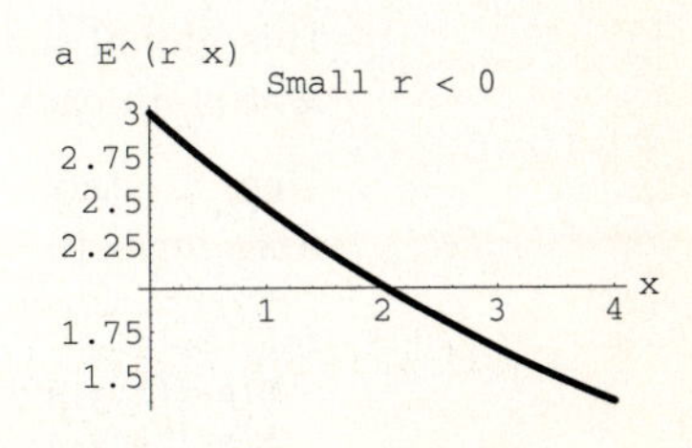

A small negative r forces small-time negative growth.

Here is what a big negative r forces:

In[24]:=
```
Clear[f,x]
a = 3; r = -7; f[x_] = a E^(r x);
Plot[f[x],{x,0,4},
PlotStyle->{{Red,Thickness[0.015]}},
AxesLabel->{"x","a E^(r x)"},
PlotLabel->"Big r < 0",
PlotRange->All];
```

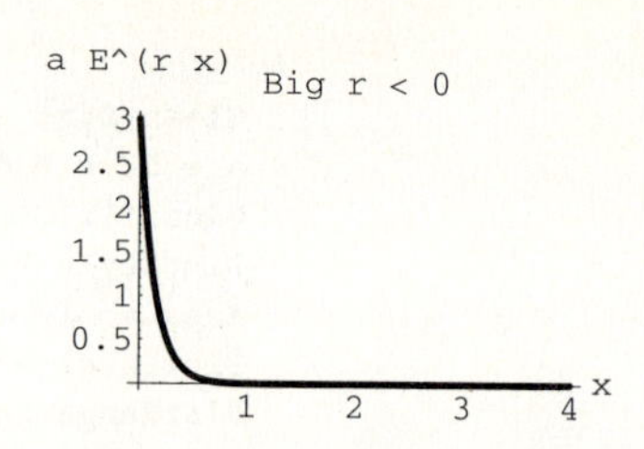

A big negative r forces big-time (negative) growth until x reaches the point at which $f[x] = ae^{rx}$ is all used up. Folks call this by the name "exponential decay."

■ B.4) Dominance in the global scale

Here are plots of x^2, $x^2 + 18\,x + 5$, and $x^2 - 50\,x + 100$ on the same axes for $-10 \leq x \leq 10$:

In[25]:=
```
Clear[f,g,h,x]
f[x_] = x^2;
g[x_] = x^2 + 18 x + 5;
h[x_] = x^2 - 50 x + 100;
Plot[{f[x],g[x],h[x]},{x,-10,10},
AxesLabel->{"x",""},PlotRange->All];
```

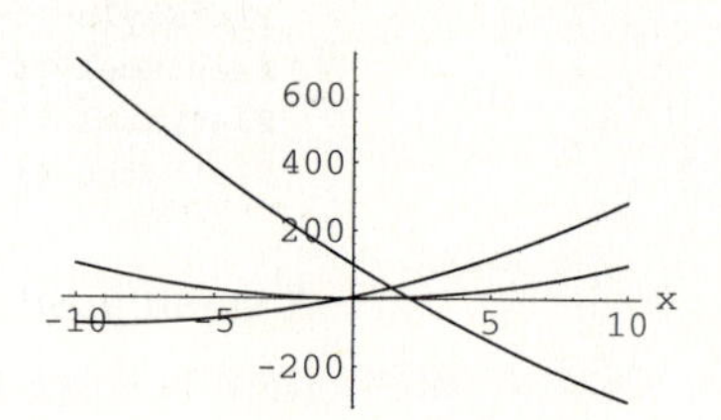

This is a mess. Now look at the same functions plotted on the huge interval $-5000 \leq x \leq 5000$:

In[26]:=
```
Plot[{f[x],g[x],h[x]},{x,-5000,5000},
AxesLabel->{"x",""},PlotRange->All];
```

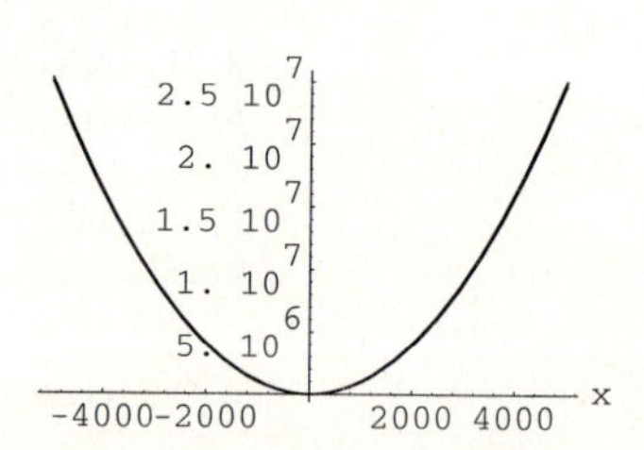

Order out of chaos. In this global scale, the functions all look the same! In fact, in this global scale, x^2, $x^2 + 18\,x + 12$, and $x^2 - 13\,x - 18$ all look like x^2. In other words, in each formula x^2 is the dominant term in the global scale.

Here are plots of x^3, $x^3 - 3\,x^2 - 61\,x + 65$, and $x^3 - 18\,x^2 + 30\,x - 980$ on the same axes for $-10 \leq x \leq 10$:

In[27]:=
```
Clear[f,g,h,x]
f[x_] = x^3;
g[x_] = x^3 - 3 x^2 - 61 x + 65 ;
h[x_] = x^3 - 18 x^2 + 30 x - 980 ;
Plot[{f[x],g[x],h[x]},{x,-10,10},
AxesLabel->{"x",""},PlotRange->All];
```

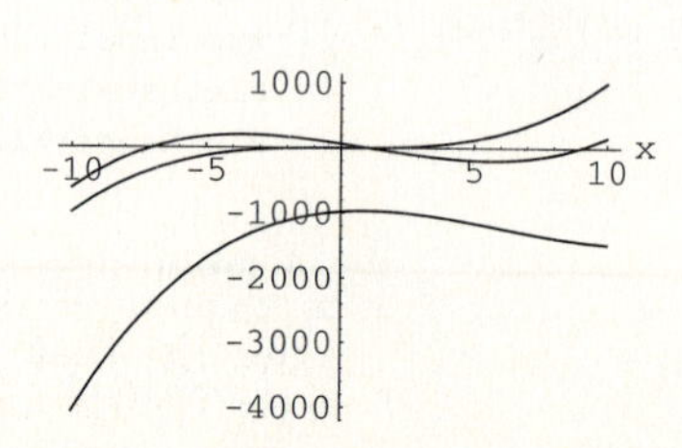

Another mess. Now look at the same functions plotted on the huge interval $-10000 \leq x \leq 10000$:

In[28]:=
```
Plot[{f[x],g[x],h[x]},{x,-10000,10000},
AxesLabel->{"x",""},PlotRange->All];
```

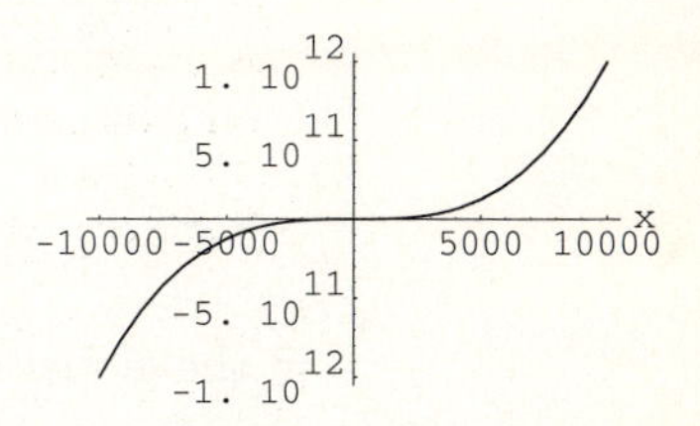

Order out of chaos. In this global scale, the functions all look the same! In this global scale, x^3, $x^3 - 3\,x^2 - 61\,x + 65$, and $x^3 - 18\,x^2 + 30\,x - 12$ all look like x^3. In other words, in each formula x^3 is the dominant term in the global scale.

B.4.a) When you plot x^4, $x^4 + 8\,x^3 - 73\,x^2 - 512\,x + 571$, and $x^4 - 12\,x^3 - 73\,x^2 + 958\,x - 864$ on a really huge interval, what do the plots look like?

Answer: They all look like the dominant term x^4. Try it and see:

In[29]:=
```
Clear[x]
Plot[{x^4,
x^4 + 8 x^3 - 73 x^2 -512 x + 571,
x^4 - 12 x^3 -73 x^2 + 958 x - 864 },
{x,-5000,5000}, AxesLabel->{"x","y"},
PlotRange->All];
```

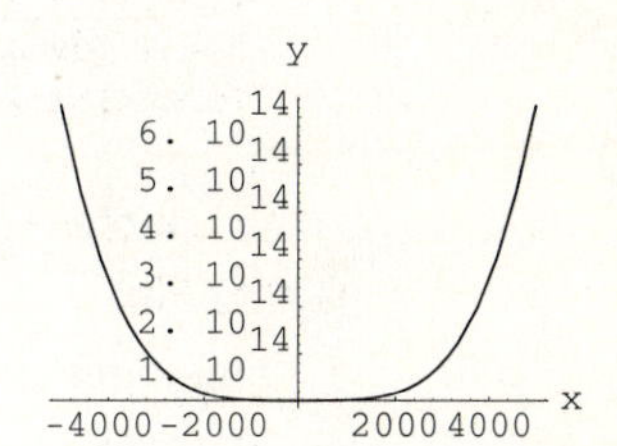

Yep; just like x^4.

B.4.b) Here is a global scale plot of

$$\frac{2\,x^4 + 5\,x^2 + 4\,x + 1}{x^3 + 6\,x^2 + 1}:$$

In[30]:=
```
globalscale =
Plot[(2 x^4 + 5 x^2 + 4 x + 1)/(x^3 + 6 x^2 + 1),
{x,-1000,1000},AxesLabel->{"x",""} ];
```

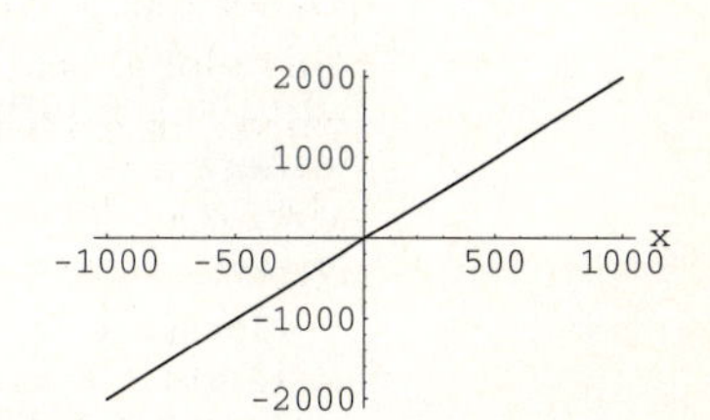

The plot looks like a straight line.

Account for this outcome.

Answer: The reason for this is that if you ignore all but the dominant terms in

$$\frac{2\,x^4 + 5\,x^2 + 4\,x + 1}{x^3 + 6\,x^2 + 1},$$

then you get

$$\frac{2\,x^4}{x^3} = 2\,x$$

in the global scale. The straight line you see is $f[x] = 2\,x$. Check it out:

In[31]:=
```
lineplot = Plot[2 x,{x,-1000,1000},
PlotStyle->{{Thickness[0.03],Red}},
AxesLabel->{"x",""},
DisplayFunction->Identity];
Show[lineplot,globalscale,
DisplayFunction->$DisplayFunction];
```

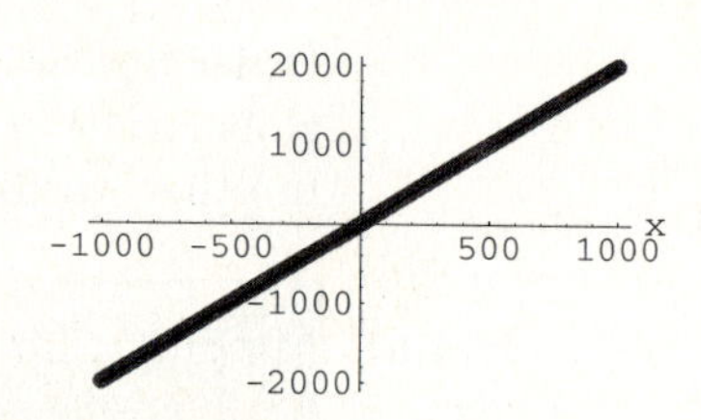

Yes ma'am. The global scale of

$$\frac{2\,x^4 + 5\,x^2 + 4\,x + 1}{x^3 + 6\,x^2 + 1}$$

is

$$\frac{2\,x^4}{x^3} = 2\,x.$$

This does not mean that these two are even close on small intervals:

In[32]:=
```
Plot[{2 x,
(2 x^4+5 x^2+4 x+1)/(x^3+6 x^2+1)},
{x,-3,3},
PlotStyle->
{{Thickness[0.03],Red},{Thickness[0.01]}},
AxesLabel->{"x",""} ];
```

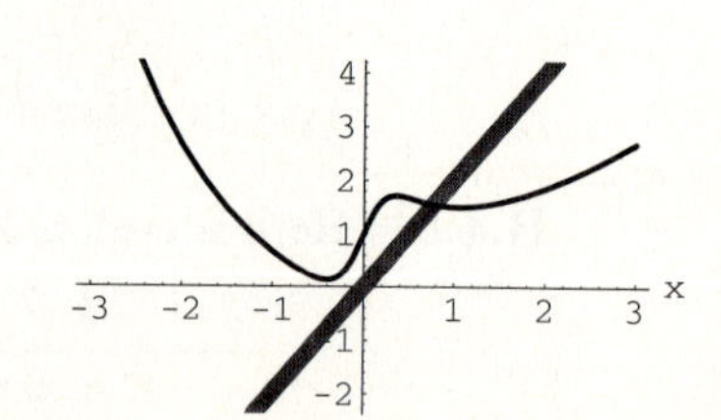

But as the intervals get bigger and bigger, the plots look more and more like each other:

In[33]:=
```
Plot[{2 x,
(2 x^4 + 5 x^2 + 4 x + 1)/(x^3 + 6 x^2 + 1)},
{x,-10,10},
PlotStyle->
{{Thickness[0.03],Red},{Thickness[0.01]}},
AxesLabel->{"x",""} ];
```

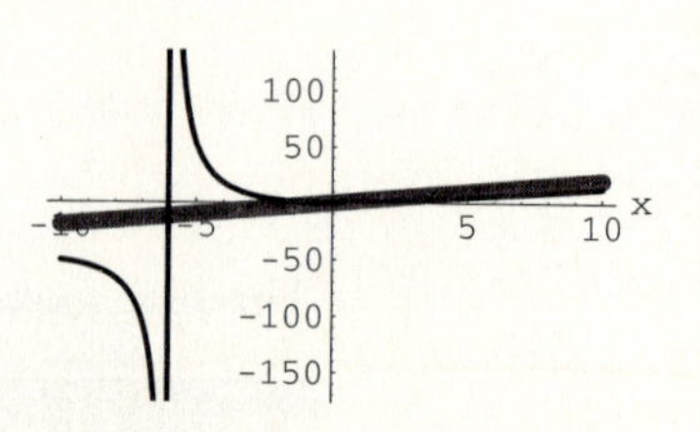

In[34]:=

```
Plot[{2 x,
(2 x^4 + 5 x^2 + 4 x + 1)/(x^3 + 6 x^2 + 1)},
{x,-50,50},
PlotStyle->
{{Thickness[0.03],Red},{Thickness[0.01]}},
AxesLabel->{"x",""} ];
```

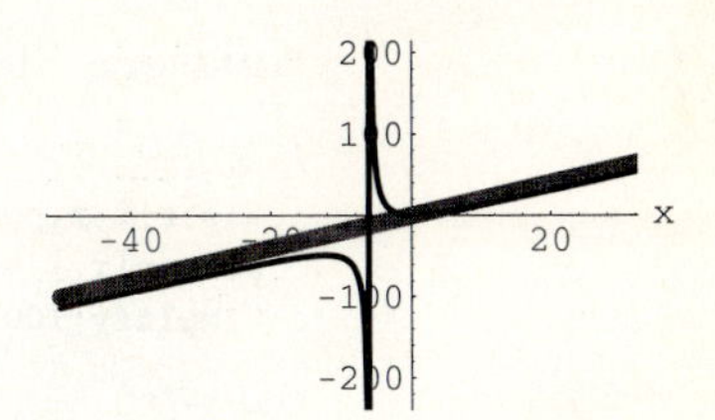

In[35]:=

```
Plot[{2 x,
(2 x^4 + 5 x^2 + 4 x + 1)/(x^3 + 6 x^2 + 1)},
{x,-500,500},
PlotStyle->
{{Thickness[0.03],Red},{Thickness[0.01]}},
AxesLabel->{"x",""} ];
```

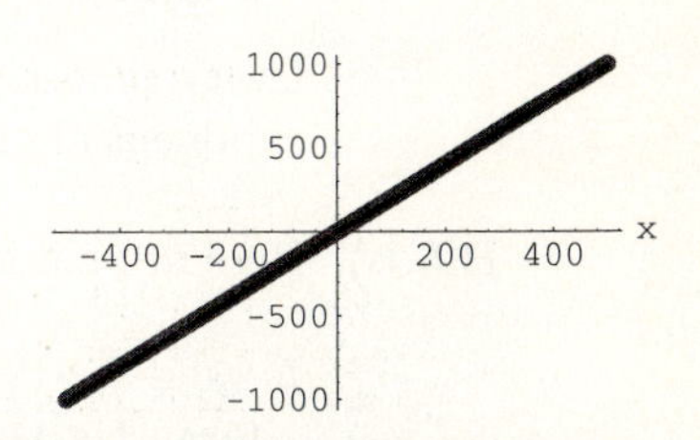

B.4.c) What's the moral of all the plots done above?

Answer: Only the dominant terms influence the global scale behavior.

■ B.5) Percentage growth rate and dominance in the global scale

B.5.a.i) Given a function $f[x]$, what does the function percen$[x]$ specified by

$$f[x+1] = f[x] + f[x]\,\text{percen}[x]$$

measure?

Answer: 100 percen$[x]$ measures the average percent growth rate of $f[x]$ as x advances by one unit.

B.5.a.ii) Here is a formula for percen$[x]$:

In[36]:=

```
Clear[f,percen,x]; equation = f[x + 1] == f[x] + f[x] percen[x];
Simplify[Solve[equation,percen[x]]]
```

Out[36]=

```
                     f[1 + x]
{{percen[x] -> -1 + --------}}
                       f[x]
```

This tells you that percen$[x] = (f[x+1]/f[x]) - 1$.

Use this formula to measure the average percent growth rate of $f[x] = 3.8\,e^{0.5x}$ as x advances by one unit.

Answer: Try it out:

In[37]:=
```
Clear[f,x,percen]; f[x_] = 3.8 E^(0.50 x);
percen[x_] = (f[x + 1]/f[x]) - 1;
Simplify[100 percen[x]]
```

Out[37]=
```
64.8721
```

Every time x goes up by 1, $f[x] = 3.8\, e^{0.5x}$ goes up by about 65%. That's big-time exponential growth.

B.5.b) Here are plots of $f[x] = x^7$ and $g[x] = e^{0.3x}$ for $0 \leq x \leq 9$.

In[38]:=
```
Clear[f,g,x]
f[x_] = x^7; g[x_] = E^(0.3 x);
Plot[{f[x],g[x]},{x,0,20},
PlotStyle->{{Thickness[0.01],GrayLevel[0.5]},
{Thickness[0.02],Red}},
PlotRange->All,AxesLabel->{"x",""}];
```

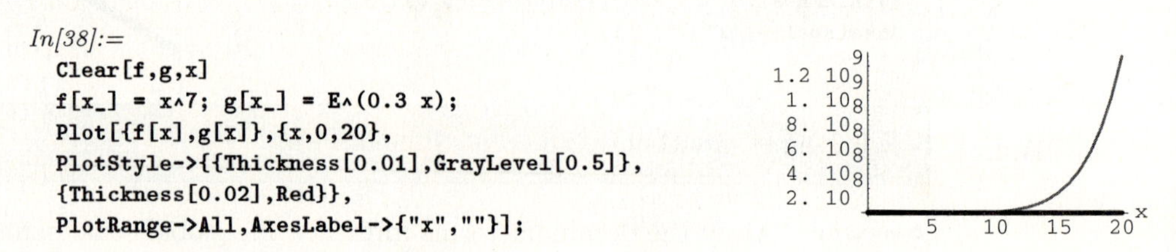

(The plot of $f[x] = x^7$ is the gray one.)

That's $f[x] = x^7$ riding high above $g[x] = e^{0.3x}$. Now look at what happens as x grows from 0 to 50:

In[39]:=
```
Plot[{f[x],g[x]},{x,0,50},
PlotStyle->{{Thickness[0.01],GrayLevel[0.5]},
{Thickness[0.02],Red}},
PlotRange->All,
AxesLabel->{"x",""}];
```

8. 10^11
6. 10^11
4. 10^11
2. 10^11
x
10 20 30 40 50

That's $f[x] = x^7$ still riding high above $g[x] = e^{0.3x}$. Now look at what happens as x grows from 0 to 100:

In[40]:=
```
Plot[{f[x],g[x]},{x,0,100},
PlotStyle->{{Thickness[0.01],GrayLevel[0.5]},
{Thickness[0.02],Red}},
PlotRange->All,AxesLabel->{"x",""}];
```

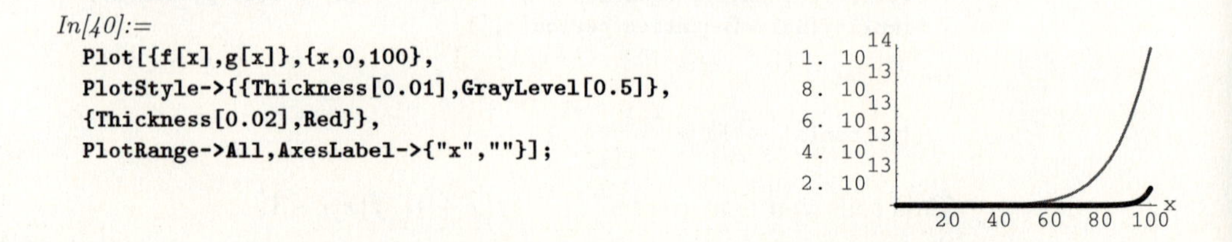

That's $g[x] = e^{0.3x}$ beginning to make its move. Now look at what happens as x grows from 0 to 125:

In[41]:=
```
Plot[{f[x],g[x]},{x,0,125},
PlotStyle->{{Thickness[0.01],GrayLevel[0.5]},
{Thickness[0.02],Red}},
PlotRange->All,AxesLabel->{"x",""}];
```

The positions are now reversed; that's $g[x] = e^{0.3x}$ pulling way above $f[x] = x^7$.

> Use the idea of percentage growth to explain why it was inevitable that the exponential function $g[x] = e^{0.3x}$ would eventually pull ahead of the power function $f[x] = x^7$.

Answer:

In[42]:=
```
Clear[f,g,x]; f[x_] = x^7; g[x_] = E^(0.3 x);
fpercen[x_] = (f[x + 1]/f[x]) - 1;
Simplify[100 fpercen[x]]
```

Out[42]=
```
                   7
            (1 + x)
100 (-1 + ---------)
               7
              x
```

In[43]:=
```
gpercen[x_] = (g[x + 1]/g[x]) - 1;
Simplify[100 gpercen[x]]
```

Out[43]=
```
34.9859
```

Plot the percentage growth rates of both $f[x]$ and $g[x]$:

In[44]:=
```
Plot[{100 fpercen[x],100 gpercen[x]},{x,1,400},
PlotStyle->{{Thickness[0.01],GrayLevel[0.5]},
{Thickness[0.02],Red}},
PlotRange->{0,100},AxesLabel->{"x","% growth"}];
```

The plot of the percentage growth of $f[x] = x^7$ is the gray plot.

Now you can see why $g[x] = e^{0.3x}$ must pull ahead of $f[x] = x^7$. As x goes up by 1, the percentage growth of $f[x] = x^7$ dies off to 0%, but the percentage growth of $g[x] = e^{0.3x}$ maintains itself at a constant rate of about 35%.

B.5.c)

> Explain why if you take any positive numbers a, b, k, and r, then the exponential function $g[x] = a\,e^{rx}$ must pull ahead of the power function $f[x] = b\,x^k$. In short, explain why exponential growth always dominates power growth.

Answer:

In[45]:=

```
Clear[f,g,b,a,r,k,x]; f[x_] = b x^k; g[x_] = a E^(r x);
fpercen[x_] = (f[x + 1]/f[x]) - 1; Simplify[100 fpercen[x]]
```

Out[45]=

```
                 k
          (1 + x)
100 (-1 + --------)
              k
             x
```

As x goes up by 1, the percentage growth rate of $f[x] = b\,x^k$ is

$$100\left(\left(\frac{1+x}{x}\right)^k - 1\right)\%.$$

In the global scale, this is

$$100\left(\left(\frac{x}{x}\right)^k - 1\right)\% = 100\,(1^k - 1)\,\% = 0\%.$$

The percentage growth rate of $f[x] = b\,x^k$ dies off to nothing. Now look at the percentage growth rate of $g[x] = b\,e^{rx}$:

In[46]:=

```
gpercen[x_] = (g[x + 1]/g[x]) - 1; Simplify[100 gpercen[x]]
```

Out[46]=

```
             r
100 (-1 + E )
```

As x goes up by 1, the percentage growth rate of $g[x] = a\,e^{rx}$ maintains itself at a constant $100\,(e^r - 1)\,\%$.

Because $e^r - 1 > 0$ for $r > 0$, the percentage growth of $g[x] = a\,e^{rx}$ does not die off and this is why the exponential function $a\,e^{rx}$ dominates the power function $b\,x^k$ no matter what positive a, b, k, and r you go with.

Tutorials

■ T.1) Global scale

T.1.a) Look at:

In[1]:=

```
Clear[x,f]
f[x_] =
x^5 - 1251 x^4 - 28750 x^3 + 50040 x^2 - 50001 x;
Plot[f[x],{x,-1000,1000},AxesLabel->{"x","f[x]"},
PlotRange->All];
```

f[x]
x
-1000
-500
500
1000
$-5.\ 10^{14}$
$-1.\ 10^{15}$
$-1.5\ 10^{15}$
$-2.\ 10^{15}$

> Is this a good global scale plot of
>
> $$f[x] = x^5 - 1251\,x^4 - 28750\,x^3 + 50040\,x^2 - 50001\,x?$$
>
> Why or why not? If it is not a good global scale plot of $f[x]$, then give a good global scale plot of $f[x]$.

Answer: A good global scale plot of

$$f[x] = x^5 - 1251\,x^4 - 28750\,x^3 + 50040\,x^2 - 50001\,x$$

looks like a plot of x^5, the dominant term for large values of x. Check:

In[2]:=
```
Plot[{f[x],x^5},{x,-1000,1000},
AxesLabel->{"x","y"}];
```

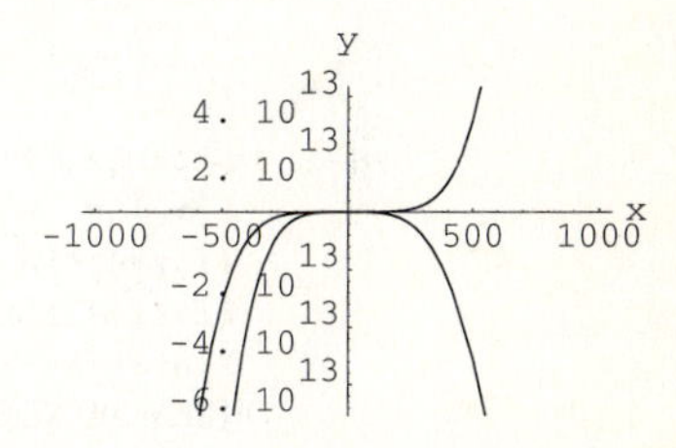

These curves don't look a bit alike. So the plot for $-1000 \leq x \leq 1000$ is not a good global scale plot of $f[x]$. Try a longer interval:

In[3]:=
```
Plot[{f[x],x^5},{x,-4000,4000},
AxesLabel->{"x",""},PlotRange->All];
```

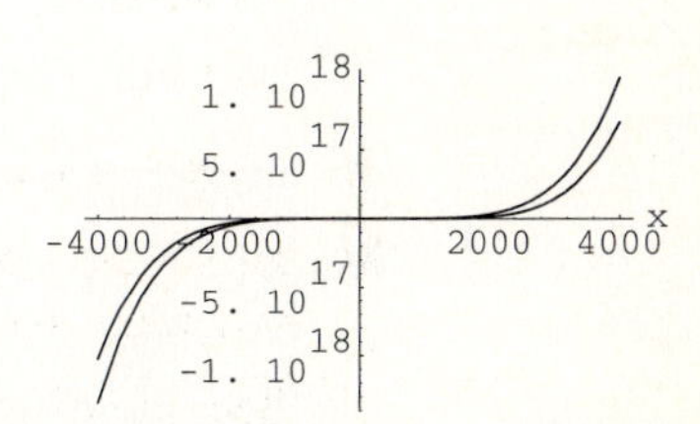

Now they have the same shape; so a good global scale plot of

$$f[x] = x^5 - 1251\,x^4 - 28750\,x^3 + 50040\,x^2 - 50001\,x$$

is:

In[4]:=
```
Plot[f[x],{x,-4000,4000},
AxesLabel->{"x","f[x]"},PlotRange->All];
```

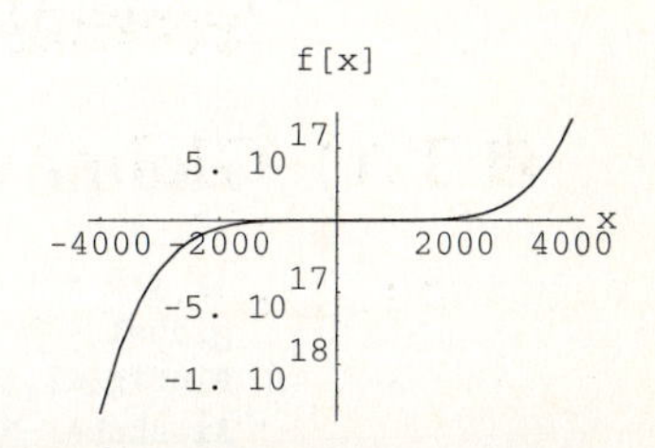

T.1.b)

> Put
>
> $$f[x] = \frac{7\,x^4 + 5\,x^2 + 4\,x + 1}{x^4 + 8\,x^2 - 17}.$$

What do you say are the limiting values

$$\lim_{x\to\infty} f[x] \qquad \text{and} \qquad \lim_{x\to -\infty} f[x]?$$

Answer: By ignoring all but the dominant terms, you can see that the global scale behavior of

$$f[x] = \frac{7\,x^4 + 5\,x^2 + 4\,x + 1}{x^4 + 8\,x^2 - 17}$$

is the same as the global scale behavior of $7\,x^4/x^4 = 7$. Here is a plot:

In[5]:=
```
Clear[x,f]
f[x_] =
(7 x^4 + 5 x^2 + 4 x +1)/(x^4 + 8 x^2 - 17);
Plot[{7,f[x]},{x,-1000,1000},
PlotStyle->{{GrayLevel[0.5],Thickness[0.03]},
{Thickness[0.01]}},
AxesLabel->{"x",""},PlotRange->{0,8}];
```

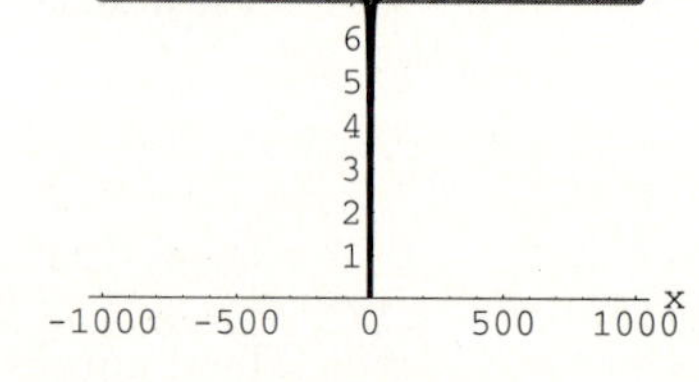

Yep; the global scale behavior is the constant 7. There is no choice but to say

$$\lim_{x\to\infty} f[x] = 7 \qquad \text{and} \qquad \lim_{x\to -\infty} f[x] = 7.$$

T.1.c) Plot

$$f[x] = \frac{2\,x^5 - 4\,x + 1}{x^4 + 6\,x^2 + 12}$$

in global scale. What simpler function mimics the global scale behavior of $f[x]$? Give a number b so that $f[x]$ is in its global scale behavior for $|x| > b$.

Answer: Open the door and take a look:

In[6]:=
```
Clear[x,f]
f[x_] =
(2 x^5 - 4 x + 1)/(x^4 + 6 x^2 + 12);
global =
Plot[f[x],{x,-1000,1000},
AxesLabel->{"x","y"}];
```

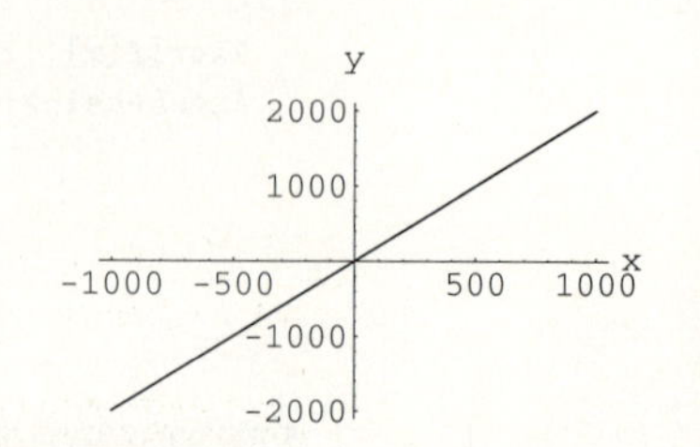

Sure looks like a line in the global scale. Just for the heck of it, look at a local scale plot:

In[7]:=
```
Plot[f[x],{x,-2,2},
AxesLabel->{"x","y"}];
```

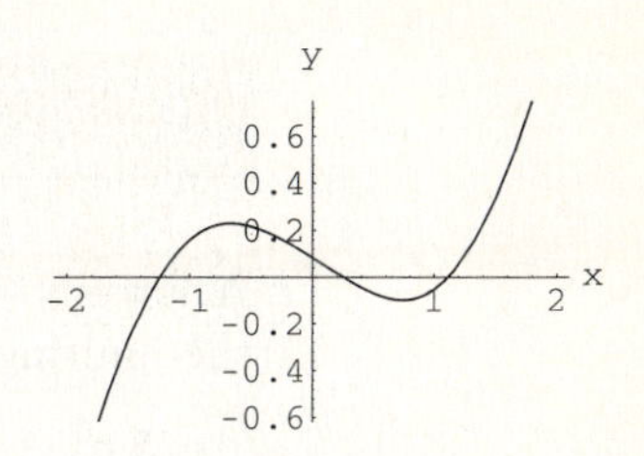

This local action is obscured in the global plot. Now look at the dominant terms. The function is

$$f[x] = \frac{2\,x^5 - 4\,x + 1}{x^4 + 6\,x^2 + 12}.$$

The dominant term in the numerator is $2\,x^5$. The dominant term in the denominator is x^4.

The global scale behavior of

$$\frac{2\,x^5 - 4\,x + 1}{x^4 + 6\,x^2 + 12}$$

is the same as the global scale behavior of $2\,x^5/x^4 = 2\,x$. You gotta say that $2\,x$ mimics the global scale behavior of f[x]. Check:

In[8]:=
```
Plot[{2 x,f[x]},
{x,-1000,1000},
PlotStyle->{{Red,Thickness[0.03]},
{Thickness[0.01]}},
AxesLabel->{"x",""}];
```

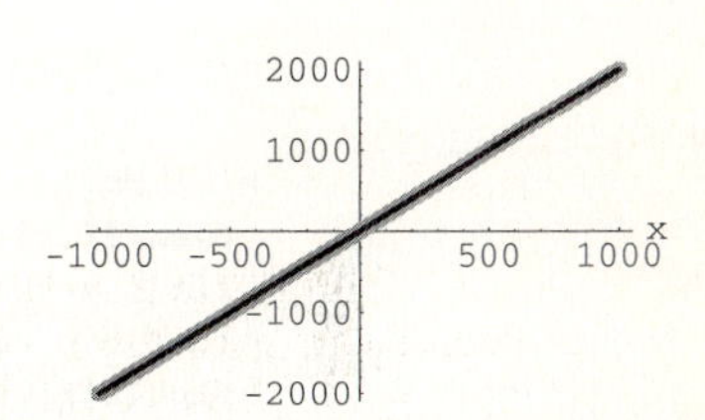

Good. In this global scale, the plots of the functions $f[x]$ and $2\,x$ share the same ink. Let's see when $f[x]$ goes into its global scale behavior:

In[9]:=
```
Plot[{2 x,f[x]}, {x,-50,50},
PlotStyle->{{Red,Thickness[0.01]},
{Thickness[0.01]}}, AxesLabel->{"x",""}];
```

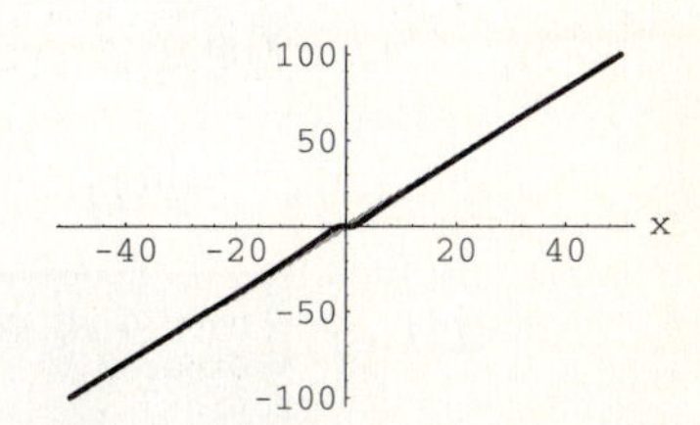

$f[x]$ seems to be into its global scale behavior for $|x| > 20$. Take $b = 20$ and you're out of here.

T.1.d) What do you say is the limiting value

$$\lim_{x \to \infty} \frac{5\,x^6 - \cos[x] - 6\,x^3}{e^{0.7\,x} + 8\sin[x]}?$$

Illustrate with a plot.

Answer: By ignoring all but the dominant terms as x goes to ∞, you see that the limiting behavior of

$$\frac{5\,x^6 - \cos[x] - 6\,x^3}{e^{0.7\,x} + 8\sin[x]}$$

is the same as the limiting behavior of

$$\frac{5\,x^6}{e^{0.7x}}.$$

And because exponential growth dominates power growth,

$$\lim_{x\to\infty} \frac{5\,x^6}{e^{0.7x}} = 0.$$

So

$$\lim_{x\to\infty} \frac{5\,x^6 - \cos[x] - 6\,x^3}{e^{0.7x} + 8\sin[x]} = 0.$$

See it happen:

In[10]:=
```
Clear[f,x]
f[x_] =
(5 x^6 - Cos[x] - 6 x^3)/(E^(0.7 x) + 8 Sin[x]);
Plot[f[x],{x,0,300},
PlotStyle->{{Thickness[0.01],Blue}},
PlotRange->All,
AxesLabel->{"x","f[x]"}];
```

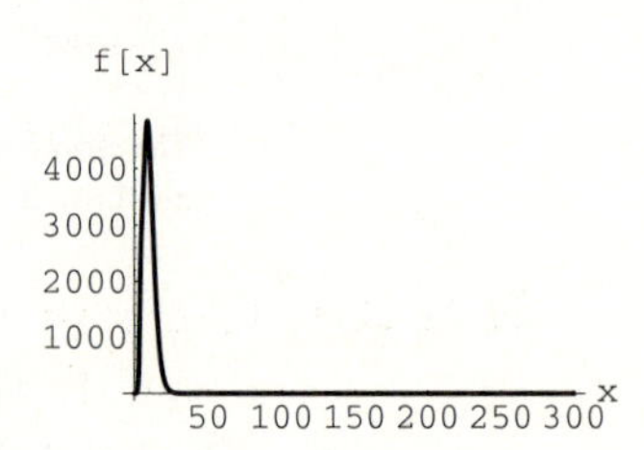

Yep;

$$\lim_{x\to\infty} \frac{5\,x^6 - \cos[x] - 6\,x^3}{e^{0.7x} + 8\sin[x]} = 0.$$

T.1.e) What do you say is the limiting value

$$\lim_{x\to\infty} \frac{3\,e^{-x} - e^{-4x}}{e^{-x} + e^{-2x}\cos[x]^2}?$$

Answer: Take

$$f[x] = \frac{3\,e^{-x} - e^{-4x}}{e^{-x} + e^{-2x}\cos[x]^2}$$

and multiply top and bottom by e^{4x} to get

$$f[x] = \frac{3\,e^{3x} - 1}{e^{3x} + e^{2x}\cos[x]^2}.$$

The quotient of the dominant terms is $3\,e^{3x}/e^{3x} = 3$. So

$$\lim_{x\to\infty} \frac{3\,e^{-x} - e^{-4x}}{e^{-x} + e^{-2x}\cos[x]^2} = 3.$$

T.1.f) Rank the following functions in order of dominance as $x \to \infty$:

$0.0005\,x^{112}$, $e^{0.1x}$, $0.6\,x^2$,
$\sqrt{x}$, $12\,x$, $0.4\,x^3$,
$0.0000017\,e^{2x}$, $100\,x^{0.3}$.

Answer: For really large x's, the functions line up in the order:

$0.0000017\,e^{2x}$,
$e^{0.1x}$,
$0.0005\,x^{112}$,
$0.4\,x^3$,
$0.6\,x^2$,
$12\,x$,
$\sqrt{x} = x^{0.5}$,
$100\,x^{0.3}$.

In the long run, exponential growth always beats power growth.

T.2) Linear models

T.2.a) Water freezes at 32 degrees Fahrenheit and at 0 degrees Celsius. Water boils at 212 degrees Fahrenheit and at 100 degrees Celsius. Find the line function that takes Celsius degrees as input and spits out the corresponding Fahrenheit reading.

Give a plot depicting Celsius readings versus Fahrenheit readings for the range $-30 \leq$ Celsius reading ≤ 50.

Answer: Use the line function

$$\text{fahr}[\text{celsius}] = b + r\,\text{celsius}$$

that runs through $\{0, 32\}$ and $\{100, 212\}$:

In[11]:=

```
Clear[fahr,celsius]
fahr[celsius_] = Fit[{{0,32},{100,212}},{1,celsius},celsius]
```

Out[11]=

```
32. + 1.8 celsius
```

Here is the plot:

In[12]:=
```
Plot[fahr[celsius],{celsius,-30,50},
AxesLabel->{Celsius, Fahrenheit}];
```

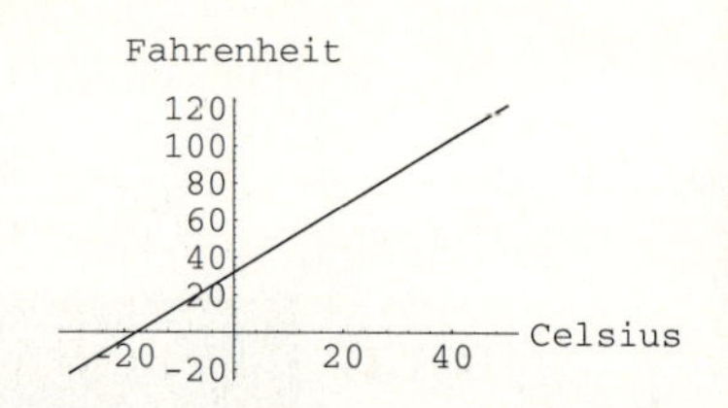

Fifty degrees Celsius is darn hot.

T.2.b) A new John Deere tractor costs \$27,000. After 10 years of heavy use, it is predicted to be worth \$12,000 (in today's dollars). Give a reasonable estimate of its value in five years.

Answer: Run a line through $\{0, 27000\}$ and $\{10, 12000\}$.

In[13]:=
```
Clear[f,t]
f[t_] = Fit[{{0,27000},{10,12000}},{1,t},t]
```

Out[13]=
```
27000. - 1500. t
```

Note:

In[14]:=
```
{f[0],f[10]}
```

Out[14]=
```
{27000., 12000.}
```

Good; $f[0]$ and $f[10]$ are right on the money. Here is a plot of the linear model for 10 years:

In[15]:=
```
Plot[f[t],{t,0,10},
PlotStyle->{{Thickness[0.01],DarkGreen}},
AxesLabel->{"years","value"}];
```

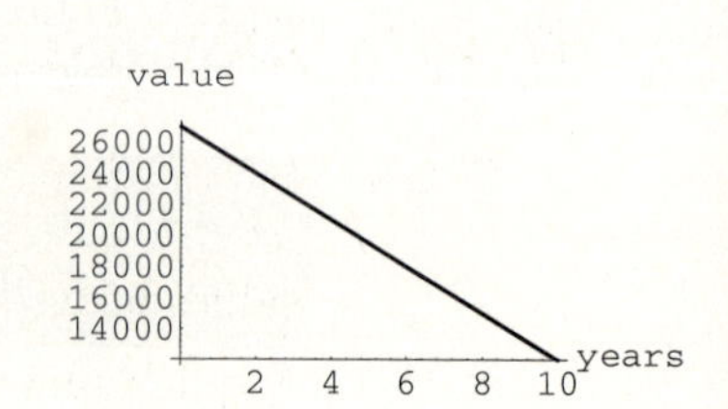

The predicted value in dollars after five years is:

In[16]:=
```
f[5]
```

Out[16]=
```
19500.
```

Here is another way of doing the same thing: The new tractor costs \$27,000, and its value falls to \$12,000 in 10 years. This means that its value fell \$15,000 in 10 years. So it is reasonable to say that its value fell by \$1500 dollars per year.

This tells us that its value in t years is given by $\$(27000 - 1500\,t)$. This agrees with the earlier function produced above.

In[17]:=
```
f[t]
```
Out[17]=
```
27000. - 1500. t
```

T.2.c) Hertz offers you a Ford Taurus for 34 dollars plus 29 cents per mile. Avis offers a Chevrolet Lumina for 29 dollars plus 32 cents per mile. How does the length of your trip tell you which deal is better?

Answer: Take a look:

In[18]:=
```
Clear[hertz, avis,x];
hertz[x_] = 34 + 0.29 x;
avis[x_] = 29 + 0.32 x;
Plot[{hertz[x],avis[x]},{x,0,300},
PlotStyle->{{Thickness[0.006]},Red}];
```

120 100 80 60
50 100 150 200 250 300

That's the Avis line starting under the Hertz line and crossing the Hertz line somewhere near $x = 160$. To find the exact point where they cross, look at:

In[19]:=
```
Solve[hertz[x] == avis[x],x]
```
Out[19]=
```
{{x -> 166.667}}
```

So on trips of fewer than 167 miles, rent Avis's Lumina; for longer trips go for Hertz's Taurus.

■ T.3) Data analysis and compromise lines

T.3.a) Information relating the variable x and the function $f[x]$ may consist of nothing more than some measurements indicating the observed value of $f[x]$ for various values of x.

A quick plot of the points usually reveals to the alert eye whether the functional relationship giving $f[x]$ in terms of x is of the form $f[x] = b + a\,x$. Look at this data plot:

In[20]:=
```
data =
{{-0.8, -1.3},{-0.4, 0.08},{-0.2, 0.73},
{0.2, 1.46},{0.4, 1.92},{0.8, 3.35},{1.4, 4.01},
{1.8, 5.14},{2.6, 6.98},{3.2, 7.58},{3.6, 9.28}};
dataplot = ListPlot[data,AxesLabel->{"x","f[x]"},
PlotStyle->{Red,PointSize[.015]}];
```

f[x]
8 6 4 2
1 2 3 x

No one line function can run through all these points, but your eye tells you that they are strung out more or less in a straight line. Probably the irregularities grew out of errors in measurements.

> How do you use *Mathematica* to get a good compromise line function that does a good job of going with the flow of these data points?

Answer: Try this:

In[21]:=

```
Clear[compromisef]
compromisef[x_] = Fit[data,{1,x},x]
```

Out[21]=

```
0.992799 + 2.24676 x
```

In[22]:=

```
compromiseplot = Plot[compromisef[x],{x,-1,4},
PlotStyle->{{Thickness[0.01],GrayLevel[0.5]}},
DisplayFunction->Identity];
Show[compromiseplot,dataplot,
DisplayFunction->$DisplayFunction];
```

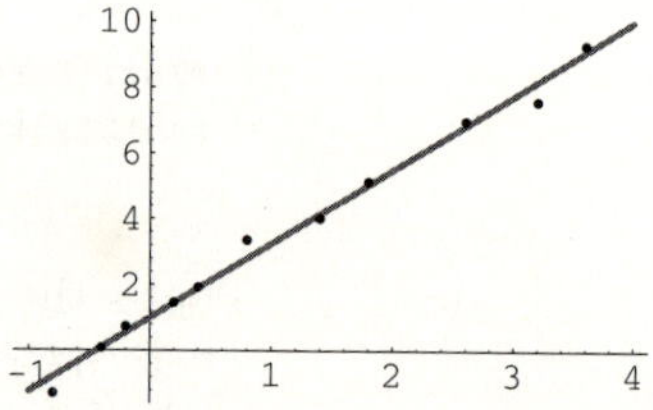

Nice compromise line. Later in the course you'll learn how *Mathematica* did this.

■ T.4) Functions given by data lists

Many functions, especially those arising out of experimental data in the lab, do not come with formulas but come intially as a data list. One such function is the function that measures the United States national debt.

Here's a plot of all the data you might want in the form $\{x, y\}$ where x is a year and y is the national debt in billions of dollars for year x for reasonably selected years starting with 1900:

In[23]:=

```
nationaldebt =
{{1900, 1.2}, {1910, 1.1}, {1920, 24.2}, {1930, 16.1},
 {1940, 43}, {1945, 258.7}, {1950, 256.1},
 {1955, 272.8}, {1960, 284.1}, {1965, 313.8},
 {1970, 370.1}, {1975, 533.2}, {1976, 620.4},
 {1977, 698.8}, {1978, 771.5}, {1979, 826.5},
 {1980, 907.7}, {1981, 997.9}, {1982, 1142},
 {1983, 1377.2}, {1984, 1572.3}, {1985, 1823.1},
 {1986, 2125.3}, {1987, 2350.3}, {1988, 2602.3},
 {1989, 2857.4}, {1990, 3233.3}, {1991, 3502.}};
```

In[24]:=
```
dataplot =
ListPlot[nationaldebt,
AxesLabel->{"x","y"},
PlotStyle->{Red,PointSize[0.02]}];
```

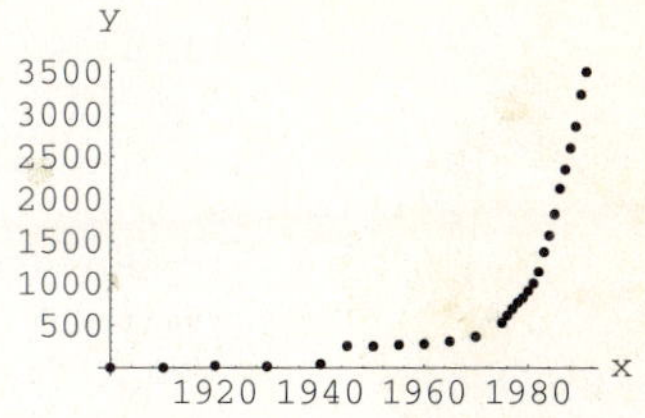

It doesn't make sense to try to put a compromise line through these points as was done in an earlier tutorial. But there is something almost as good and maybe even better. *Mathematica* can put a function right through all these points. Here it is:

In[25]:=
```
Clear[debt]
debt[t_] = Interpolation[nationaldebt][t]
```

Out[25]=
```
InterpolatingFunction[{1900, 1991}, <>][t]
```

The formula for this function is so long and detailed that *Mathematica* spares you the pain of seeing the formula, but you can plot the function:

In[26]:=
```
functplot = Plot[debt[t],{t,1900,1990},
PlotStyle->{{Thickness[0.01],GrayLevel[0.5]}},
PlotRange->All,DisplayFunction->Identity];
Show[functplot,dataplot,
AxesLabel->{"t","Debt"},
DisplayFunction->$DisplayFunction];
```

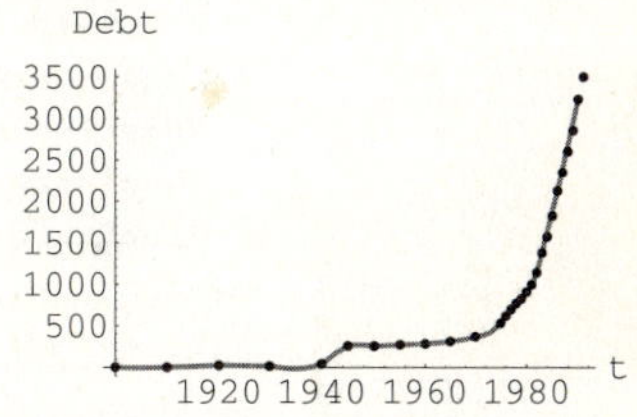

A mighty fine job. Having this function at your fingertip control allows you to analyze the heck out of the national debt function.

T.4.a) Here is a plot of debt$[t+1]-$ debt$[t]$ for $1900 \leq t \leq 1989$:

In[27]:=
```
Clear[growth]
growth[t_] = debt[t + 1] - debt[t];
growthplot =
Plot[growth[t],{t,1900,1989},
PlotStyle->{{Thickness[0.01],Red}},
PlotRange->All,
AxesLabel->{"t","growth"}];
```

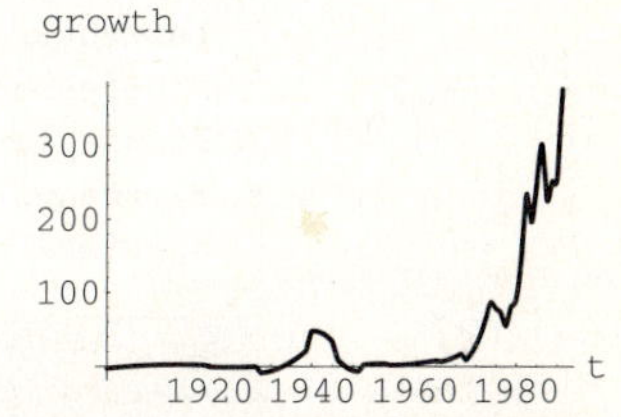

Discuss what information the plot exhibits. Use your knowledge of twentieth-century American history to comment on the rises and falls in that plot.

Answer: The last function plotted is growth$[t]=$ debt$[t+1]-$ debt$[t]$, which is nothing but the net growth in the national debt for the year starting at t and ending at $t+1$.

One feature certainly worth a comment is the crest near the middle of the year 1940–1941:

In[28]:=
```
twar = 1940.5;
warpoint = Graphics[{PointSize[0.04],
Point[{twar,growth[twar]}]}];
Show[warpoint,growthplot,PlotRange->All,
Axes->True];
```

growth

300
200
100

1920 1940 1960 1980 t

This is the time of the beginning of the buildup for World War II. It shows that the Roosevelt government was probably gearing up for war well in advance of the Japanese bombing of Pearl Harbor on December 7, 1941.

Ronald Reagan was President from 1980 through 1988:

In[29]:=
```
reagan1 = 1980;
reagan2 = 1988;
reaganpoint1 =Graphics[{PointSize[0.04],
Point[{reagan1,growth[reagan1]}]}];
reaganpoint2 =Graphics[{PointSize[0.04],
Point[{reagan2,growth[reagan2]}]}];
Show[reaganpoint1,reaganpoint2,growthplot,
PlotRange->All,Axes->True];
```

growth

300
200
100

1920 1940 1960 1980 t

Reagan almost lost all control as the growth increased until late in his term. He succeeded in driving the growth down, but he lost it at the end of his term.

T.4.b) Here is a plot of $100(\text{debt}[t+1]/\text{debt}[t]-1)$ for $1953 \leq t \leq 1989$:

In[30]:=
```
Clear[percentgrowth]
percentgrowth[t_] = 100 (debt[t + 1]/debt[t] - 1);
percentgrowthplot =
Plot[percentgrowth[t],{t,1953,1989},
PlotStyle->{{Thickness[0.01],Red}},
PlotRange->All,
AxesLabel->{"t","percentgrowth"}];
```

percentgrowth

20
15
10
5

1960 1970 1980 t

Interpret the plot.

Answer: The last function plotted is $\text{growth}[t] = 100(\text{debt}[t+1]/\text{debt}[t]-1)$, which is nothing but the net percent growth in the national debt for the year starting at t and ending at $t+1$.

You gotta say that over the last 15 years the percent growth per year is averaging better than 10%, and this presents the haunting spectre of dreaded exponential growth. The politicians better do something soon.

Eisenhower, Kennedy, and Johnson (1952–1969) did a better job of holding down the percentage increase in the national debt than any of the more recent presidents. Nixon (1969–1973) began to let things get out of control. You can see that Ford and Carter (1973–1980) dealt with the problem rather successfully, but Reagan (1981–1988) lost it.

■ T.5) The trig functions sin[x] and cos[x]

T.5.a) Here is a plot of sin[x] for $-2\pi \leq x \leq 4\pi$:

In[31]:=

```
Clear[f,x]
f[x_] = Sin[x];
sinplot = Plot[f[x],{x,-2 Pi,4 Pi},
PlotStyle->{{Red,Thickness[0.01]}},
AxesLabel->{"x","Sin[x]"}];
```

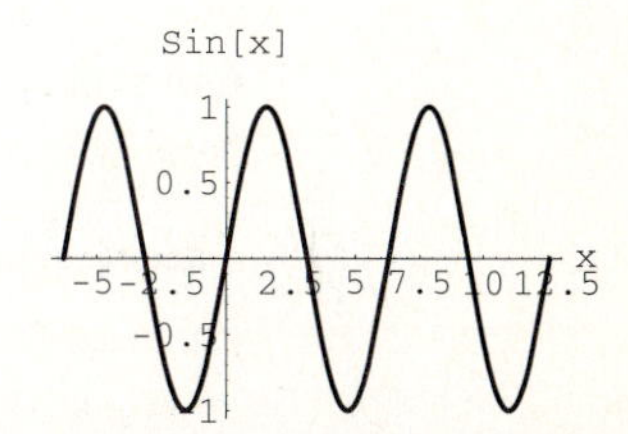

What's worth knowing about sin[x]?

Answer: From the point of view of global dominance, sin[x] just doesn't stack up. Sin[x] is so busy oscillating that it has no time to get big in the global scale. The main features you should know are all exhibited nicely in the plot.

The main features to remember are that sin[x] oscillates between -1 and 1 and sin[0] $= 0$.

If you're hazy about the association between sin[x] and right triangles, don't worry. You can do a whale of a lot with sin[x] if you recognize the plot above as a plot of sin[x].

T.5.b) Here is a plot of cos[x] for $-2\pi \leq x \leq 4\pi$:

In[32]:=

```
Clear[f,x]
f[x_] = Cos[x];
cosplot = Plot[f[x],{x,-2 Pi,4 Pi},
PlotStyle->{{Blue,Thickness[0.01]}},
AxesLabel->{"x","Cos[x]"}];
```

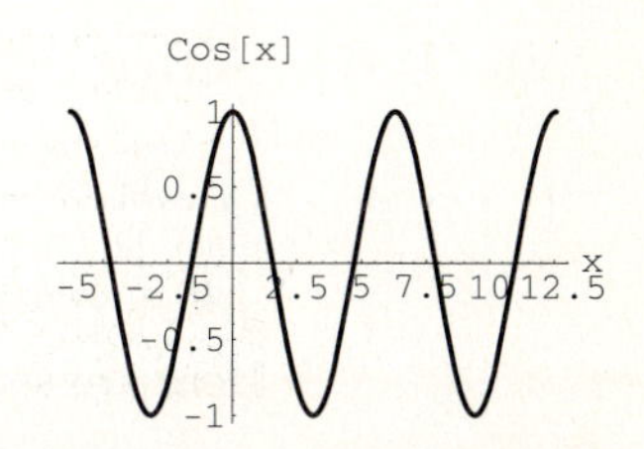

What's worth knowing about cos[x]?

Answer: From the point of view of global dominance, cos[x] just doesn't stack up. Cos[x] is so busy oscillating that it has no time to get big in the global scale. The main features you should know are all exhibited nicely in the plot.

The main features to remember are that cos[x] oscillates between -1 and 1 and cos[0] $= 1$.

If you're hazy about the association between cos[x] and right triangles, don't worry. You can do a whale of a lot with cos[x] if you recognize the last plot as a plot of cos[x].

T.5.c) Hey! The write-up on cos[x] is obviously a copy, paste, and edit job of the write-up on sin[x]. What gives?

Answer: Good question. Take a look at both plots together.

In[33]:=
```
Show[cosplot,sinplot,AxesLabel->{"x",""}];
```

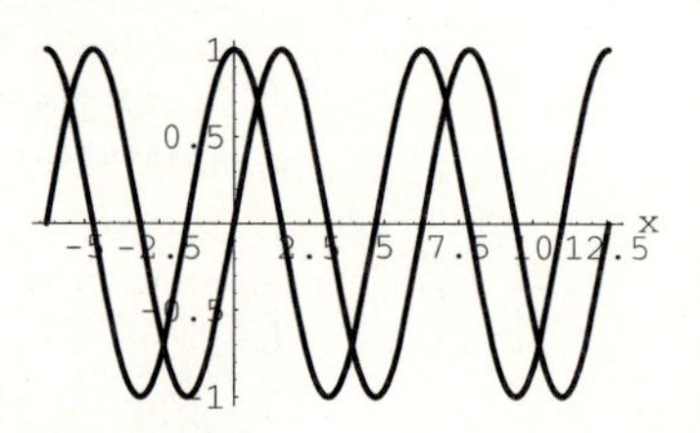

In shape, the two curves are identical because they are just shifted versions of each other. This is explained by the fact that cos[x] = sin[$x + \pi/2$].

In[34]:=
```
Clear[x]
Cos[x] == Sin[x + Pi/2]
```

Out[34]=
```
True
```

To get the cos[x] curve from the sin[x] curve, you just push the sin[x] curve $\pi/2$ units to the left.

■ T.6) Another linear model: Drinking and driving

Calculus&*Mathematica* thanks pharmacokineticist Professor Al Staubus of the College of Pharmacy at Ohio State University and Medical Doctor Jim Peterson of Urbana, Illinois for help on this problem.

Here are some basic data:

Alcohol content of various beverages: Here is a chart revealing how many grams of alcohol are found in the following beverages:

12	ounce regular beer	13.6 grams of alcohol
12	ounce light beer	11.3
4	ounce port wine	16.4
4	ounce burgundy wine	10.9

4 ounce rose wine	10.0
1.5 ounce 100-proof vodka	16.7
1.5 ounce 100-proof bourbon	16.7
1.5 ounce 80-proof vodka	13.4
1.5 ounce 80-proof bourbon	13.4

Body fluid supply: A college-age male in good shape weighing K kilograms has about 0.68 K liters of fluid in his body. Males in poor shape have less.

A college-age female in good shape weighing K kilograms has about 0.65 K liters of fluid in her body. Females in poor shape have less.

Threshold for legal driving: In all states, if your body fluids contain more than 1 gram of alcohol per liter of body fluids, then you are too drunk to drive legally.

In some states, if your body fluids contain more than 0.8 grams of alcohol per liter of body fluids, then you are too drunk to drive legally.

Alcohol elimination: The typical human liver can eliminate 12 grams of alcohol per hour.

Jennifer is a party-loving student who weighs 136 pounds. Her weight in kilograms is:

In[35]:=
```
Convert[136 PoundWeight,KilogramWeight]
```

Out[35]=
```
61.6885 KilogramWeight
```

According to the data above, her supply of body fluids in liters meausures out to:

In[36]:=
```
Jenniferfluids = 0.65 61.6885 Liters
```

Out[36]=
```
40.0975 Liters
```

With no alcohol in her system, she quickly downs a vodka and tonic made with four 1.5-ounce shots of 80-proof vodka. The alcohol in her body fluids measures out to:

In[37]:=
```
Jenniferalcohol = 4 13.4 Grams
```

Out[37]=
```
53.6 Grams
```

Her alcohol concentration is:

In[38]:=
```
Jenniferalcohol/Jenniferfluids
```

Out[38]=
```
1.33674 Grams
-------------
   Liters
```

Jennifer is not a legal driver in any state.

The blood test often administered in DUI cases measures the alcohol concentration of the blood. The alcohol concentration of the blood is the same as the alcohol concentration of the body fluids as a whole.

T.6.a) Assuming Jennifer drinks no more alcohol, report on how long she has to wait until her body fluids contain the relatively safe alcohol concentration of 0.5 grams/liter. When will she become legal for driving?

Answer: The typical human liver can eliminate 12 grams of alcohol per hour. If you measure t in hours, the function that describes the alcohol content (in grams) in her body is a line function whose growth rate is -12. And this function must go through:

In[39]:=
```
{0,Jenniferalcohol}
```
Out[39]=
```
{0, 53.6 Grams}
```

The function that measures the alcohol content (in grams) in her blood t hours after she took the drink is:

In[40]:=
```
Clear[alcohol,t]
alcohol[t_] = 53.6 - 12 ( t - 0)
```
Out[40]=
```
53.6 - 12 t
```

Her total body fluids is:

In[41]:=
```
Jenniferfluids
```
Out[41]=
```
40.0975 Liters
```

The concentration of alcohol in her body fluids in grams/liter is measured by:

In[42]:=
```
Clear[concentration]
concentration[t_] = alcohol[t]/40.0975
```
Out[42]=
```
0.0249392 (53.6 - 12 t)
```

A plot:

In[43]:=
```
Plot[{concentration[t],0.5},{t,0,4},
PlotStyle->{{Thickness[0.01]},{Red}},
PlotRange->{0,concentration[0]},
AxesLabel->{"hours","concentration"}];
```

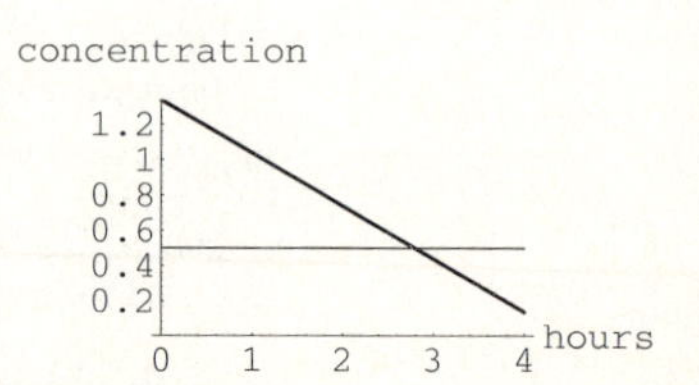

She has to wait almost three hours until her body fluids contain the relatively safe alcohol concentration of 0.5 grams/liter.

To see when she becomes legal to drive in all states, look at:

In[44]:=

```
Solve[concentration[t] == 0.8,t]
```

Out[44]=

```
{{t -> 1.7935}}
```

Almost two hours.

To see when she becomes legal to drive in some states, look at:

In[45]:=

```
Solve[concentration[t] == 1.0,t]
```

Out[45]=

```
{{t -> 1.12521}}
```

A little more than an hour.

T.6.b) Bubba, a student at a large Midwestern university, is going to a party. He plans to be cool. More than that, though, he plans to drink plenty of beer at the party; after all he's got a reputation to uphold. But he is also worried about his driver's license. So he comes up with the scheme: He'll drink beer gradually at the rate of one 12-ounce beer every 10 minutes for the first hour; then he'll nurse one 12-ounce beer every half hour for the next two hours. After all this beer, he'll stop drinking and just sit around looking cool until he's sure his body fluid alcohol concentration is down to 0.5 grams/liter, a safe level for driving.

The only hitch is that Bubba can't figure out how long he has to wait from the time of his last beer until it's safe for him to drive, so he calls you up and asks you to figure it out for him.

> Help out this poor devil.

Answer: The first question you ask is how much Bubba weighs. He says that he weighs 187 pounds. You say, "Now I can go to work."

In[46]:=

```
Convert[187 PoundWeight,KilogramWeight]
```

Out[46]=

```
84.8217 KilogramWeight
```

According to the data above, his body fluids in liters measures out to:

In[47]:=

```
Bubbafluids = (0.68) 84.8217 Liters
```

Out[47]=

```
57.6788 Liters
```

During the first hour he plans to drink six 12-ounce beers. Each typical beer contains 13.6 grams of alcohol. This means he plans to consume

In[48]:=
```
6 13.6
```
Out[48]=
```
81.6
```

grams of alcohol during the first hour. This is the same as saying that the alcohol in his body is growing at a rate of 81.6 grams per hour during the first hour. But his liver is eliminating the alcohol at a rate of 12 grams per hour. This means that the alcohol in his body during the first hour is measured in grams by:

In[49]:=
```
Clear[alcohol1,t]
alcohol1[t_] = 0 + (81.6 - 12) t
```
Out[49]=
```
69.6 t
```

His body fluid supply is:

In[50]:=
```
Bubbafluids
```
Out[50]=
```
57.6788 Liters
```

His alcohol concentration during the first hour is measured by:

In[51]:=
```
Clear[concentration1]
concentration1[t_] = alcohol1[t]/57.6788
```
Out[51]=
```
1.20668 t
```

In[52]:=
```
firsthour =
Plot[{concentration1[t],0.5},{t,0,1},
PlotStyle->{{Thickness[0.01]},{Red}},
PlotRange->{0,concentration1[1]},
AxesLabel->{"hours","concentration"}];
```


Bubba will be feeling no pain at the end of the first hour.

During the second and third hours he plans to drink four 12-ounce beers. Each beer contains 13.6 grams of alcohol. This means he plans to consume

In[53]:=
```
4 13.6
```
Out[53]=
```
54.4
```

grams of alcohol during the second and third hours. This is the same as saying that his alcohol concentration is growing at a rate of

In[54]:=
```
54.4/2
```
Out[54]=
```
27.2
```

grams per hour during the second and third hours. But his liver is eliminating the alcohol at a rate of 12 grams per hour. This means that the alcohol in his body during the second and third hours is measured in grams by:

In[55]:=
```
Clear[alcohol23,t]
alcohol23[t_] = alcohol1[1] + (27.2 - 12) (t - 1)
```
Out[55]=
```
69.6 + 15.2 (-1 + t)
```

The reason for the presence of the alcohol1[1] term is that when $t = 1$,

$$\text{alcohol23}[t] = \text{alcohol1}[t].$$

His alcohol concentration during the second and third hours is measured in grams per liter by:

In[56]:=
```
Clear[concentration23]
concentration23[t_] = alcohol23[t]/57.6788
```
Out[56]=
```
0.0173374 (69.6 + 15.2 (-1 + t))
```

In[57]:=
```
secondandthirdhours =
Plot[{concentration23[t],0.5},{t,1,3},
PlotStyle->{{Thickness[0.01]},{Red}},
PlotRange->{0,concentration23[3]},
AxesLabel->{"hours","concentration"}];
```

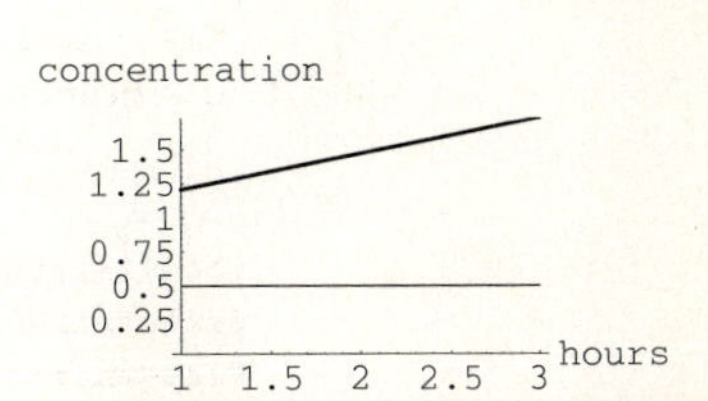

Here are both plots combined into one:

In[58]:=
```
Show[firsthour,secondandthirdhours,PlotRange->All];
```

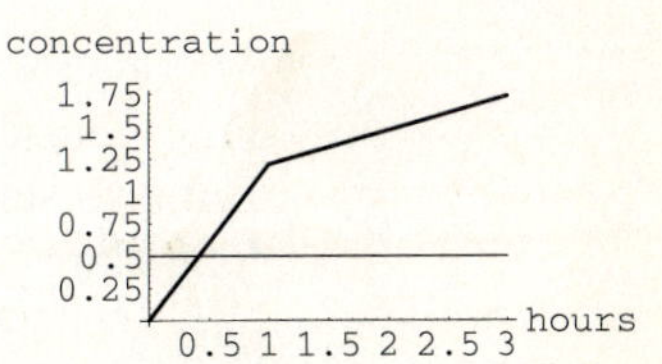

At the end of the first three hours, Bubba will be stinko.

After the third hour, Bubba will drink nothing but his liver will continue to eliminate alcohol at a rate of 12 grams per hour. This means that the alcohol in his body after the third hour is measured in grams by:

In[59]:=

```
Clear[alcoholafter3]
alcoholafter3[t_] = alcohol23[3]  - 12 ( t - 3)
```

Out[59]=

```
100. - 12 (-3 + t)
```

The reason for the presence of the alcohol23[3] term is that when $t = 3$,

$$\text{alcoholafter3}[t] = \text{alcohol23}[t].$$

His alcohol concentration in grams per liter after the third hour is measured by:

In[60]:=

```
Clear[concentrationafter3]
concentrationafter3[t_] = alcoholafter3[t]/57.6788
```

Out[60]=

```
0.0173374 (100. - 12 (-3 + t))
```

A plot:

In[61]:=

```
afterthirdhour =
Plot[{concentrationafter3[t],0.5},{t,3,10},
PlotStyle->{{Thickness[0.01]},{Red}},
PlotRange->{0,concentration23[3]},
AxesLabel->{"hours","concentration"}];
```

The whole story:

In[62]:=

```
Show[firsthour,secondandthirdhours,
afterthirdhour,
PlotRange->All];
```

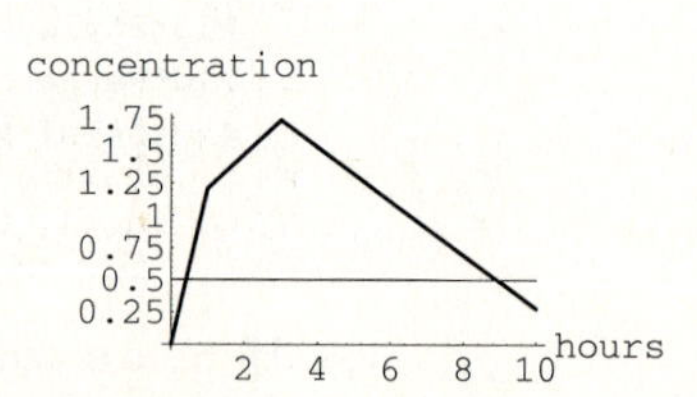

You look at this and you tell Bubba that he should plan to stay at the party about nine hours, drinking for the first three hours and sobering up for six hours. For Bubba's sake, you hope the others like Bubba's company. He's going to be there a long time.

Give It a Try

Experience with the starred ($\star$) problems will be especially beneficial for understanding later lessons.

■ G.1) Line fundamentals★

G.1.a) Here are five line functions all with the same growth rate:

In[1]:=
```
Clear[f1,f2,f3,f4,f5,x]
a = 0.86;
f1[x_] = -1 + a x; f2[x_] =  0 + a x;
f3[x_] =  1 + a x; f4[x_] =  2 + a x;
f5[x_] =  3 + a x;
Plot[{f1[x],f2[x],f3[x],f4[x],f5[x]},{x,-1,4},
AxesLabel->{"x",""}];
```

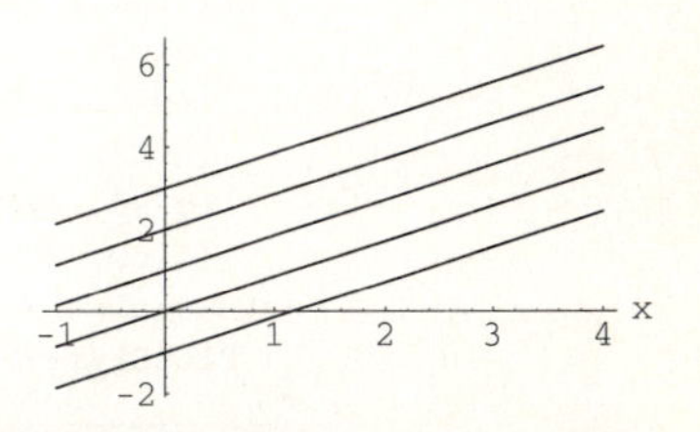

Look at the plot and fill in the blank: The plots of two line functions with the same growth rate are ____________ to each other.

G.1.b) A function $f[x]$ starts out at $x = 0$ with a value of 7.9 and goes up at a constant rate of 0.3 units on the y-axis for each unit on the x-axis.

Give a formula for this function and plot it.

G.1.c) A function $f[x]$ starts out at $x = 1.1$ with a value of 7.3 and goes up at a rate of -0.3 units on the y-axis for each unit on the x-axis.

Give a formula for this function and plot it.

G.1.d) Given a constant b, what happens to $f[x] = -2x + b$ every time x goes up by one unit?

G.1.e) Here are the two points $\{1, 3\}$ and $\{2, 2\}$:

In[2]:=
```
point1 =
Graphics[{Blue,PointSize[0.03], Point[{1,3}]}];
point2 =
Graphics[{Blue,PointSize[0.03], Point[{3,2}]}];
twopoints =
Show[point1,point2,PlotRange->{{0,5},{0,5}},
AxesOrigin->{0,0},Axes->True,AxesLabel->{"x","y"},
AspectRatio->Automatic];
```

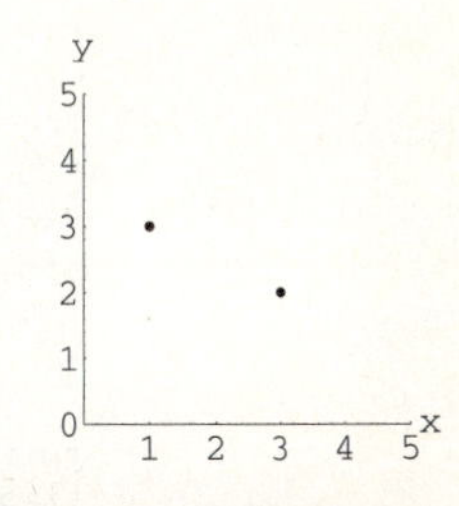

Find a formula for the line function whose plot hits both points. Confirm your answer with a plot.

G.1.e.i) Here are some points:

In[3]:=
```
data =
{{-2.01, -2.59}, {-1.74, -2.17},{-1.5, -1.65},
 {-1.27, -1.31}, {-1.04, -1.20}, {-0.84, -1.08},
 {-0.45, -0.74}, {-0.35, -0.22}, {0., 0.2},
 {0.27, 0.37}, {0.51, 0.45}, {0.73, 0.71},
 {0.98, 1.19}, {1.23, 1.68},{1.58, 1.94},
 {1.69, 2.02}, {2.01, 2.20}};
dataplot = ListPlot[data,AxesLabel->{"x","y"},
PlotStyle->{Red,PointSize[0.03]}];
```

Explain why it is impossible to get a single line function whose plot runs through all of these points.

G.1.e.ii) Use the *Mathematica* Fit instruction to find a compromise line function whose plot flows with the data. Confirm with a plot.

G.1.e.iii) Use the *Mathematica* Interpolation instruction to come up with a function whose plot runs through all these points. Confirm with a plot. What value would you predict at $x = 1.77$?

■ G.2) Global scale★

G.2.a) Look at:

In[4]:=
```
Clear[x,f]
f[x_] = x^4 - 10000000 x^2;
Plot[f[x],{x,-1000,1000},
AxesLabel->{"x","f[x]"},
PlotRange->All];
```

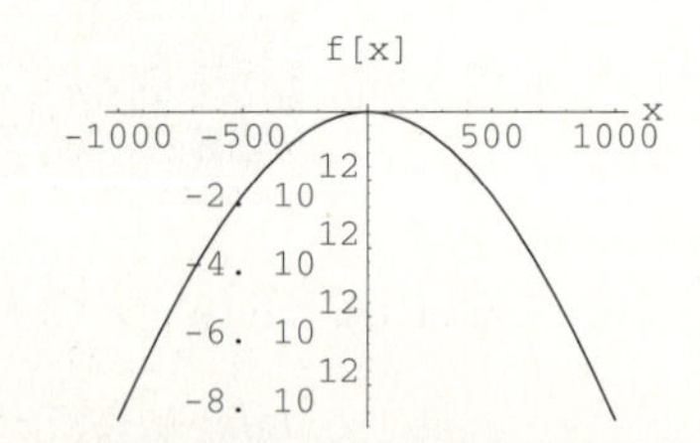

Is this a good global scale plot of $f[x] = x^4 - 10000000\, x^2$? Why or why not? If it is not a good global scale plot of $f[x]$, then give a good global scale plot of $f[x]$.

G.2.b) Put

$$f[x] = \frac{2\,x^6 + 50\,x^2}{x^6 + 3\,x^2 + 1}.$$

What do you say are the limiting values

$$\lim_{x \to \infty} f[x] \qquad \text{and} \qquad \lim_{x \to -\infty} f[x]?$$

G.2.c) What do you say is the limiting value

$$\lim_{x \to \infty} \frac{x^9 + 4\,e^{0.6x}}{3\,x^{12} + 2\,e^{0.6x}}?$$

Illustrate with a plot.

G.2.d) What do you say is the limiting value

$$\lim_{x \to \infty} \frac{3\,x^8 - 123\cos[x] - 6\,x^2}{e^{0.4x}}?$$

Illustrate with a plot.

G.2.e) What do you say is the limiting value

$$\lim_{x \to \infty} e^{-0.8x}\,(1 + 5\,x^6)?$$

Illustrate with a plot.

G.2.f) What do you say is the limiting value

$$\lim_{x \to \infty} \frac{3\,e^{-x} - 5\,e^{-3x}}{e^{-3x} + e^{-x}}?$$

Illustrate with a plot.

G.2.g) Rank the following functions in order of dominance as $x \to \infty$:

$0.0001\,x^{24}$, $0.0004\,e^{0.01x}$, $89\,x^2$,
$\sqrt{x}$, $17\,x$, $0.08\,x^3$,
$0.0000013\,e^{2x}$, $100\,x^{0.4}$.

G.2.h) Plot

$$f[x] = \frac{2\,x^4 - 40\,x + 1}{x^2 + x + 12}$$

in global scale. What simpler function mimics the global scale behavior of $f[x]$? Give a number b so that $f[x]$ is in its global scale behavior for $|x| > b$.

G.3) Linear models★

G.3.a) A new Maytag 17.1 cubic feet refrigerator costs \$875 and is projected to cost \$71 per year to operate. Assuming the Maytag people are truthful in saying that their appliances never break down, give a plot showing the total projected costs of buying and operating the Maytag at any time during the first 10 years of ownership.

> How much does the Maytag owner have to shell out for 10 years of cold soda and beer? How much would it have cost after $7\frac{1}{2}$ years?

G.3.b) You have a chance to buy a used Chevrolet Caprice Classic V8 for \$3150 and you have a chance to buy a used Honda Accord LX for \$4375. Your mechanic has assured you that both cars are in tip-top shape and should give many repair-free miles. The Chevy delivers 17 miles per gallon and the Honda delivers 32 miles per gallon. Assume that gasoline costs \$1.19 per gallon. Put

chevy[x] = the cost of buying and fueling the Chevy for x miles.

honda[x] = the cost of buying and fueling the Honda for x miles.

> Plot chevy[x] and honda[x] on the same axes for $0 \leq x \leq 50{,}000$ miles.
>
> How many miles would you have to drive before the cost of buying and fueling the Chevy would become greater than the cost of buying and fueling the Honda?

G.3.c) Water freezes at 32 degrees Fahrenheit and at 0 degrees Celsius. Water boils at 212 degrees Fahrenheit and at 100 degrees Celsius. Find the line function that takes Fahrenheit degrees as input and spits out the corresponding Celsius reading.

> Give a plot depicting Fahrenheit readings versus Celsius readings for the range $-25 \leq$ Fahrenheit reading ≤ 120.

G.3.d) A certain airplane uses k gallons of fuel for each combined take-off and landing. If this airplane uses 1081 gallons of fuel on a 270-mile trip and uses 1575 gallons on a 435-mile trip, then:

> Estimate the miles per gallon delivered by the airplane and estimate the value of k.

G.4) Compromise lines through data★

G.4.a) This problem was adapted from Edward Tufte's book *The Visual Display of Quantitative Information*, Graphics Press, 1983. It is based on data from a paper by R. Doll, "Etiology of Lung Cancer," *Advances in Cancer Research*, 3 (1955), 1–50.

Here are data in the form $\{x, y\}$ for various countries. In each case, x stands for the per capita cigarette consumption of the country in 1930 and y stands for the death rate of males for lung cancer in 1950 measured in deaths per million population of the same country. In all cases x and y are rounded to the nearest 50.

In[5]:=
```
Iceland = {200,50}; Norway = {250,100}; Sweden = {300,100};
Denmark = {350,150}; Australia = {450,150}; Holland =  {450,250};
Canada =  {500,150}; Switzerland = {550,250}; Finland = {1150,350};
Britain = {1150,450}; America = {1300,200};
data = {Iceland,Norway,Sweden,Denmark,Australia,Holland,Canada,
Switzerland,Finland,Britain,America};
```

And a plot:

In[6]:=
```
dataplot =
ListPlot[data,
PlotStyle->{Red,PointSize[.03]}];
```

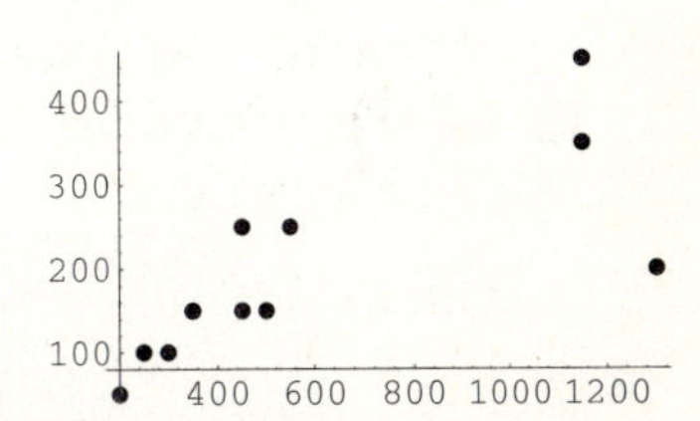

Use *Mathematica* to get a compromise line function that does a good job of going with the flow of these data points. Add the plot of the compromise line to the plot. Discuss the message the plot tries to convey.

G.4.b.i) Each of the following measurements $\{x, y\}$ stands for the pull x (in pounds) needed to lift the corresponding weight y (also in pounds) by means of a certain pulley block:

In[7]:=
```
data = {{14.2,28}, {26.6,60}, {38.1,86}, {50.3,114}, {62.8,146}, {75.1,172}};
```

And a plot:

In[8]:=

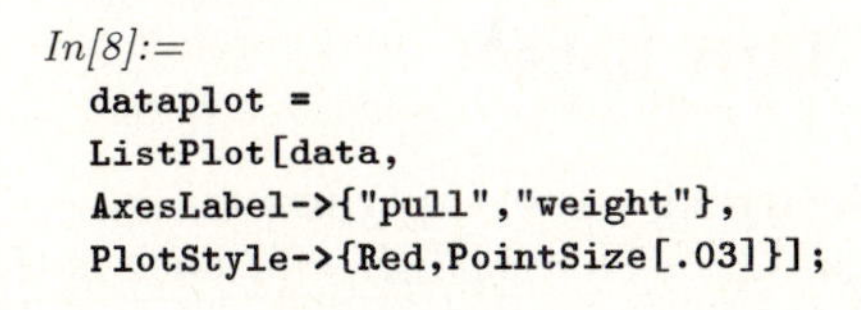

```
dataplot =
ListPlot[data,
AxesLabel->{"pull","weight"},
PlotStyle->{Red,PointSize[.03]}];
```

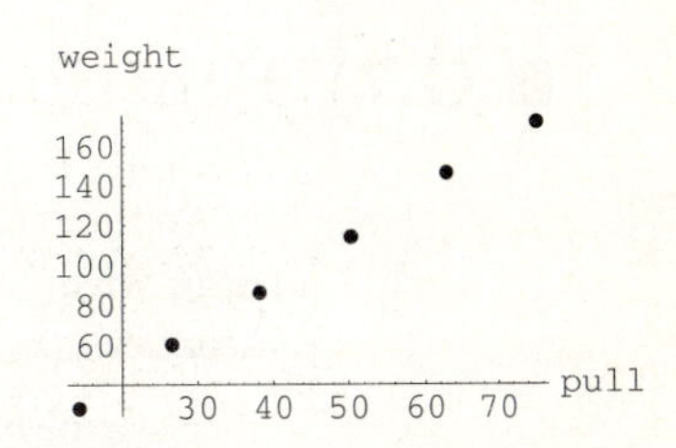

The data contain built-in measurement errors.

Use *Mathematica* to get a compromise line function that does a good job of going with the flow of these data points. Add the plot of the compromise line to the plot.

G.4.b.ii) Take the line function that you found in part G.4.b.i) above and use it to estimate the weight that can be lifted by pulls of 60 pounds and of 100 pounds.

G.4.b.iii) Take the formula for the line function that you found in part G.4.b.i) above and use it to try to answer the question: Each extra pound of pull results in about how many extra pounds of lift?

■ G.5) Globs

This problem appears only in the electronic version.

■ G.6) Percentage growth★

G.6.a) Measure the average percent growth rate of $f[x] = 4.8\,e^{2x}$ as x advances by one unit. Does your result depend on where x starts and stops as it advances by one unit?

G.6.b) Measure the average percent growth rate of $f[x] = 5.1\,e^{-3x}$ as x advances by one unit. Does your result depend on where x starts and stops as it advances by one unit?

G.6.c) Measure the average percent growth rate of $f[x] = 5.1\,x^3$ as x advances by one unit. Does your result depend on where x starts and stops as it advances by one unit?

■ G.7) Functions given by data lists★

Many functions, especially those arising out of experimental data in the lab, do not come with formulas but come intially as a data list. One such is the function that measures the United States population.

Here's a plot of all the results of all the censuses ever taken by the United States government in the form $\{x, y\}$ where x is a year and y is the U.S. population in millions for year x:

In[9]:=
```
uspop =
{{1790,3.9},{1800,5.3},{1810,7.24},{1820,9.6},
 {1830,12.86},{1840,17},{1850,23.2},{1860,31.4},
 {1870,38.5},{1880,50.2},{1890,63},{1900,76.2},
 {1910,92.2},{1920,106},{1930,123.2},{1940,132.2},
 {1950,151.3},{1960,179.3},{1970,203.3},
 {1980,226.5},{1990,248.7}};
popdataplot = ListPlot[uspop,
AxesLabel->{"year","population"},
PlotStyle->{Red,PointSize[0.03]}];
```

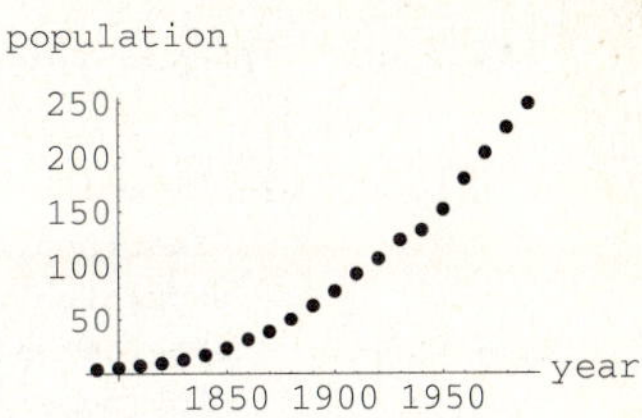

It doesn't make sense to try to put a compromise line through these points. But there is something almost as good and maybe even better. *Mathematica* can put a function right through all these points.

Here it is:

In[10]:=
```
Clear[pop]
pop[t_] = Interpolation[uspop][t]
```

Out[10]=
```
InterpolatingFunction[{1790, 1990}, <>][t]
```

The formula for this function is so long and detailed that *Mathematica* spares you the pain of seeing the formula, but you can plot the function:

In[11]:=
```
functplot =
Plot[pop[t],{t,1790,1990},
PlotStyle->{{Thickness[0.01],Blue}},
PlotRange->All,
DisplayFunction->Identity];
Show[functplot,popdataplot,
AxesLabel->{"t",""},
DisplayFunction->$DisplayFunction];
```

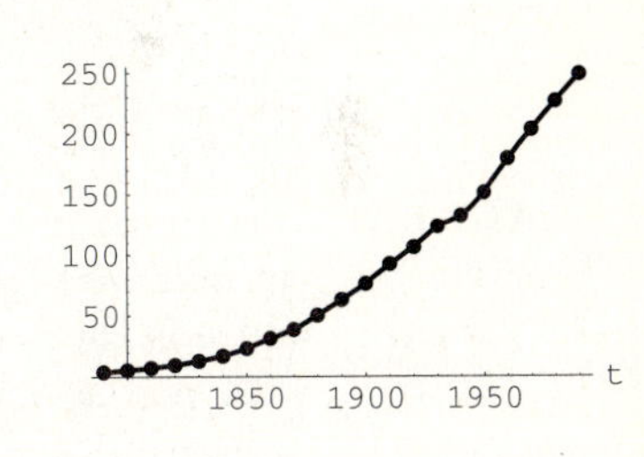

Totally acceptable. Having this function at your fingertip control allows you to analyze the heck out of the population function.

For instance, if you want to get a decent estimate of the population in millions of the United States in 1957, the time of hot Chevy V-8's and Mickey Mantle, just ask for it:

In[12]:=
```
pop[1957]
```

Out[12]=
```
170.733
```

G.7.a) Give the year of your birth and estimate the U.S. population for that year.

G.7.a.i) Here is a plot of pop$[t+1]-$ pop$[t]$ for $1790 \leq t \leq 1989$:

In[13]:=

```
Clear[popgrowth]
popgrowth[t_] = pop[t + 1] - pop[t];
popgrowthplot =
Plot[popgrowth[t],{t,1790,1989},
PlotStyle->{{Thickness[0.01],Blue}},
PlotRange->All,AxesLabel->
{"t",""}];
```

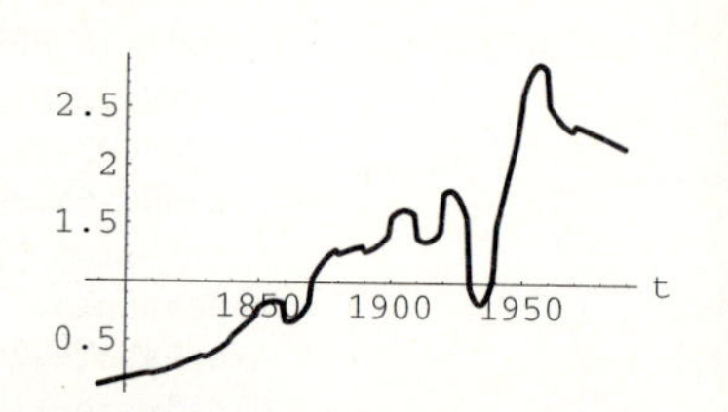

Discuss what information the plot exhibits. Use your knowledge of American history to comment on the crests and dips.

G.7.a.ii) Here is a plot of $100\,(\text{pop}[t+1]/\text{pop}[t]-1)$ for $1790 \leq t \leq 1989$:

In[14]:=

```
Clear[poppercentgrowth]
poppercentgrowth[t_] = 100 (pop[t + 1]/pop[t] - 1);
poppercentgrowthplot =
Plot[poppercentgrowth[t],{t,1790,1989},
PlotStyle->{{Thickness[0.01],Blue}},
PlotRange->All,AxesLabel->{"t",""}];
```

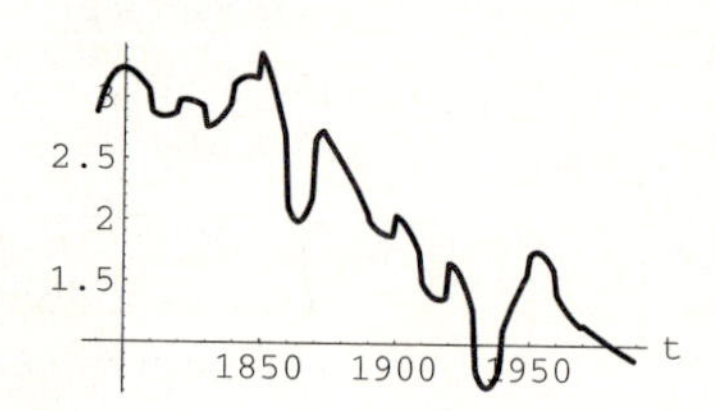

Interpret the plot.

G.7.b.i) Just to really foul things up, here's a table and a plot of some immigration numbers (in millions) by decade from 1820 until 1985. The form is $\{x, y\}$ where x is a year and y is the U.S. average number of immigrants per year for the 10 years starting with year $x-5$ and ending with $x+5$.

In[15]:=

```
immigrationdata =
{{1825,0.015},{1835,0.060},{1845,0.171},
 {1855,0.260},{1865,0.215},{1875,0.281},{1885,0.525},
 {1895,0.369},{1905,0.880},{1915,0.574},{1925,0.411},
 {1935,.053},{1945,0.104},{1955,0.252},{1965,0.332},
 {1975,0.449},{1985,0.734}};
immigrationdataplot = ListPlot[immigrationdata,
AxesLabel->{"year","immigrants"},
PlotStyle->{DarkGreen,PointSize[0.03]}];
```

immigrants
0.8
0.6
0.4
0.2
1850 1900 1950
year

Use *Mathematica* to run a function through the data and then analyze the data the way you did in part G.7.a). Decide whether or not you think any of this information is relevant to the population analysis you did in part G.7.a). If so, use it. If not, explain why in some detail.

■ G.8) Another linear model: Drinking and driving

Calculus&*Mathematica* thanks pharmacokineticist Professor Al Staubus of the College of Pharmacy at Ohio State University and Jim Peterson, M.D. of Urbana, Illinois for help on this problem.

Here are some basic data:

Alcohol content of various beverages: Here is a chart revealing how many grams of alcohol are found in the following beverages:

12	ounce regular beer	13.6 grams of alcohol
12	ounce light beer	11.3
4	ounce port wine	16.4
4	ounce burgundy wine	10.9
4	ounce rosé wine	10.0
1.5	ounce 100-proof vodka	16.7
1.5	ounce 100-proof bourbon	16.7
1.5	ounce 80-proof vodka	13.4
1.5	ounce 80-proof bourbon	13.4

Body fluid supply: A college-age male in good shape weighing K kilograms has about 0.68 K liters of body fluids. Males in poor shape have less.

A college-age female in good shape weighing K kilograms has about 0.65 K liters of body fluids. Females in poor shape have less.

For instance, if you are a typical college-age female and you weigh 120 pounds, then your weight in kilograms is:

In[16]:=
```
Convert[120 PoundWeight,KilogramWeight]
```
Out[16]=
```
83.461 KilogramWeight
```

And your body fluids measure out to about:

In[17]:=
```
0.65 54.4311 Liters
```
Out[17]=
```
35.3802 Liters
```

But if you are a typical college-age male and you weigh 184 pounds, then your weight in kilograms is:

In[18]:=
```
Convert[184 PoundWeight,KilogramWeight]
```
Out[18]=
```
83.461 KilogramWeight
```

And your body fluids in liters measure out to about:

In[19]:=
```
0.65 83.461 Liters
```
Out[19]=
```
54.2497 Liters
```

Threshold for legal driving: In all states, if your body fluids contain more than 1 gram of alcohol per liter, then you are too drunk to drive legally.

In some states, if your body fluids contain more than 0.8 grams of alcohol per liter, then you are too drunk to drive legally.

Alcohol elimination: The typical human liver can eliminate 12 grams of alcohol per hour.

G.8.a) Your gang is going to a party and they are riding in your car. Your strategy is to nurse one 12-ounce beer each hour you are at the party.

> If you start with no alcohol in your system and you stay for five hours, will your blood alcohol concentration ever exceed 0.8 grams/liter? Use your own weight.

G.8.b) Your gang is going to a party and they are riding in your car. Your strategy is to figure out how many 4-ounce glasses of burgundy you can nurse per hour so that your body fluids never contain more than 0.8 grams/liter for the whole six hours.

> Do it.

G.8.c) Chip, a male student weighing 135 pounds, drives to the party with the idea of nursing two 12-ounce beers each hour for five hours.

> If he starts with no alcohol in his system, how many grams of alcohol are there in each liter of his body fluids at the end of the five hours?
>
> After the five hours, about how long must he go without drinking alcohol in order to reduce his body fluid alcohol concentration to a legal driving level of under 0.8 grams/liter?

G.8.d) The chairman of the Champaign City Council Task Force on Alcohol has a good record for establishing programs to deal with excessive drinking by party-loving students at the University of Illinois at Urbana-Champaign.

Unfortunately on Saturday August 1, 1992, this poor fellow was hoisted by his own petard by being arrested for driving with a concentration of 1.6 grams of alcohol per liter of his body fluids. At the time of his arrest, he said he had drunk only a couple of glasses of wine and believed that his driving was not impaired.

> Assuming that he weighs about 160 pounds and that he is in pretty good shape and assuming that he drank nothing but burgundy, what is the minimum

number of 4-ounce glasses of burgundy he could have consumed prior to his arrest?

■ G.9) Interpolation and approximation★

Here is a plot of $f[x] = \sqrt{x}$ for $0 \leq x \leq 10$:

In[20]:=
```
Clear[f,x]
f[x_] = Sqrt[x];
fplot = Plot[f[x],{x,0,10},
PlotStyle->{{Blue,Thickness[0.01]}},
AxesLabel->{"x","f[x]"}];
```

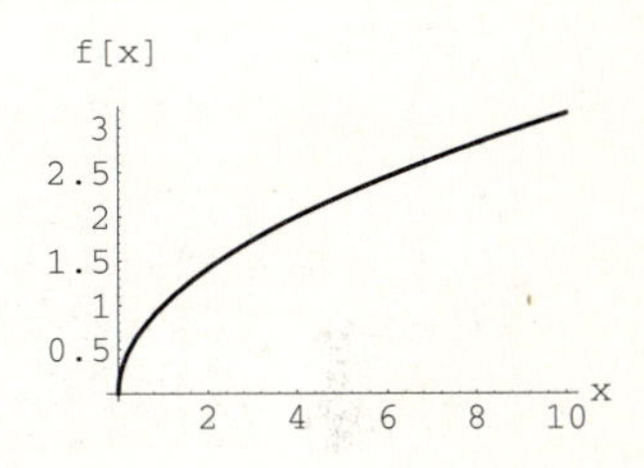

Here is the same plot shown with the points $\{0, f[0]\}$, $\{10, f[10]\}$ and the plot of the *Mathematica* interpolation function that runs through these two points:

In[21]:=
```
points = {{0,f[0]},{10,f[10]}};
pointplot = ListPlot[points,
PlotStyle->PointSize[0.03],
DisplayFunction->Identity]; Clear[runner]
runner[x_] = Interpolation[points][x];
runnerplot = Plot[runner[x],{x,0,10},
PlotStyle->{{Red,Thickness[0.01]}},
DisplayFunction->Identity];
Show[fplot,pointplot,runnerplot,
DisplayFunction->$DisplayFunction];
```

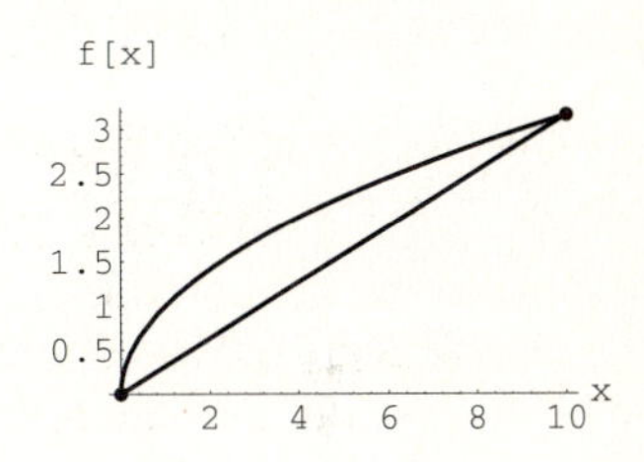

Here's what happens when you add the point $\{5, f[5]\}$:

In[22]:=
```
points = {{0,f[0]},{5,f[5]},{10,f[10]}};
pointplot = ListPlot[points,
PlotStyle->PointSize[0.03],
DisplayFunction->Identity]; Clear[runner]
runner[x_] = Interpolation[points][x];
runnerplot = Plot[runner[x],{x,0,10},
PlotStyle->{{Red,Thickness[0.01]}},
DisplayFunction->Identity];
Show[fplot,pointplot,runnerplot,
DisplayFunction->$DisplayFunction];
```

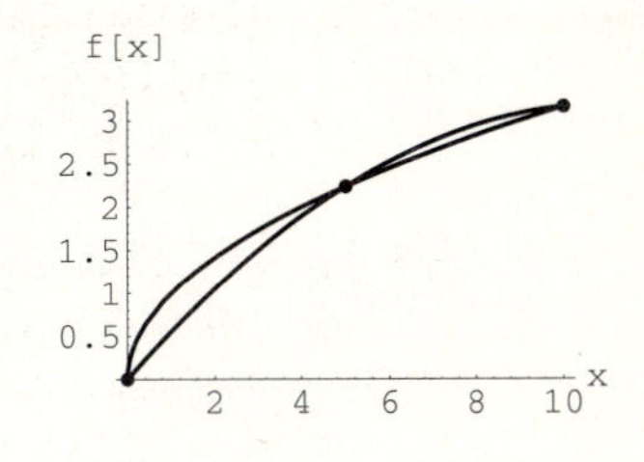

G.9.a) Your job is to come up with three new points $\{a, f[a]\}$, $\{b, f[b]\}$, and $\{c, f[c]\}$ so that when you add the three new points to the list above, the plot of runner$[x]$ will be as close as you can make it to the plot of $f[x]$.

LESSON 1.02

Natural Logs and Exponentials

Basics

■ B.1) The natural base e and the natural logarithm

After 0, 1, and π, the most important number in calculus is e. Some folks would argue this point, saying that e is the most important number in calculus. As you progress through the course, you can decide for yourself.

Here is e to 10 accurate digits:

In[1]:=

```
N[E,10]
```

Out[1]=

```
2.718281828
```

Lots of folks like to say that

$$e = 2.7\text{AndrewJacksonAndrewJackson}$$

because Andrew Jackson was elected President of the United States in the year 1828. *Mathematica* can slam out the decimals of e. Here is e to 50 accurate digits (49 accurate decimals):

In[2]:=

```
N[E,50]
```

Out[2]=

```
2.7182818284590452353602874713526624977572470937
```

One hundred accurate digits:

In[3]:=

```
N[E,100]
```

Out[3]=

```
2.71828182845904523536028747135266249775724709369995957496696762772407\
  6630353547594571382178525166427
```

For reasons that you will see after just a few more lessons, e is the natural base for exponentials and logarithms.

B.1.a) Plot e^x and then plot e^{-x} and describe what you see.

Answer: Here is a plot of e^x:

In[4]:=

```
Clear[x]
Plot[E^x,{x,0,7},
PlotStyle->{{Blue,Thickness[0.015]}},
AxesLabel->{"x","E^x"}];
```

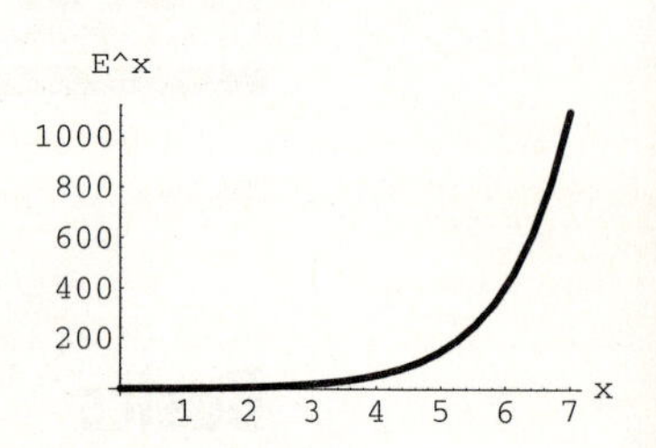

Pristine exponential growth.

Here is a plot of e^{-x}:

In[5]:=

```
Clear[x]
Plot[E^(-x),{x,0,7},
PlotStyle->{{Blue,Thickness[0.015]}},
AxesLabel->{"x","E^(-x)"}];
```

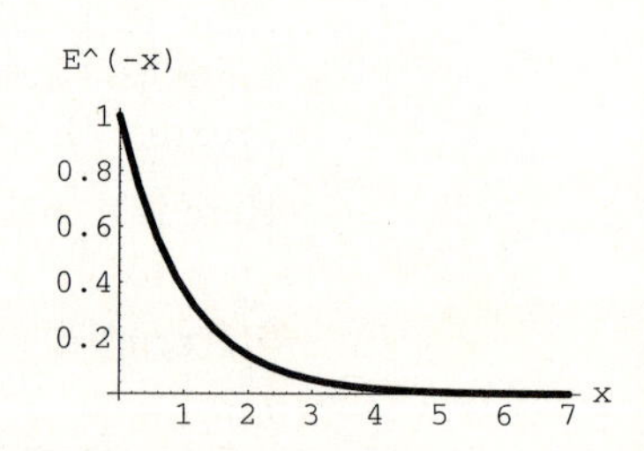

Pristine exponential decay.

B.1.b) The natural logarithm is the logarithm function whose base is e. Every scientific calculator has a button that activates this function; so you know it's a pretty big deal in science. Saying that $y = \log[x]$ is the same as saying $e^y = x$. Try it out:

In[6]:=

```
x = 12;
y = N[Log[x]];
```

In[7]:=

```
x == E^y
```

Out[7]=

```
True
```

Play with other x's and rerun.

B.1.b.i) Plot $\log[x]$ and describe what you see.

Answer:

In[8]:=
```
Clear[x]
Plot[Log[x],{x,0,8},
PlotStyle->{{Blue,Thickness[0.015]}},
AxesLabel->{"x","Log[x]"}];
```

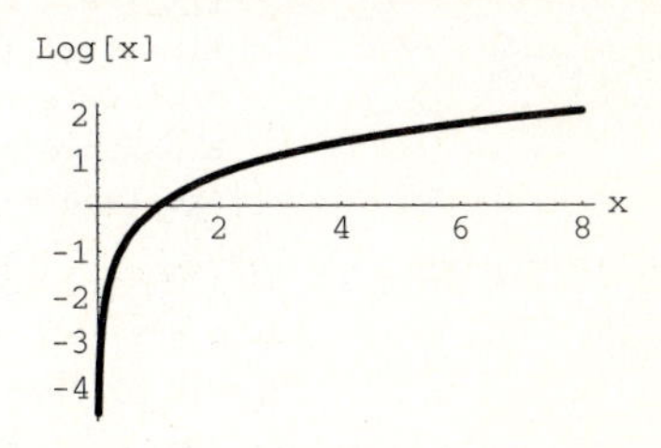

Logarithmic growth is fast at first and then very slow later, just the opposite of exponential growth.

B.1.b.ii) The main laws of exponents are

$$e^{a+b} = e^a e^b$$
$$e^{a-b} = \frac{e^a}{e^b}$$
$$\left(e^a\right)^b = e^{ab}.$$

You've seen these before and in previous math courses you saw how these laws of exponents are related to the main properties of $\log[x]$:

$$\log[a\,b] = \log[a] + \log[b]$$
$$\log\left[\frac{a}{b}\right] = \log[a] - \log[b]$$
$$\log[a^b] = b\log[a].$$

Check them out:

In[9]:=
```
a = 5.1; b = 3.9; {N[Log[a b]],N[Log[a] + Log[b]]}
```

Out[9]=
```
{2.99022, 2.99022}
```

In[10]:=
```
{N[Log[a/b]],N[Log[a] - Log[b]]}
```

Out[10]=
```
{0.268264, 0.268264}
```

In[11]:=
```
{N[Log[a^b]],N[b Log[a]]}
```

Out[11]=
```
{6.35404, 6.35404}
```

Change a and b and rerun. (Be careful not to try negative a's and b's. Reason: Within the realm of real numbers, $\log[x]$ makes no sense for $x \leq 0$.)

Here is a plot of the exponential function $f[x] = 2\,e^{0.4x}$:

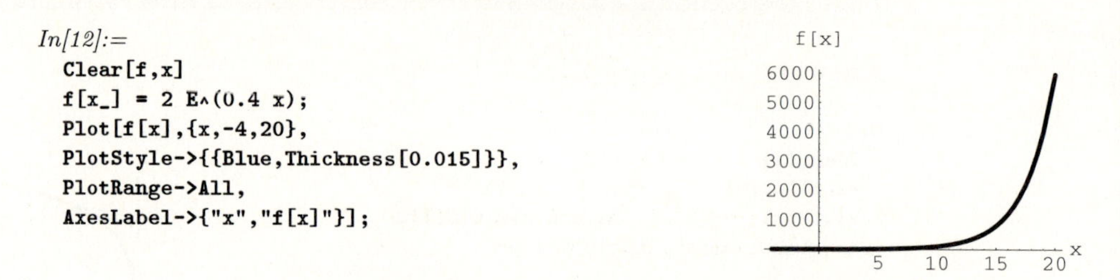

In[12]:=
```
Clear[f,x]
f[x_] = 2 E^(0.4 x);
Plot[f[x],{x,-4,20},
PlotStyle->{{Blue,Thickness[0.015]}},
PlotRange->All,
AxesLabel->{"x","f[x]"}];
```


Look at that rascal grow! And here is a plot of $\log[f[x]]$ over the same interval:

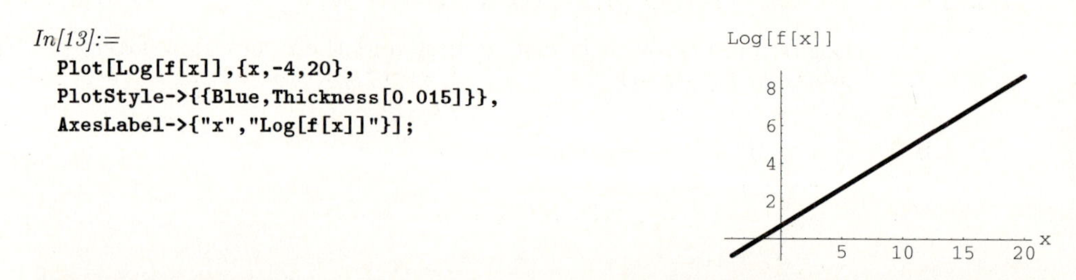

In[13]:=
```
Plot[Log[f[x]],{x,-4,20},
PlotStyle->{{Blue,Thickness[0.015]}},
AxesLabel->{"x","Log[f[x]]"}];
```


> Use the properties of logarithms to explain why this plot turned out to be a straight line.

Answer: Well, $f[x] = 2\,e^{0.4x}$; so

$$\begin{aligned}\log[f[x]] &= \log[2\,e^{0.4x}] \\ &= \log[2] + \log[e^{0.4x}] \\ &= \log[2] + 0.4\,x\log[e] \\ &= \log[2] + 0.4\,x\end{aligned}$$

because $\log[e] = 1$:

In[14]:=
```
Log[E] == 1
```

Out[14]=
```
True
```

This tells you and everyone else that $\log[f[x]]$ is the line function that passes through $\{0, \log[2]\}$ with constant growth rate 0.4.

B.1.b.iii)

> If you take any exponential function $f[x] = a\,e^{rx}$, where a and r are constants with $a > 0$, and then you plot $\log[f[x]]$, why are you sure to get the plot of a line function?

Answer: It's the same as above. $f[x] = a\,e^{r x}$ so

$$\begin{aligned}\log[f[x]] &= \log[a\,e^{r x}]\\ &= \log[a] + \log[e^{r x}]\\ &= \log[a] + r\,x\log[e]\\ &= \log[a] + r\,x.\end{aligned}$$

This tells you and everyone else that $\log[f[x]]$ is the line function that goes through $\{0, \log[a]\}$ with constant growth rate r.

■ B.2) Percentage growth, doubling time and half life

Line functions post a constant growth rate. Every time x goes up by 1 unit, $f[x] = a\,x + b$ goes up by a units.

In[15]:=

```
Clear[f,a,b,x]
f[x_] = a x + b; Expand[f[x + 1] - f[x]]
```

Out[15]=

```
a
```

Exponential functions do not do this:

In[16]:=

```
Clear[f,x]
f[x_] = 2 E^(0.6 x); Simplify[f[x + 1] - f[x]]
```

Out[16]=

```
          0.6 x
1.64424 E
```

The bigger x is, the faster $f[x] = 2\,e^{0.6x}$ grows. On the other hand, exponential functions do post a constant percentage growth rate as x goes up by 1:

In[17]:=

```
Simplify[100 ((f[x + 1]/f[x]) - 1)]
```

Out[17]=

```
82.2119
```

This tells you that $f[x] = 2\,e^{0.6x}$ goes up by about 82.2% every time x goes up by 1. In fact if you take any exponential function $f[x] = a\,e^{r x}$ and look at its percentage growth rate as x goes up by 1, then you get:

In[18]:=

```
Clear[f,a,r,x]
f[x_] = a E^(r x);
Simplify[100 (f[x + 1]/ f[x] - 1)]
```

Out[18]=

```
             r
100 (-1 + E )
```

This tells you that every time x goes up by 1, $f[x] = a\,e^{rx}$ goes up by $100\,(e^r - 1)\,\%$. Now look at this:

In[19]:=
```
Clear[f,a,r,x,h]; f[x_] = a E^(r x);
Simplify[100 (f[x + h]/ f[x] - 1)]
```

Out[19]=
```
            h r
100 (-1 + E    )
```

What does this tell you?

Answer: This tells you that every time x goes up by h units, $f[x] = a\,e^{rx}$ goes up by $100\,(e^{hr} - 1)\,\%$.

B.2.b.i) How do you make an exponential function going through $\{0, 5\}$ that grows 30% every time x goes up by one unit?

Answer: Start with:

In[20]:=
```
Clear[f,r,x]; f[x_] = 5 E^(r x)
```

Out[20]=
```
   r x
5 E
```

And look at:

In[21]:=
```
f[0]
```

Out[21]=
```
5
```

No sweat; $f[x]$ goes right through $\{0, 5\}$. Now set r to achieve the growth of 30% every time x goes up by one unit:

In[22]:=
```
equation = Simplify[100 (f[x + 1]/ f[x] - 1)] == 30
```

Out[22]=
```
            r
100 (-1 + E ) == 30
```

This is the same as

$$-1 + e^r = 0.3;$$
$$e^r = 1.3;$$
$$r = \log[1.3].$$

Try it out:

In[23]:=
```
r = Log[1.3]; f[x]
```

Out[23]=
```
     0.262364 x
5 E
```

In[24]:=
```
Simplify[100 (f[x + 1]/ f[x] - 1)]
```

Out[24]=
```
30.
```

Yep; this $f[x]$ goes up 30% every time x goes up by 1.

B.2.b.ii) How do you make an exponential function going through $\{0,5\}$ that grows 10% every time x goes up by 1/3 unit?

Answer: Again start with:

In[25]:=
```
Clear[f,a,r,x]; f[x_] = 5 E^(r x)
```

Out[25]=
```
   r x
5 E
```

And look at:

In[26]:=
```
f[0]
```

Out[26]=
```
5
```

No sweat; $f[x]$ goes right through $\{0,5\}$. Now set r to achieve the growth of 10% every time x goes up by 1/3 unit:

In[27]:=
```
equation = Simplify[100 (f[x + 1/3]/ f[x] - 1)] == 10
```

Out[27]=
```
               r/3
100 (-1 + E      ) == 10
```

This is the same as

$$-1 + e^{r/3} = 0.1;$$
$$e^{r/3} = 1.1;$$
$$\frac{r}{3} = \log[1.1];$$
$$r = 3\log[1.1].$$

Try it out:

In[28]:=
```
r = 3 Log[1.1]; f[x]
```

Out[28]=

```
  0.285931 x
5 E
```

In[29]:=

```
Simplify[100 (f[x + 1/3]/ f[x] - 1)]
```

Out[29]=

```
10.
```

Yep; this $f[x]$ goes up 10% every time x goes up by 1/3.

B.2.c.i) A certain measurement increases exponentially with time. This means that the measurement is made at time t by a function $f[t] = a\,e^{rt}$ where a and r are positive. Here's an example:

In[30]:=

```
Clear[f,a,r,t]
f[t_] = a E^(r t);
a = 2.3;
r = 0.7;;
Plot[f[t],{t,0,10},
PlotStyle->{{Red,Thickness[0.01]}},
AxesLabel->{"t", "f[t]"}];
```


> Given an exponential function $f[t] = a\,e^{rt}$ where a is positive and r is positive, what is the value of h that makes $f[t + h] = 2\,f[t]$? Most folks call this h by the name "doubling time."

Answer: Your job is to find a time measurement h so that $f[t+h] = 2\,f[t]$. This is the same as saying that $f[t]$ goes up by 100% every time t goes up by h:

In[31]:=

```
Clear[f,a,r,h,t]
f[t_] = a E^(r t);
equation = Simplify[100 (f[t + h]/ f[t] - 1)] == 100
```

Out[31]=

```
            h r
100 (-1 + E    ) == 100
```

This is the same as $-1 + e^{hr} = 1$ or

$$e^{hr} = 2,$$

$$h\,r = \log[2],$$

$$h = \frac{\log[2]}{r}.$$

See it happen for $f[t] = 0.125\,e^{0.91t}$:

In[32]:=

```
Clear[f,a,r,h,t]; f[t_] = a E^(r t); a = 0.125; r = 0.91;
h = Log[2]/r; Table[f[t],{t,0,10 h,h}]
```

Out[32]=

```
{0.125, 0.25, 0.5, 1., 2., 4., 8., 16., 32., 64., 128.}
```

Or via a plot:

In[33]:=

```
fplot = Plot[f[t],{t,0,10 h},
PlotStyle->Thickness[0.01],DisplayFunction->Identity];
doubles = Table[Graphics[{Red,Line[{{t,0},
{t,f[t]},{0,f[t]}}]}],{t,0,10 h,h}];
Show[fplot,doubles,PlotRange->All,
AxesLabel->{"t","f[t]"},
DisplayFunction->$DisplayFunction];
```

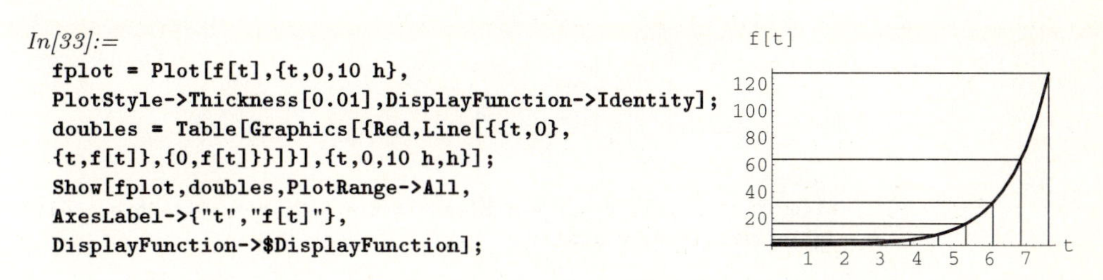

Perfecto. Every time t goes up by $h = \log[2]/r$, the function $f[t] = a\,e^{rt}$ doubles. This is why most folks say that the doubling time of $f[t] = a\,e^{rt}$ (with $a > 0$ and $r > 0$) is $\log[2]/r$ time units.

B.2.c.ii) A certain measurement decays exponentially with time. This means that the measurement is made at time t by a function $f[t] = a\,e^{rt}$ where a is positive and r is negative.

Here's an example:

In[34]:=

```
Clear[f,a,r,t]
f[t_] = a E^(r t); a = 15.3; r = -1.4;;
Plot[f[t],{t,0,10},
PlotStyle->{{Red,Thickness[0.01]}},
AxesLabel->{"t", "f[t]"},PlotRange->All];
```

f[t]
15
12.5
10
7.5
5
2.5
t
2 4 6 8 10

> Given an exponential function $f[t] = a\,e^{rt}$ where a is positive and r is negative, what is the value of h that makes $f[t+h] = f[t]/2$? Most folks call this h by the name "half life."

Answer: Your job is to find a time measurement h so that $f[t+h] = f[t]/2$. This is the same as saying that $f[t]$ goes down by 50% every time t goes up by h:

In[35]:=

```
Clear[f,a,r,h,t]; f[t_] = a E^(r t);
equation = Simplify[100 (f[t + h]/ f[t] - 1)] == -50
```

Out[35]=

```
               h r
100 (-1 + E       ) == -50
```

This is the same as $-1 + e^{hr} = -1/2$ or

$$e^{hr} = \frac{1}{2}$$
$$h\,r = \log\left[\frac{1}{2}\right] = -\log[2]$$
$$h = -\frac{\log[2]}{r}.$$

Check it out for $f[t] = 512\,e^{-0.1t}$:

In[36]:=
```
Clear[f,a,r,h,t]; f[t_] = a E^(r t); a = 512; r = -0.1; h = -Log[2]/r;
Table[f[t],{t,0,10 h,h}]
```

Out[36]=
```
{512, 256., 128., 64., 32., 16., 8., 4., 2., 1., 0.5}
```

Or via a plot:

In[37]:=
```
fplot = Plot[f[t],{t,0,10 h},
PlotStyle->Thickness[0.01],
DisplayFunction->Identity];
halves = Table[
Graphics[{Red,Line[{{t,0},{t,f[t]},{0,f[t]}}]}],
{t,0,10 h,h}]; Show[fplot,halves,PlotRange->All,
AxesLabel->{"t","f[t]"},
DisplayFunction->$DisplayFunction];
```

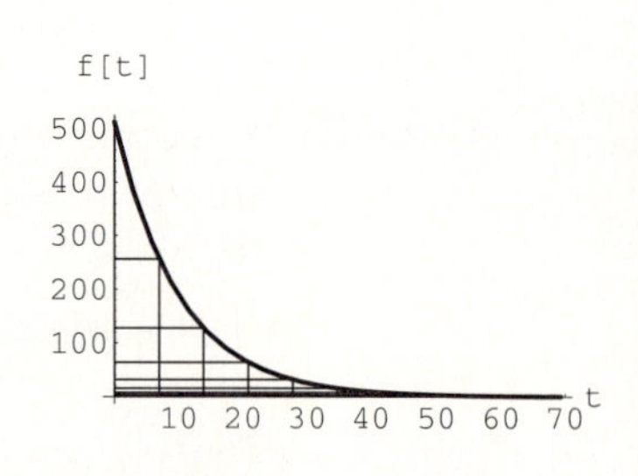

Coolness.

Every time t goes up by $h = -\log[2]/r$, then the function $f[t] = a\,e^{rt}$ halves itself. This is why most folks say that the half life of $f[t] = a\,e^{rt}$ (with $a > 0$ and $r < 0$) is $-\log[2]/r$ time units.

■ B.3) Unnatural bases

B.3.a.i) Look at this:

In[38]:=
```
Clear[a]; E^Log[a] == a
```

Out[38]=
```
True
```

> Apparently *Mathematica* is programmed to think $e^{\log[a]} = a$. Explain why this is OK.

Answer: Saying that $b = \log[a]$ is the same as saying that $e^b = a$.

B.3.a.ii) Look at plots of $f[x] = 3(4^{0.7x})$ and $g[x] = 3\,e^{(0.7\log[4])x}$ on the same axes:

In[39]:=
```
Clear[f,g,x]
f[x_] = 3 4^(0.7 x);
g[x_] = 3 E^((0.7 Log[4]) x);
Plot[{f[x],g[x]},{x,0,5},AxesLabel->{"x",""}];
```

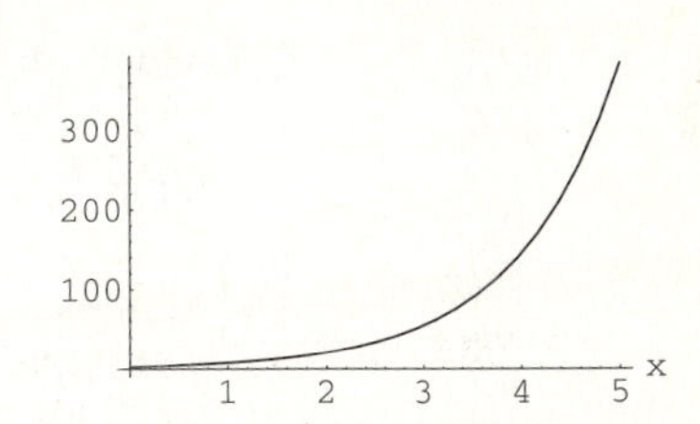

Two functions were plotted but only one graph came out. Is this an accident?

Answer: No way! In mathematics, there are no accidents.

Look at the formulas for the functions $f[x] = 3(4^{0.7x})$ and

$$g[x] = 3\,e^{(0.7\log[4])x} = 3\left(e^{(\log[4])}\right)^{0.7x}.$$

These formulas are disguised versions of the same thing because $e^{\log[4]} = 4$:

In[40]:=
```
E^Log[4]
```

Out[40]=
```
4
```

B.3.a.iii) What's the moral?

Answer: If you start with an unnatural exponential function $f[x] = a\,b^{cx}$ whose base is b and not e, then you can rewrite the formula as $f[x] = a\,e^{\log[b]cx}$ so that the base is the natural base e. In other words, any exponential function involving an unnatural base can be written as an exponential involving the natural base e. This is the way nearly all scientists do it.

B.3.b) Given any positive base b, saying that $y = \log[b, x]$ is the same as saying $b^y = x$. Look at the output:

In[41]:=
```
Clear[b,x]
Log[b,x]
```

Out[41]=
```
Log[x]
------
Log[b]
```

Apparently *Mathematica* is programmed to think that $\log[b, x] = \log[x]/\log[b]$. Explain why this is OK.

Answer: Saying that $y = \log[b, x]$ is the same as saying $b^y = x$. This is the same as saying $e^{\log[b] y} = x$. This is the same as saying $\log[x] = \log[b]\, y$. This is the same as saying $y = \log[x]/\log[b]$. (Whew!)

The upshot: Saying that

$$y = \log[b, x]$$

is the same as saying

$$y = \frac{\log[x]}{\log[b]}.$$

B.3.b.ii) What's the moral?

Answer: All log functions are multiples of $\log[x]$, the logarithm with base e.

If you want the base 10 logarithm, $\log[10, x]$, you can get it directly or you can get it from the natural $\log[x]$ function through the simple formula $\log[x]/\log[10]$.

Try it out:

In[42]:=

```
ColumnForm[Table[N[{x,Log[10,x],Log[x]/Log[10]}],{x,1,6,0.5}]]
```

Out[42]=

```
{1., 0, 0}
{1.5, 0.176091, 0.176091}
{2., 0.30103, 0.30103}
{2.5, 0.39794, 0.39794}
{3., 0.477121, 0.477121}
{3.5, 0.544068, 0.544068}
{4., 0.60206, 0.60206}
{4.5, 0.653213, 0.653213}
{5., 0.69897, 0.69897}
{5.5, 0.740363, 0.740363}
{6., 0.778151, 0.778151}
```

The first slot is x, the second is $\log[10, x]$, and the third is $\log[x]/\log[10]$ with x running from 1 to 6 in increments of 0.5.

Tutorials

■ T.1) Exponential models

T.1.a) Living tissue contains two kinds of carbon. One kind, carbon-14 $\left({}^{14}_{6}C\right)$, is radioactive with a half life of about 5750 years; the other is not radioactive. In living tissue the ratio of the two is always constant, but when the tissue dies, the radioactive carbon begins to decay exponentially while the other carbon remains.

To date a fossil, scientists measure the amount of each kind of carbon present to determine how much of the radioactive carbon has decayed.

If measurements show that the radioactive carbon in a fossil has decayed by 73%, then how old is the fossil?

Answer: When you mention half life, you are talking about exponential decay. Measuring t (= time) in years with $t = 0$ at the time of death of the animal that became the fossil, you know that the percentage of radioactive carbon left in the fossil t years after the death of the animal is given by

$$\text{radiocarbon}[t] = a\, e^{rt}$$

where r is negative. You also know that

$$100 = \text{radiocarbon}[0] = a\, e^{r0} = a.$$

In addition you can determine r through the half life formula discussed in one of the Basics.

In[1]:=

```
halflife = 5750; Clear[r]; Solve[halflife == -Log[2]/r,r]
```

Out[1]=

```
         -Log[2]
{{r -> ---------}}
          5750
```

Here comes the function.

In[2]:=

```
Clear[radiocarbon,t]; a = 100; r = N[-Log[2]/5750];
radiocarbon[t_] = a E^(r t)
```

Out[2]=

```
       100
-----------------
 0.000120547 t
E
```

The radioactive carbon in a fossil has decayed by 73%; so its age is determined by:

In[3]:=

```
N[Solve[radiocarbon[t] == ( 100 - 73),t]]
```

Out[3]=

```
{{t -> 10861.6}}
```

The fossil is about 11,000 years old.

T.1.b.i) This problem appears only in the electronic version.

■ T.2) Exponential data

T.2.a) Here are the two points {1.2, 2.4} and {3.1, 5.7} and the line function going through them.

In[4]:=

```
data = {{1.2,2.4},{3.1,5.7}};
dataplot =
ListPlot[data,PlotStyle->{PointSize[0.03],Blue},
DisplayFunction->Identity]; Clear[line]
line[x_] = Fit[data,{1,x},x];
lineplot = Plot[line[x],{x,-1,6},
DisplayFunction->Identity];
Show[dataplot,lineplot,
DisplayFunction->$DisplayFunction];
```

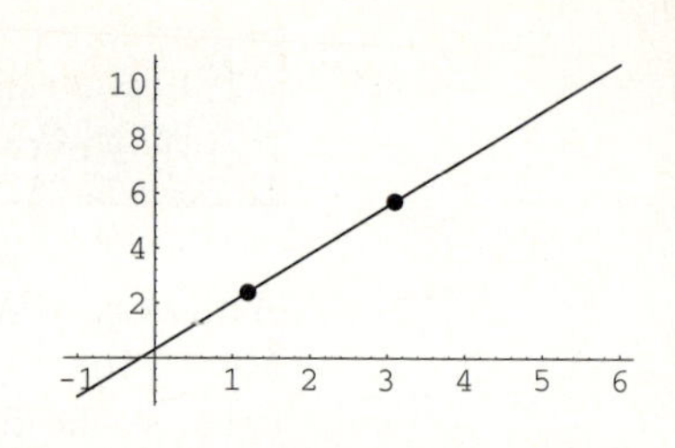

> Add to this plot the plot of the exponential function $f[x] = a\,e^{rx}$ that goes through the same two points.

Answer: Call $f[x]$ the exponential function that goes through these two points:

In[5]:=

```
data = {{1.2,2.4},{3.1,5.7}};
```

$\log[f[x]]$ is a line function that goes through:

In[6]:=

```
logdata = {{1.2,Log[2.4]},{3.1,Log[5.7]}}
```

Out[6]=

```
{{1.2, 0.875469}, {3.1, 1.74047}}
```

So $\log[f[x]]$ is given by:

In[7]:=

```
Clear[logf]; logf[x_] = Fit[logdata,{1,x},x]
```

Out[7]=

```
0.329155 + 0.455262 x
```

But $f[x] = e^{\log[f[x]]}$, so $f[x]$ is given by:

In[8]:=

```
Clear[f]; f[x_] = E^logf[x]
```

Out[8]=

```
 0.329155 + 0.455262 x
E
```

This is the exponential function $f[x] = a\,e^{rx}$ with $a = e^{0.329155}$ and $r = 0.455262$. Here comes the plot:

In[9]:=

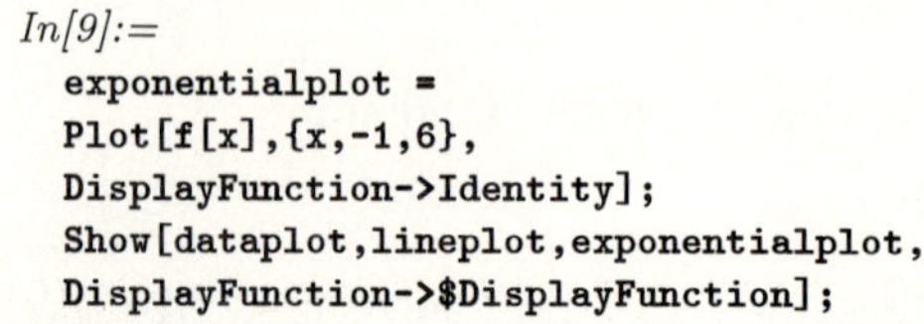

```
exponentialplot =
Plot[f[x],{x,-1,6},
DisplayFunction->Identity];
Show[dataplot,lineplot,exponentialplot,
DisplayFunction->$DisplayFunction];
```

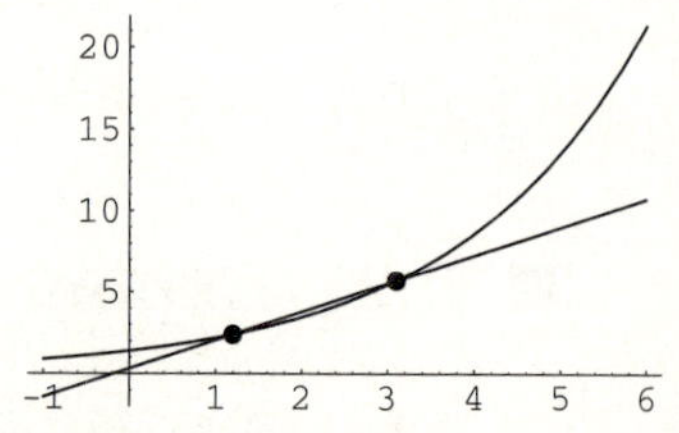

Lookin' fine.

T.2.b.i) Information between the variable x and the function y may consist of nothing more than some measurements indicating the observed value of y for various values of x. A quick plot of the points usually reveals to the alert eye whether the functional relationship of y in terms of x is of the form $y = b + a\,x$.

If the plot of the points looks like this:

In[10]:=

```
data = {{-0.8, -0.84},{-0.4, 0.29},{-0.2, 0.73},
{0.2, 1.45},{0.4, 1.75},{0.8, 2.84},{1.4, 4.01},
{1.8, 4.78},{2.6, 6.85},
{3.2, 8.23},{3.6, 9.39}};
dataplot = ListPlot[data,AxesLabel->{"x","y"},
PlotStyle->{Red,PointSize[0.015]}];
```

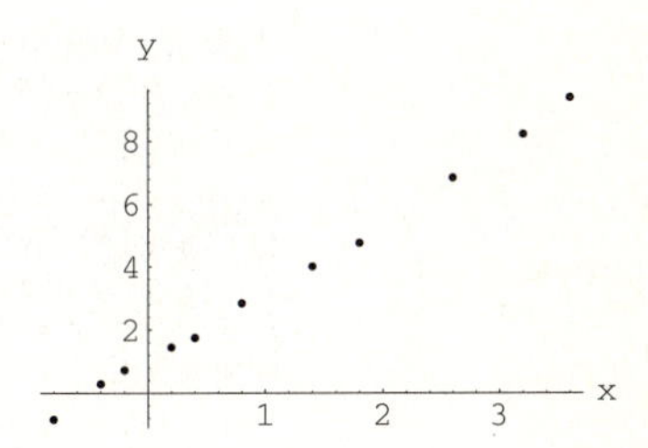

Then you are happy fitting these data with a line:

In[11]:=

```
Clear[fitter,x]; fitter[x_] = Fit[data,{1,x},x]
```

Out[11]=

```
1.00154 + 2.25897 x
```

In[12]:=

```
fitplot =
Plot[fitter[x],{x,-1,4},
PlotStyle->{{Blue,Thickness[0.01]}},
DisplayFunction->Identity];

Show[fitplot,dataplot,AxesLabel->{"x","y"},
DisplayFunction->$DisplayFunction];
```

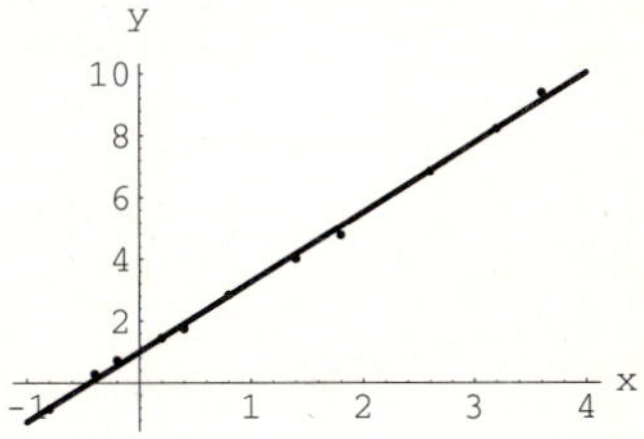

Not bad. Sometimes the points do not line up in a straight line:

In[13]:=

```
data = {{-1.00, 2.41}, {-0.45, 3.38}, {0.11, 3.80}, {0.52, 4.53},
{1.45, 6.74}, {2.02, 8.31}, {2.54, 10.23}, {3.14, 14.01}, {3.49, 15.00},
{3.97, 18.32}, {4.54, 22.38}, {5.02, 27.47}};
```

In[14]:=

```
dataplot =
ListPlot[data,AxesLabel->{"x","y"},
PlotStyle->{Red,PointSize[0.015]}];
```

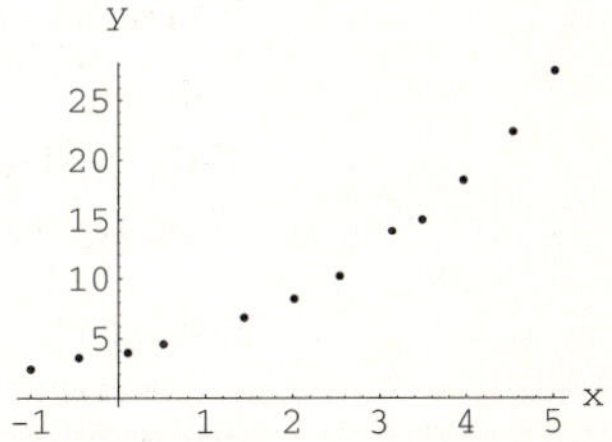

This looks suspiciously exponential.

> How do you better eyeball an exponential relationship? How do you come up with a compromise exponential function $f[x] = a\,e^{rx}$ whose plot runs through or near these data points?

Answer: Here are the data points in the form $\{x, y\}$:

In[15]:=
```
data = {{-1.00, 2.41}, {-0.45, 3.38}, {0.11, 3.80}, {0.52, 4.53},
{1.45, 6.74}, {2.02, 8.31}, {2.54, 10.23}, {3.14, 14.01},{3.49, 15.00},
{3.97, 18.32}, {4.54, 22.38}, {5.02, 27.47}};
```

The following instruction keeps the first slot fixed but replaces the corresponding second slots by its base e logarithm:

In[16]:=
```
Clear[j]
logdata = Table[{data[[j,1]],Log[data[[j,2]]]},{j,1,Length[data]}]
```

Out[16]=
```
{{-1., 0.879627}, {-0.45, 1.21788}, {0.11, 1.335}, {0.52, 1.51072},
 {1.45, 1.90806}, {2.02, 2.11746}, {2.54, 2.32532}, {3.14, 2.63977},
 {3.49, 2.70805}, {3.97, 2.90799}, {4.54, 3.10817}, {5.02, 3.31309}}
```

Now plot the logdata points:

In[17]:=
```
logdataPlot =
ListPlot[logdata,
PlotStyle->{Red,PointSize[.015]}];
```

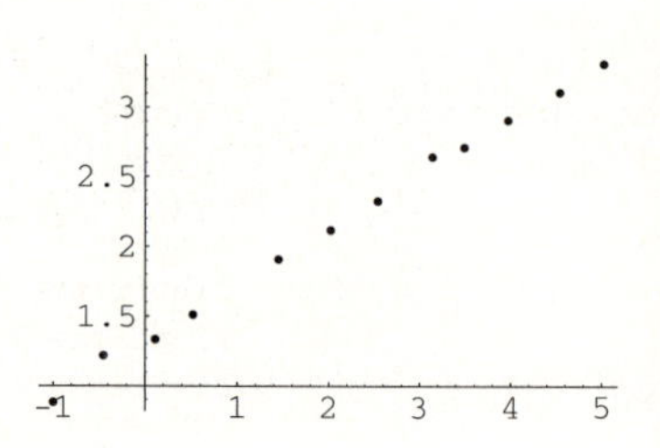

These points nearly line up in a perfect straight line. This is a dead give-away that the data display a strong exponential relationship.

Let's agree that $f[x] = a\, e^{rx}$ is a compromise exponential function whose plot does a good job of going with the flow of these data points. Then $\log[f[x]]$ is a compromise line function that flows with:

In[18]:=
```
logdata
```

Out[18]=
```
{{-1., 0.879627}, {-0.45, 1.21788}, {0.11, 1.335}, {0.52, 1.51072},
 {1.45, 1.90806}, {2.02, 2.11746}, {2.54, 2.32532}, {3.14, 2.63977},
 {3.49, 2.70805}, {3.97, 2.90799}, {4.54, 3.10817}, {5.02, 3.31309}}
```

So $\log[f[x]]$ is given by:

In[19]:=
```
Clear[logf]; logf[x_] = Fit[logdata,{1,x},x]
```

Out[19]=
```
1.32182 + 0.398787 x
```

But $f[x] = e^{\log[f[x]]}$, so $f[x]$ is given by:

In[20]:=

```
Clear[f]; f[x_] = E^logf[x]
```

Out[20]=

```
 1.32182 + 0.398787 x
E
```

This is the exponential function $f[x] = a\,e^{rx}$ with $a = e^{1.32182}$ and $r = 0.398787$. Here comes the plot:

In[21]:=

```
exponentialplot = Plot[f[x],{x,-2,6},
DisplayFunction->Identity];
Show[dataplot,exponentialplot,
DisplayFunction->$DisplayFunction];
```

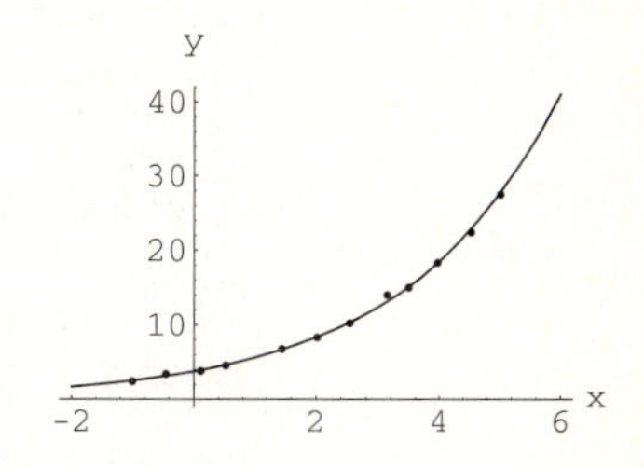

No problem at all.

T.2.b.ii) If there were an entry $\{2.35, y\}$ in the exponential data table in part T.2.b.i), what would you bet that y would be?

Answer: Go with the exponential fit $f[x]$ and plug in $x = 2.35$:

In[22]:=

```
f[2.35]
```

Out[22]=

```
9.57326
```

T.2.b.iii) If there were an entry $\{4.85, y\}$ in the exponential data table in part T.2.b.i), what would you bet that y would be?

Answer: Go with the exponential fit $f[x]$ and plug in $x = 4.85$:

In[23]:=

```
f[4.85]
```

Out[23]=

```
25.9441
```

■ T.3) The number *e* and finance

T.3.a) You put \$3500 in a certificate of deposit that pays 5.5% simple interest in a year. Use an exponential function to calculate the value of the account at the end of each of the first five years. Make some nice graphics.

Answer: Measure t in years with $t = 0$ corresponding to the time of the deposit of \$3500. Put balance[t] $= a\, e^{rt}$:

In[24]:=
```
Clear[balance,t,a,r]; balance[t_] = a E^(r t)
```

Out[24]=
```
   r t
a E
```

You know $3500 = \text{balance}[0] = a\, e^{r0} = a$. So:

In[25]:=
```
a = 3500; balance[t]
```

Out[25]=
```
      r t
3500 E
```

Now set r to give you the advertised 5.5% interest per year by solving

$$100\left(\frac{\text{balance}[t+1]}{\text{balance}[t]} - 1\right) = 5.5$$

for r:

In[26]:=
```
Simplify[100(balance[t + 1]/balance[t] - 1)] == 5.5
```

Out[26]=
```
           r
100 (-1 + E ) == 5.5
```

This tells you $e^r - 1 = 0.055$; $e^r = 1.055$:

In[27]:=
```
r = Log[1.055];
balance[t]
```

Out[27]=
```
      0.0535408 t
3500 E
```

Check it out:

In[28]:=
```
Table[100(balance[t + 1]/balance[t] - 1),{t,0,10}]
```

Out[28]=
```
{5.5, 5.5, 5.5, 5.5, 5.5, 5.5, 5.5, 5.5, 5.5, 5.5, 5.5}
```

Yep, the balance goes up by 5.5% each year. Here's a table indicating the balances in your account at the end of each of the first five years. (The reason that the := sign is used below is to tell the machine not to calculate the function until after data are fed in.)

In[29]:=

```
Clear[balancebar,balancebaredge]
balancebar[t_] := Graphics[{DarkGreen,
Polygon[{{t,0},{t + 1,0},{t + 1,balance[t]},
{t,balance[t]}}]}]
edges[t_] :=
Graphics[Line[{{t,0},{t + 1,0},{t + 1,balance[t]},
{t,balance[t]},{t,0}}]];
Show[Table[{balancebar[t],edges[t]},{t,0,4}],
Axes->True,AxesLabel->{"years","balance"}];
```

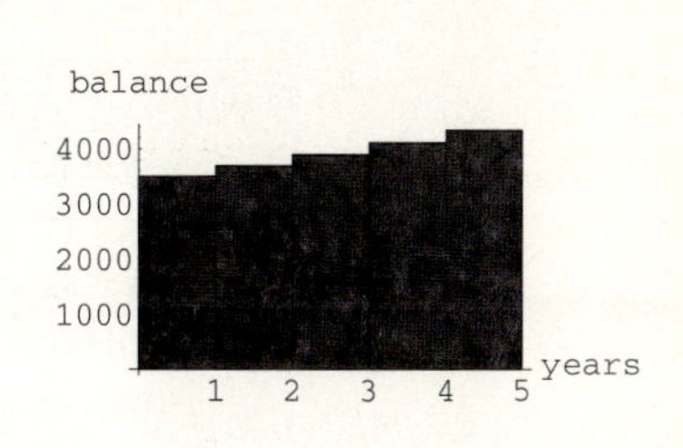

Here's what happens over the first 20 years:

In[30]:=

```
Show[Table[{balancebar[t],edges[t]},{t,0,19}],
Axes->True,AxesLabel->{"years","balance"}];
```

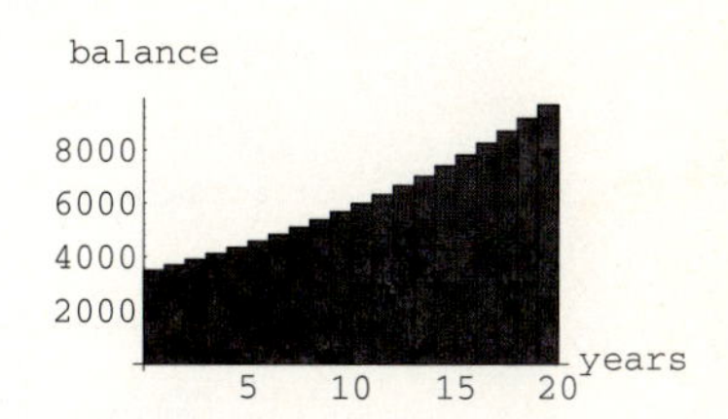

Or in cold numbers:

In[31]:=

```
Table[balance[t],{t,0,19}]
```

Out[31]=

```
{3500, 3692.5, 3895.59, 4109.84, 4335.89, 4574.36, 4825.95, 5091.38,
  5371.4, 5666.83, 5978.51, 6307.32, 6654.23, 7020.21, 7406.32,
  7813.67, 8243.42, 8696.81, 9175.13, 9679.76}
```

Over time, even a small interest rate like 5.5% can amount to something. In 50 years, that $3500 will grow to:

In[32]:=

```
balance[50]
```

Out[32]=

```
50896.9
```

Rerun with different interest rates to get a feeling for what's happening.

T.3.b) You charge $3500 on your MasterCard and make no payments for a year. The advertised interest rate is 15.3% per year. So at the end of the year you figure that you owe:

In[33]:=

```
expected = 3500 + 0.153 3500
```

Out[33]=

```
4035.5
```

But at the end of the year, when you look at the bill from the MasterCard people, you are stunned because they are billing you for more than $4035.50.

What gives?

Answer: If you read the fine print, you learn that MasterCard does not calculate simple interest as done in part T.3.a) above. They take the advertised interest rate:

In[34]:=
```
advertisedyearly = 15.3
```
Out[34]=
```
15.3
```

Then they break up the year into 12 months and then charge you (15.5/12) % per month.

In[35]:=
```
actualmonthly = advertisedyearly/12
```
Out[35]=
```
1.275
```

The financial folks call this compound interest. Here's how you calculate compound interest: Measure t in years with $t = 0$ corresponding to the time of the charge of \$3500. Put

$$\text{balance}[t] = a\, e^{rt} :$$

In[36]:=
```
Clear[balance,t,a,r]; balance[t_] = a E^(r t)
```
Out[36]=
```
   r t
a E
```

You know $3500 = \text{balance}[0] = a\, e^{r0} = a$. So:

In[37]:=
```
a = 3500; balance[t]
```
Out[37]=
```
      r t
3500 E
```

Now look at the actual monthly interest percentage:

In[38]:=
```
actualmonthly
```
Out[38]=
```
1.275
```

Set r to correspond to the actual monthly interest by solving

$$100\left(\frac{\text{balance}[t+1/12]}{\text{balance}[t]} - 1\right) = 1.275$$

for r:

In[39]:=
```
Simplify[100(balance[t + 1/12]/balance[t] - 1)] == actualmonthly
```
Out[39]=
```
             r/12
100 (-1 + E     ) == 1.275
```
This tells you that

$$e^{r/12} - 1 = 0.01275;$$

$$e^{r/12} = 1.01275;$$

$$\frac{r}{12} = \log[1.01275] :$$

In[40]:=
```
r = (12) Log[1.01275]
```
Out[40]=
```
0.152033
```
The exponential function that measures what you owe at the end of each of the 12 months of the year is:

In[41]:=
```
balance[t]
```
Out[41]=
```
      0.152033 t
3500 E
```
Because t is measured in years, here is what to expect to pay at the end of the year:

In[42]:=
```
balance[1]
```
Out[42]=
```
4074.69
```
A bit more than what you expected:

In[43]:=
```
expected
```
Out[43]=
```
4035.5
```
Compound interest works in favor of the MasterCard people. Here is a chart displaying how your balance grows month-by-month:

In[44]:=
```
jump = 1/12;
Clear[balancebar,edges]
balancebar[t_] := Graphics[{DarkGreen,
Polygon[{{t,0},{t + jump,0},
{t + jump,balance[t]},{t,balance[t]}}]}];
edges[t_] := Graphics[Line[{{t,0},{t + jump,0},
{t + jump,balance[t]},{t,balance[t]},{t,0}}]];
Show[Table[{balancebar[t],edges[t]}, {t,0,1-jump,jump}],
Axes->True,AxesLabel->{"years","balance"}];
```

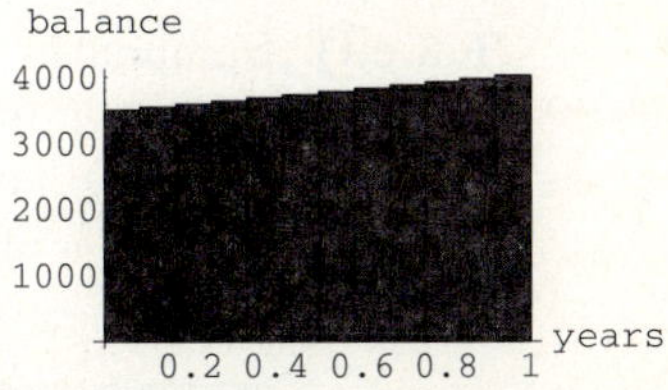

If you don't pay the bill for two years, here's what happens month-by-month:

In[45]:=
```
Show[Table[{balancebar[t],edges[t]},
{t,0,2-jump,jump}],
Axes->True,
AxesLabel->{"years","balance"}];
```

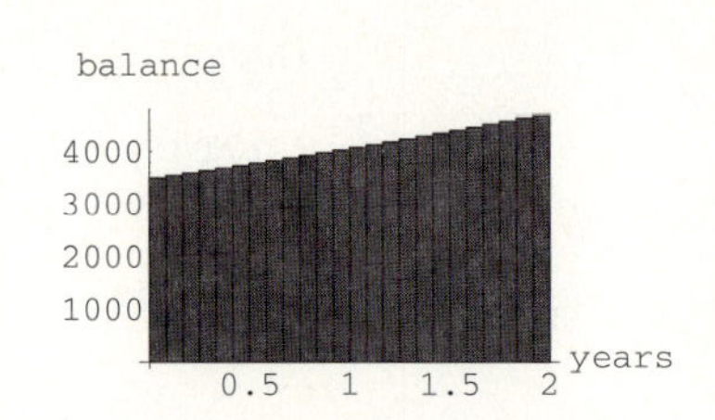

Four years:

In[46]:=
```
Show[Table[{balancebar[t],edges[t]},
{t,0,4-jump,jump}],
Axes->True,
AxesLabel->{"years","balance"}];
```

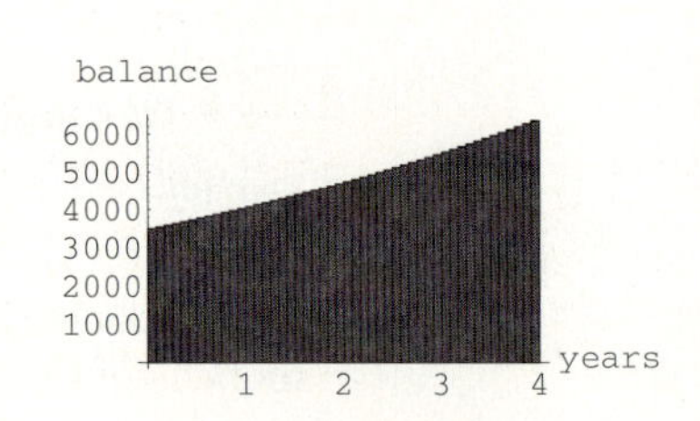

Ten years:

In[47]:=
```
Show[Table[{balancebar[t],edges[t]},
{t,0,10-jump,jump}],
Axes->True,
AxesLabel->{"years","balance"}];
```

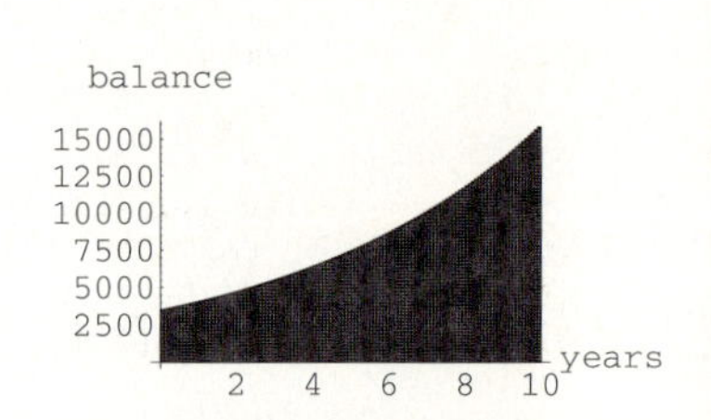

That's big-time exponential growth kicking in. No wonder financial advisors tell everyone to try to pay off their credit cards.

After 20 years, the balance will be:

In[48]:=
```
balance[20]
```

Out[48]=
```
73216.4
```

Gimme a break.

T.3.c.i) Six months ago you got married and are happy to learn that a baby is on the way. After the initial euphoria, you come back to earth when you realize that you are going to have to save for college for the little kid. You find a bank that will pay you 6.1% interest compounded daily (365 times per year) and you've got $5000 to invest.

Describe how the $5000 will grow.

Answer:

In[49]:=

```
advertisedyearly = 6.1
```

Out[49]=

```
6.1
```

Break up the year into 365 days:

In[50]:=

```
actualdaily = advertisedyearly/365
```

Out[50]=

```
0.0167123
```

Measure t in years with $t = 0$ corresponding to the time of the deposit of $5000. Put balance$[t] = a\,e^{rt}$:

In[51]:=

```
Clear[balance,t,a,r]; balance[t_] = a E^(r t)
```

Out[51]=

```
   r t
a E
```

You know $5000 = \text{balance}[0] = a\,e^{r0} = a$. So:

In[52]:=

```
a = 5000;
balance[t]
```

Out[52]=

```
      r t
5000 E
```

Now look at the actual daily interest percentage:

In[53]:=

```
actualdaily
```

Out[53]=

```
0.0167123
```

Set r to correspond to the actual daily interest by solving:

In[54]:=

```
Simplify[100 (balance[t + 1/365]/balance[t] - 1)] == actualdaily
```

Out[54]=

```
             r/365
100 (-1 + E     ) == 0.0167123
```

This tells you that

$$e^{r/365} - 1 = 0.000167123;$$

$$e^{r/365} = 1.000167123;$$

$$\frac{r}{365} = \log[1.000167123]:$$

In[55]:=
```
r = (365) Log[1.000167123]
```
Out[55]=
```
0.0609948
```

The exponential function that measures what you have at the end of each of the 12 months of the year is:

In[56]:=
```
balance[t]
```
Out[56]=
```
       0.0609948 t
5000 E
```

After 18 years, you've got:

In[57]:=
```
balance[18]
```
Out[57]=
```
14989.4
```

Watch it grow:

In[58]:=
```
Plot[balance[t],{t,0,18},
PlotStyle->{{DarkGreen,Thickness[0.015]}},
AxesLabel->{"years","balance"},
AxesOrigin->{0,balance[0]}];
```

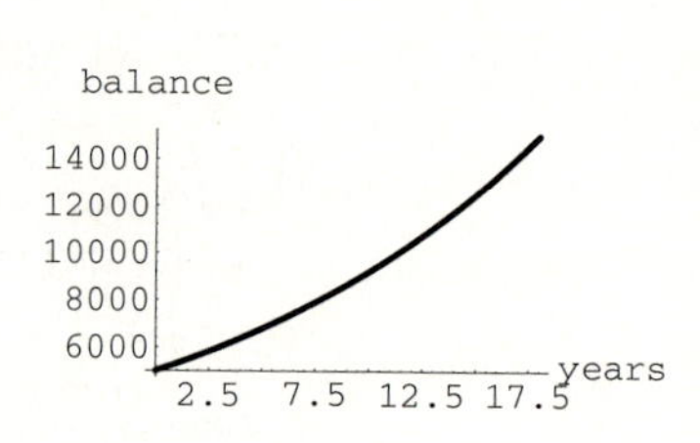

You look at this and wonder about inflation.

T.3.c.ii) You find a new bank that pays you 6.6% interest compounded monthly (12 times per year) and wonder how long it will take for your $5000 to grow to $30,000.

> Estimate how long it will take.

Answer:

In[59]:=
```
advertisedyearly = 6.6
```
Out[59]=
```
6.6
```

Break up the year into 12 months:

In[60]:=
```
actualmonthly = advertisedyearly/12
```
Out[60]=
```
0.55
```

Measure t in years with $t = 0$ corresponding to the time of the deposit of \$5000. Put balance$[t] = a\,e^{rt}$:

In[61]:=
```
Clear[balance,t,a,r]; balance[t_] = a E^(r t)
```

Out[61]=
```
     r t
a E
```

You know $5000 = \text{balance}[0] = a\,e^{r0} = a$. So:

In[62]:=
```
a = 5000;
balance[t]
```

Out[62]=
```
        r t
5000 E
```

Now look at the actual monthly interest percentage:

In[63]:=
```
actualmonthly
```

Out[63]=
```
0.55
```

Set r to correspond to the actual monthly interest by solving:

In[64]:=
```
Simplify[100 (balance[t + 1/12]/balance[t] - 1)] == actualmonthly
```

Out[64]=
```
                r/12
100 (-1 + E    ) == 0.55
```

This tells you that

$$e^{r/12} - 1 = 0.0055;$$

$$e^{r/12} = 1.0055;$$

$$\frac{r}{12} = \log[1.0055]:$$

In[65]:=
```
r = (12) Log[1.0055]
```

Out[65]=
```
0.0658192
```

The exponential function that measures what you have at the end of each of the 12 months of the year is:

In[66]:=
```
balance[t]
```

Out[66]=
```
        0.0658192 t
5000 E
```

Watch it grow over the first 50 years:

In[67]:=
```
Plot[{balance[t],30000},{t,0,50},
PlotStyle->{{DarkGreen,Thickness[0.015]},Red},
AxesLabel->{"years","balance"},
AxesOrigin->{0,balance[0]}];
```

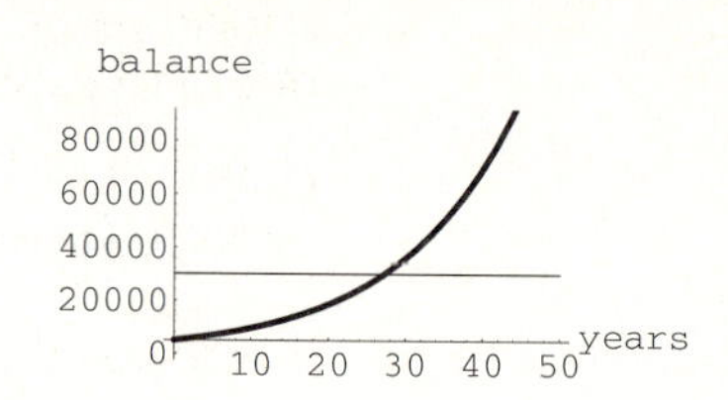

It will take about 27 or 28 years to reach the $30,000 mark. If you want the precise answer, here it is:

In[68]:=
```
N[Solve[balance[t] == 30000,t]]
```

Out[68]=
```
{{t -> 27.2225}}
```

Here is the algebra behind the answer:

Look at:

In[69]:=
```
balance[t] == 30000
```

Out[69]=
```
      0.0658192 t
5000 E              == 30000
```

This is the same as $e^{0.0658192t} = 6$. This is the same as $0.0658192\,t = \log[6]$; and this means that t is given by:

In[70]:=
```
N[Log[6]/0.0658192]
```

Out[70]=
```
27.2224
```

This is the same answer the machine gave you above.

T.3.d) Some banks advertise a passbook interest rate of 100 x percent and say that they are compounding this interest rate k times a year. Then they announce the resulting effective interest rate. As a mathematics student, you are entitled to more information.

When you start with an annouced interest rate of 100 x percent compounded k times a year, then the effective percentage interest rate is given by the formula

$$100\left(e^{k\log[1+x/k]}-1\right).$$

Try it out on an advertised interest rate of 6% compounded 12 times a year (monthly):

In[71]:=
```
Clear[effective,x,k]
effective[x_,k_] = 100 (E^(k Log[1 + x/k]) - 1); effective[0.06,12]
```

Out[71]=
```
6.16778
```

An interest rate of 6% compounded 12 times a year carries an effective interest rate of 6.16778%.

If you hold the interest rate at 6% but compound 365 times (daily), then the effective interest rate is:

In[72]:=
```
effective[0.06,365]
```

Out[72]=
```
6.18313
```

The more often the bank compounds, the higher the effective interest rate. Play with this function and answer the questions:

> What does the effective interest rate measure? Where does the formula come from?

Answer: Play: Look at 5% compounded 12 times a year.

In[73]:=
```
effective[0.05,12]
```

Out[73]=
```
5.11619
```

This says that 5% compounded 12 times a year results in an effective interest rate of 5.11619%.

In[74]:=
```
effective[0.05,365]
```

Out[74]=
```
5.12675
```

This says that 5% compounded 365 times a year results in an effective interest rate of 5.12675%. This corresponds to compounding daily. Here's what happens when they compound each hour:

In[75]:=
```
effective[0.05,(365) (24)]
```

Out[75]=
```
5.12709
```

Each minute:

In[76]:=
```
effective[0.05,(365) (24) (60)]
```

Out[76]=
```
5.12711
```

Each second:

In[77]:=
```
effective[0.05,(365) (24) (60) (60)]
```

Out[77]=
```
5.12711
```

To five accurate decimals, compounding each minute is the same as compounding each second.

Now check into what the formula

$$\text{effective}[x, k] = 100\left(e^{k\log[1+x/k]} - 1\right)$$

means and where it comes from. Start with 100 x percent compounded k times a year.

In[78]:=
```
Clear[balance,t,a,r,x,k]; balance[t_] = a E^(r t)
```
Out[78]=
```
   r t
a E
```

The interest rate is $100\,x$ percent compounded k times a year.

Set r to correspond to

$$100\left(\frac{\text{balance}[t+1/k]}{\text{balance}[t]} - 1\right) = 100\frac{x}{k}.$$

This results in:

In[79]:=
```
Simplify[balance[t + 1/k]/balance[t]] == 1 + x/k
```
Out[79]=
```
 r/k            x
E     == 1 + -
                k
```

This tells you that $r/k = \log[1 + x/k]$, so $r = k\log[1 + x/k]$, and

$$\text{balance}[t] = a\,e^{t\,k\log[1+x/k]}.$$

The effective interest rate is the simple interest rate that duplicates this growth at the end of the first year ($t = 1$).

$$a + a\,\text{effective}\,x = a\,e^{k\log[1+x/k]}$$

$$1 + \text{effective}\,x = e^{k\log[1+x/k]}$$

$$\text{effective}\,x = e^{k\log[1+x/k]} - 1$$

Finally multiply by 100 to convert this to a percentage. This gives

$$\text{effective}[x, k] = 100\left(e^{k\log[1+x/k]} - 1\right).$$

When you ask *Mathematica* for this function:

In[80]:=
```
effective[x,k]
```

Out[80]=

```
                x k
100 (-1 + (1 + -) )
                k
```

You get a different-looking formula because

$$k \log\left[1 + \frac{x}{k}\right] = \log\left[\left(1 + \frac{x}{k}\right)^k\right];$$

so

$$e^{k \log[1+x/k]} = \left(1 + \frac{x}{k}\right)^k.$$

Two different ways of writing the same thing.

Give It a Try

Experience with the starred (⋆) problems will be especially beneficial for understanding later lessons.

■ G.1) Exponential growth⋆

G.1.a) Here's a plot of $f[x] = e^{0.15x}$:

In[1]:=

```
Clear[f,x]
f[x_] = E^(0.15 x);
Plot[f[x],{x,0,100},
PlotStyle->{{Red,Thickness[0.01]}},
PlotRange->All,AxesLabel->{"x",""}];
```

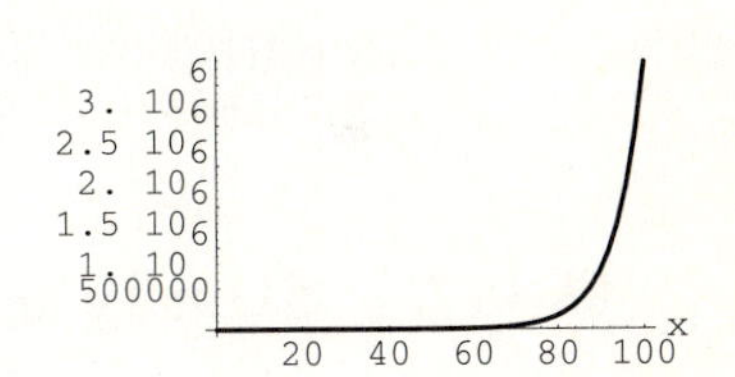

Even though 0.15 is a puny number, the function $f[x] = e^{0.15x}$ is growing huge:

In[2]:=

```
f[100]
```

Out[2]=

```
          6
3.26902 10
```

In[3]:=

```
f[1000]
```

Out[3]=

```
          65
1.39371 10
```

What do you call this spectacular growth?

G.1.b) Here's a plot of $f[x] = e^{-0.15x}$:

In[4]:=
```
Clear[f,x]
f[x_] = E^(-0.15 x);
Plot[f[x],{x,0,100},
PlotStyle->{{Red,Thickness[0.01]}},
PlotRange->All,AxesLabel->{"x",""}];
```

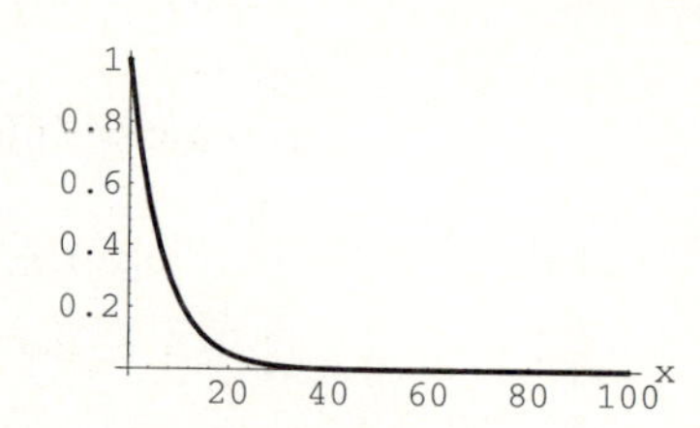

Even though 0.15 is a puny number, the function $f[x] = e^{-0.15x}$ dies off to almost nothing.

In[5]:=
```
f[100]
```
Out[5]=
```
          -7
3.05902 10
```
In[6]:=
```
f[1000]
```
Out[6]=
```
          -66
7.1751 10
```
Folks call this spectacular decay "exponential decay."

> How does the fact that $f[x] = e^{-0.15x} = 1/e^{0.15x}$ explain why the plot came out the way it did?

■ G.2) Steady growth versus steady percentage growth*

G.2.a) Line functions exhibit steady growth rates. For instance, if $f[x] = a\,x + b$, then you can be sure that $f[x]$ goes up by a units every time x goes up by one unit. Try it out:

In[7]:=
```
Clear[f,x,a,b]
f[x_] = a x + b
```
Out[7]=
```
b + a x
```
In[8]:=
```
Table[f[x],{x,0,5}]
```
Out[8]=
```
{b, a + b, 2 a + b, 3 a + b, 4 a + b, 5 a + b}
```
In[9]:=
```
Table[f[x + 1] - f[x],{x,0,5}]
```

Out[9]=

```
{a, a, a, a, a, a}
```

In[10]:=

```
Expand[f[x + 1] - f[x]]
```

Out[10]=

```
a
```

Here's a plot of the line function $f[x] = 2\,x - 3$:

In[11]:=

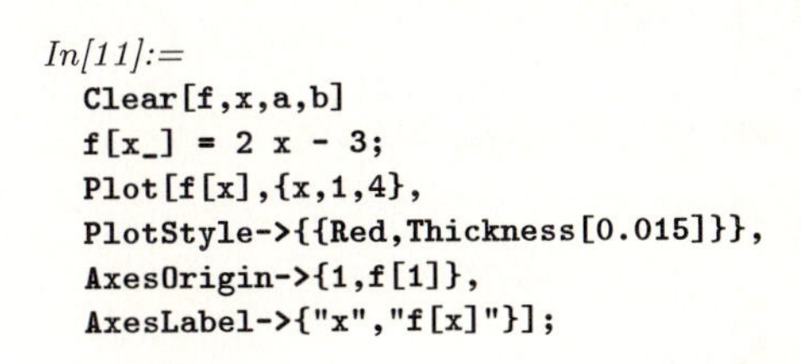

```
Clear[f,x,a,b]
f[x_] = 2 x - 3;
Plot[f[x],{x,1,4},
PlotStyle->{{Red,Thickness[0.015]}},
AxesOrigin->{1,f[1]},
AxesLabel->{"x","f[x]"}];
```

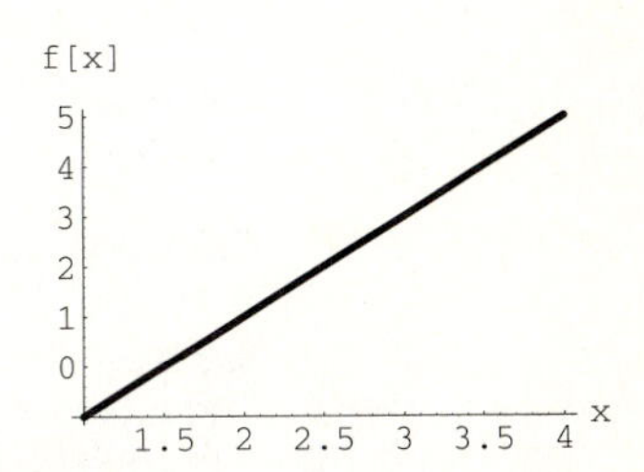

Sure enough, $f[x] = 2\,x - 3$ goes up two units every time x goes up one unit. In other words, $f[x] = 2\,x - 3$ grows exactly twice as fast as x grows.

> Fill in the blank: If you go with a line function $f[x] = a\,x + b$ and h is any number, then $f[x + h] - f[x] =$ ________.

G.2.b) Exponential functions do not exhibit a steady growth rate. Look at the exponential function $f[x] = 10\,e^{1.2x}$:

In[12]:=

```
Clear[f,x]; f[x_] = 10 E^(1.2 x)
```

Out[12]=

```
     1.2 x
10 E
```

In[13]:=

```
Factor[f[x + 1] - f[x]]
```

Out[13]=

```
          1.2 x
23.2012 E
```

The bigger x is, the faster $f[x]$ grows.

In[14]:=

```
Plot[f[x + 1] - f[x],
{x,0,6},PlotStyle->{{Red,Thickness[0.015]}},
AxesLabel->{"x","f[x]"}];
```

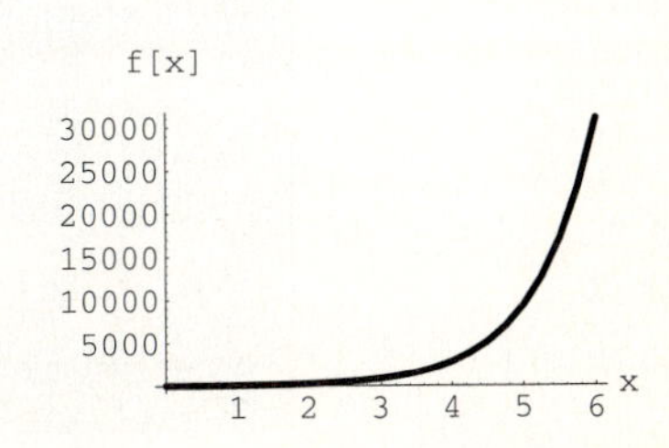

In[15]:=
```
Table[f[x + 1] - f[x],{x,0,5}]
```

Out[15]=
```
{23.2012, 77.0306, 255.751, 849.122, 2819.18, 9360.02}
```

That's big-time exponential growth.

Now look at this table:

In[16]:=
```
Table[100(f[x + 1]/f[x] - 1),{x,0,5}]
```

Out[16]=
```
{232.012, 232.012, 232.012, 232.012, 232.012, 232.012}
```

And look at this plot:

In[17]:=
```
Plot[100(f[x + 1]/f[x] - 1),{x,0,6},
PlotStyle->{{Red,Thickness[0.015]}},
AxesLabel->{"x","percentage growth"},
PlotRange->{230,235}];
```

percentage growth
235
234
233
232
231
0 1 2 3 4 5 6
x

And:

In[18]:=
```
Simplify[100(f[x + 1]/f[x] - 1)]
```

Out[18]=
```
232.012
```

The upshot is that every time x grows by 1, then $f[x]$ grows by a steady percentage of about 232%.

> Fill in the blank: Take any exponential function $f[x] = a\,e^{rx}$. Every time x grows by 1, then $f[x]$ grows by ________ percent.

G.2.c)

> Give the formula for the line function $\text{line}[x] = a\,x + b$ that goes through $\{0, 3\}$ and grows 0.2 units every time x grows by one unit.
>
> Give the formula for the exponential function
>
> $$\text{exponential}[x] = a\,e^{rx}$$
>
> that goes through $\{0, 3\}$ and grows by 20% every time x grows by one unit.
>
> Plot both functions on the same axes for $0 \le x \le 10$.
>
> Describe the difference between steady growth and steady percentage growth.

G.2.d.i) Here's an exponential function of time:

In[19]:=
```
Clear[f,a,r,t]
r = 0.05; a = 60.8; f[t_] = a E^(r t)
```
Out[19]=
```
      0.05 t
60.8 E
```

Now look at:

In[20]:=
```
Simplify[100( f[t + 1]/f[t] - 1)]
```
Out[20]=
```
5.12711
```

This tells you that $f[t]$ grows at a steady rate of about 5.127% per time unit.

> Keeping the same a, reset r so that resulting $f[t]$ grows at a steady rate of about 8% per time unit.

G.2.d.ii) Here is a general exponential function:

In[21]:=
```
Clear[f,a,r,t]
f[t_] = a E^(r t)
```
Out[21]=
```
   r t
a E
```

Now look at:

In[22]:=
```
Simplify[100(f[t + 1]/f[t] - 1)]
```
Out[22]=
```
             r
100 (-1 + E )
```

> Does the percentage growth in $f[t]$ per time unit have anything to do with the specific value of a that is used?
>
> Does the percentage growth in $f[t]$ per time unit have anything to do with the specific value of r that is used?

G.2.e.i)

> Make an exponential function $f[x] = a\,e^{rx}$ going through $\{0, 10\}$ that grows 100% every time x goes up by one unit.

G.2.e.ii) Make an exponential function $f[x] = a\,e^{rx}$ going through $\{0, 10\}$ that grows 50% every time x goes up by 1/2 unit.

G.2.e.iii) In part G.2.e.i) above, you got an exponential function that grows 100% every time x goes up by one unit.

In part G.2.e.ii) above, you got an exponential function that grows 50% every time x goes up by 1/2 unit.

Are these two functions, in fact, the very same function?

■ G.3) Exponential models*

Many of the problems here were influenced by problems in Ethan Bolker's book *Using Algebra*, Little, Brown, 1983.

G.3.a) Walter Libby won a Nobel Prize for discovering that living tissue contains two kinds of carbon. One kind, carbon-14 $\left({}^{14}_{6}C\right)$, is radioactive with a half life of about 5750 years; the other is not radioactive. In living tissue the ratio of the two is always constant, but when the tissue dies, the radioactive carbon begins to decay exponentially while the other carbon remains.

To date a fossil, scientists measure the amount of each kind of carbon present to determine how much of the radioactive carbon has decayed.

If measurements show that the radioactive carbon in an ancient wood carving has decayed by 47.3%, then how old is the wood in the carving?

G.3.b) Rocks contain two kinds of potassium. One is radioactive and the other is not. The radioactive potassium has a half life of 1.4×10^9 years.

In[23]:=

```
halflife = 1.4 10^9
```

Out[23]=

```
      9
1.4 10
```

In the oldest known rocks, geologists have determined that radioactive potassium has decayed by 82%.

Measure the age of these rocks. What does your answer tell you about the age of the planet Earth?

G.3.c) The source data for this problem come from E. Batschelet's book *Introduction to Mathematics for Life Scientists*, Springer-Verlag, New York, 1979.

When an underwater light beam is turned on in sea water, the resulting intensity of the light decays exponentially along the direction of the beam.

In fact, in clear sea water, the percentage of the original intensity x meters from the source measured along the beam is given by intensity$[x] = 100\, e^{-1.4x}$.

> Plot intensity$[x]$ for $0 \leq x \leq 12$.
>
> Calculate intensity[10] and use your result to explain why in any ocean almost no plants can grow at a depth of more than 10 meters.

G.3.d) In 1980, about 170 million motor vehicles were registered in the United States. Census figures indicate a steady growth at a rate of 4% per year.

> How many registrations are projected for the year 2000? How about 2030?

G.3.e.i) When the government annouces an inflation rate of x% per year, this means that if the inflation rate never changes, then the price of an item that costs $100 today is given by

$$P[t] = 100\, e^{rt}$$

dollars, where r is set so that

$$100\left(\frac{P[t+1]}{P[t]} - 1\right) = x.$$

Take the current year for $t = 0$ and assume that the Federal Reserve Board holds inflation at 3.5% per year for the next 50 years.

> Give a plot of the projected price over the next 50 years of an item that costs $100 today.

G.3.e.ii) Take the current year for $t = 0$ and assume that the Federal Reserve Board holds inflation at 5.5% per year for the next 50 years.

> Give a plot of the projected price over the next 50 years of an item that costs $100 today.

G.3.e.iii) Take the current year for $t = 0$ and assume that the Federal Reserve Board holds inflation at 7.5% per year for the next 50 years.

> Give a plot of the projected price over the next 50 years of an item that costs \$100 today.

G.3.e.iv) Take the current year for $t = 0$ and assume that the Federal Reserve Board holds inflation at 9.5% per year for the next 50 years.

> Give a plot of the projected price over the next 50 years of an item that costs \$100 today.

G.3.e.v)

> Discuss how you expect the Federal Reserve Board's policies on inflation to influence your future. Are a couple of extra points on the inflation rate anything to worry about?
>
> What is it about the function
>
> $$f[x] = e^{rx} \qquad (\text{for } r > 0)$$
>
> that makes long-term high inflation rates really scary?

■ G.4) Exponential data analysis★

G.4.a) The cost of a day in the hospital in the U.S. has caused lots of weeping, wailing, and gnashing of teeth. The following table gives figures in the form $\{t, H[t]\}$ where t is measured in years since 1970 and $H[t]$ is the average dollar cost per patient day in a hospital in the U.S.

Source: *Statistical Abstracts*, 1986.

In[24]:=
```
bedcost = {{0,74},{2,95},{3,102},{4,114},{5,134},
{6,153},{7,174},{8,194},{9,217},{10,245},
{11,284},{12,327},{13,369}};
```

And a plot:

In[25]:=
```
bedplot =
ListPlot[bedcost,
PlotStyle->{Blue,PointSize[0.02]}];
```

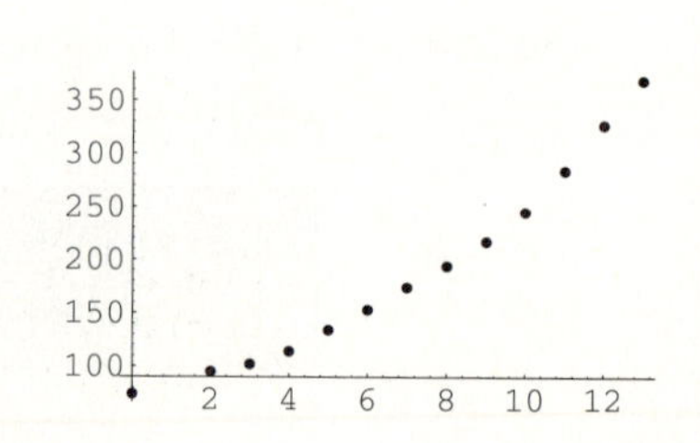

The following instruction exhibits the same data but in the form $\{t, \log[H[t]]\}$:

In[26]:=

```
logbed = Table[{bedcost[[j,1]],N[Log[bedcost[[j,2]]]]},
{j,1 Length[bedcost]}]
```

Out[26]=

```
{{0, 4.30407}, {2, 4.55388}, {3, 4.62497}, {4, 4.7362}, {5, 4.89784},
 {6, 5.03044}, {7, 5.15906}, {8, 5.26786}, {9, 5.3799},
 {10, 5.50126}, {11, 5.64897}, {12, 5.78996}, {13, 5.9108}}
```

Here is a plot:

In[27]:=

```
ListPlot[logbed,
PlotStyle->{Blue,PointSize[0.02]},
PlotLabel->"Semi-log paper plot"];
```

Semi-log paper plot

5.75
5.5
5.25
4.75
4.5
2 4 6 8 10 12

G.4.a.i) Over the years indicated, do you believe that the average dollar cost per patient day in a hospital in the U.S. is growing approximately exponentially? Explain your response.

If you believe that it is, then find a compromise exponential function $f[t] = a\,e^{rt}$ whose plot runs through or near the data. Confirm with a plot.

G.4.a.ii) Use your function to predict when it will cost \$1000 dollars to stay in a hospital for one day. How about \$2000?

G.4.b) Thomas Malthus, in his 1803 book *An Essay on the Principle of Population*, made quite a name for himself by arguing that the world population will grow exponentially which will lead to ultimate disaster.

Here are some world population data $\{t, P[t]\}$ where t is the year, $t = 0$ corresponds to 1950, and $P[t]$ is the estimated world population in millions for the corresponding year t. Here is a plot:

In[28]:=

```
worldpop =
{{0,2513},{10,3027},{20,3678},{30,4478},
 {35,4865},{36,4942},{37,5026},{38,5128},
 {39,5234},{40,5321}};
popdataplot = ListPlot[worldpop,
AxesLabel->{"year","World population"},
PlotStyle->{Red,PointSize[0.03]}];
```

World population

5000
4500
4000
3500
3000
year
10 20 30 40

(Source: *Information Please* almanac.)

This plot covers the years 1950 to 1990. Here are the same data on semi-log paper:

In[29]:=

```
Clear[k]
logworldpop =
Table[{worldpop[[k,1]],N[Log[worldpop[[k,2]]]]},
{k,1,Length[worldpop]}];
ListPlot[logworldpop,
PlotStyle->{Red,PointSize[0.03]},
PlotLabel->"Semi-log paper plot"];
```

> Explain why this plot tells you that since 1950, the world population is showing definite signs of exponential growth.
>
> Find a compromise exponential function $f[t] = a\, e^{rt}$ whose plot runs through or near the data. Confirm with a plot.
>
> Use your function to make predictions about the world population over the years 1990 to 2040.

G.4.c) Here are the United States population data $\{t, P[t]\}$ where t is the year and $P[t]$ is the official census figure in millions for the corresponding year t. Here is a plot:

In[30]:=

```
uspop =
{{0,3.9},{1800,5.3},{1810,7.24},{1820,9.6},
 {1830,12.86},{1840,17},{1850,23.2},{1860,31.4},
 {1870,38.5},{1880,50.2},{1890,63},{1900,76.2},
 {1910,92.2},{1920,106},{1930,123.2},
 {1940,132.2},{1950,151.3},{1960,179.3},
 {1970,203.3},{1980,226.5},{1990,248.7}};
popdataplot = ListPlot[uspop,
AxesLabel->{"year","population"},
PlotStyle->{Red,PointSize[0.03]}];
```

Here are the same data on semi-log paper:

In[31]:=

```
Clear[k]
logpop = Table[{uspop[[k,1]],N[Log[uspop[[k,2]]]]},
{k,1,Length[uspop]}];
ListPlot[logpop,PlotStyle->{Red,PointSize[0.03]},
PlotLabel->"Semi-log paper plot"];
```

> Is the overall growth of the United States population greater or slower than exponential growth?

G.4.d) Here are United States national debt data $\{t, \text{nationaldebt}[t]\}$ where t is the year and nationaldebt$[t]$ is the officially reported national debt in millions for the corresponding year t. Here is a plot:

In[32]:=

```
nationaldebt =
{{1980, 907.7}, {1981, 997.9},
 {1982, 1142}, {1983, 1377.2}, {1984, 1572.3},
 {1985, 1823.1}, {1986, 2125.3}, {1987, 2305.3},
 {1988, 2602.3}, {1989, 2857.4}, {1990, 3233.3},
 {1991, 3502.}};
dataplot = ListPlot[nationaldebt,
AxesLabel->{"x","y"},
PlotStyle->{Red,PointSize[0.02]}];
```

And on semi-log paper:

In[33]:=

```
Clear[k]
logdebt = Table[{nationaldebt[[k,1]],
N[Log[nationaldebt[[k,2]]]]},
{k,1,Length[nationaldebt]}];
ListPlot[logdebt,PlotStyle->{Red,PointSize[0.03]},
PlotLabel->"Semi-log paper plot"];
```

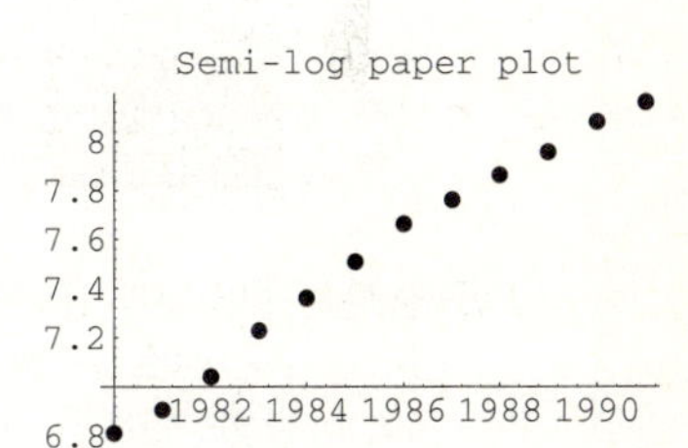

Sheesh!

> Is the overall growth of the United States national debt since 1980 approximately exponential?

G.4.e) This problem appears only in the electronic version.

■ G.5) Your money

G.5.a) On December 12, 1991, American Savings, a division of Citizens Federal Bank of Miami, Florida advertised in the *Leader* newspaper their 12-month certificates of deposit with a 5.59% nominal rate and an annual yield of 5.75%. In fine print the advertisement mentions that the 5.59% rate is being compounded daily (about 360 times per year).

> Was their advertisement truthful? How do you know?

G.5.b) The First National Bank of Snowshoe offers passbook savings accounts at 6% interest per year compounded twice a year.

The State Bank of Flatville offers passbook savings accounts at 5.9% interest per year compounded monthly.

The Bernoulli Bank of Bern offers passbook savings accounts at 5.89% per year compounded daily.

> Which bank is offering the best deal?
>
> Which bank is offering the worst deal?

G.5.c) From the Money section of the newspaper *USA Today*, June 18, 1992:

"Rates on credit card applications don't take into account monthly, and sometimes daily, compounding of finance charges. The average rate on credit cards is 18.5%, but most consumers pay effective rates of 20% or more because of compounding methods."

> What does this mean? Is this information correct? How do you know?

G.5.d) Suppose that you know that in t years you need g dollars and you know the prevailing interest rate is 100 x percent per year. For example, upon birth of a child you might want to know how much money you need to put into an account that will grow to a sum large enough to finance college expenses in 18 years.

> Announce what your goal in dollars is. Announce how long you are willing to wait for it. Assume a constant interest rate of 5% compounded 12 times a year and calculate how much money you need to put into an account today to achieve your goal in your specified time.

■ G.6) Compounding every instant★

Take an advertised interest rate 100 x percent. Saying that the interest is compounded k times a year means that if $1000 is deposited in the account, then the balance in the account after t years is given by

In[34]:=

```
Clear[balance,t,k,r]
balance[t_] = 1000 E^(r t)
```

Out[34]=

```
       r t
1000 E
```

where r is set so that the balance increases at a steady rate of

$$\left(100\frac{x}{k}\right) \text{ \% per } \left(\frac{1}{k}\right) \text{ years.}$$

Given x, you calculate r by looking at the equation

$$100\left(\left(\frac{\text{balance}[t+1/k]}{\text{balance}[t]}\right)-1\right)=100\,\frac{x}{k}:$$

In[35]:=

```
Clear[x]
Simplify[100(balance[t + 1/k]/balance[t] - 1)] == 100 x/k
```

Out[35]=

```
            r/k        100 x
100 (-1 + E   ) == -----
                         k
```

This tells you that

$$e^{r/k}-1=\frac{x}{k};$$
$$e^{r/k}=1+\frac{x}{k}$$
$$\frac{r}{k}=\log\left[1+\frac{x}{k}\right].$$

G.6.a.i) Explain why this also tells you that

$$r=\log\left[\left(1+\frac{x}{k}\right)^{k}\right].$$

G.6.a.ii) Continuing with part G.6.a.i) above, go with

$$r=\log\left[\left(1+\frac{x}{k}\right)^{k}\right].$$

This is the r so that balance$[t] = 1000\,e^{rt}$ measures the balance in the account after t years under an advertised interest rate of $100\,x$ percent compounded k times a year. The goal of this problem is to try to recognize the approximate relationship between r and x when the number k is very, very large.

Play with x and k and try to determine the relationship.

G.6.a.iii) Some banks advertise monthly compounding. Others advertise daily compounding. Still others say that they are compounding every instant. (Some banks use the term *continous compounding* instead of the term *compounding every instant.*)

Banks that advertise 100 x percent interest compounded every instant calculate the balance in the account growing from an initial deposit of \$1000 by the formula

$$\text{balance}[t]=1000\,e^{xt}.$$

Explain where you think this formula comes from.

For a given interest rate of 100 x percent, why does compounding every instant give the best deal to the customer?

G.6.a.iv) Is there a significant difference between daily compounding and compounding every instant?

G.6.a.v) If you get 100 x percent interest compounded every instant on an initial deposit of $\$K$, then the balance in the account t years after the initial deposit is

$$\text{balance}[t] = K\, e^{xt}.$$

To find how long it takes for the account to double, look at:

$$\begin{aligned} \text{balance}[t+d] &= 2\,\text{balance}[t]; \\ K\, e^{x(t+d)} &= 2\,K\, e^{xt}; \\ e^{x(t+d)} &= 2\, e^{xt}; \\ e^{xt} e^{xd} &= 2\, e^{xt}; \\ e^{xd} &= 2; \\ \log[e^{xd}] &= \log[2]; \\ x\, d &= \log[2]; \\ d &= \frac{\log[2]}{x}. \end{aligned}$$

Now look at:

In[36]:=
```
N[Log[2]]
```

Out[36]=
```
0.693147
```

and then:

Explain why financial officers often make a rough estimate of the time it takes for an investment made at an interest rate 100 x percent to double by calculating $(72/100x)$ in their heads.

For instance, a rough estimate of how long it will take money invested at 6% to double is $72/6 = 12$ years. A rough estimate of how long it will take money invested at 4% to double is $72/4 = 18$ years. This is called the Rule of 72.

■ G.7) Law and order

This problem appears only in the electronic version.

■ G.8) Unnatural bases

G.8.a) Here's a plot of 2^x:

In[37]:=

```
Clear[f,x]
f[x_] = 2^x;
fplot = Plot[f[x],{x,-1,6},
    PlotStyle->{{Blue,Thickness[0.01]}}];
```

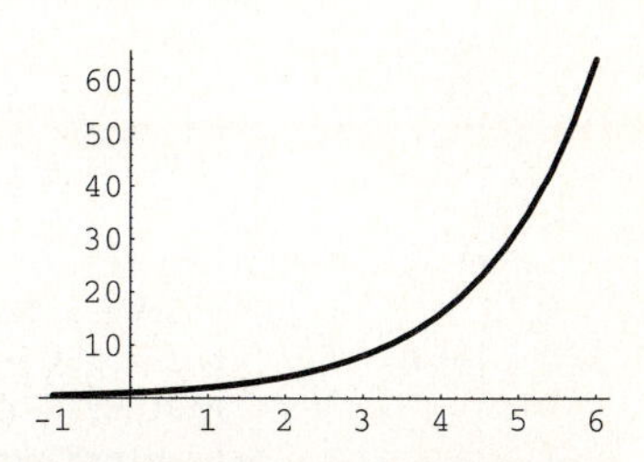

> Produce the number r so that $e^{rx} = 2^x$ and plot the resulting e^{rx} and show it with the plot of 2^x to confirm your answer.

G.8.b) When you ask *Mathematica* to give you the base 10 logarithm, you type:

In[38]:=

```
Clear[x]
Log[10,x]
```

Out[38]=

```
Log[x]
-------
Log[10]
```

> Explain why the output is correct.

■ G.9) Reflecting patterns and wandering points

Here is a true scale plot showing:

→ e^x for $-2 \leq x \leq 2$

→ x for $-2 \leq x \leq e^2$

→ $\log[x]$ for $e^{-2} \leq x \leq e^2$:

In[39]:=

```
Clear[x]; end = 2; start = -2;
plot1 = Plot[E^x,{x,start,end},
PlotStyle->{{Red,Thickness[0.01]}},
DisplayFunction-> Identity];
plot2 =  Plot[x,{x,start,E^end},
PlotStyle->{{Blue,Thickness[0.01]}},
DisplayFunction-> Identity];
plot3 =  Plot[Log[x],{x,E^(start),E^end},
PlotStyle->{{Red,Thickness[0.01]}},
DisplayFunction-> Identity];
Show[plot1,plot2,plot3,AspectRatio->Automatic,
PlotRange->All, AxesLabel->{"x",""},
DisplayFunction->$DisplayFunction];
```

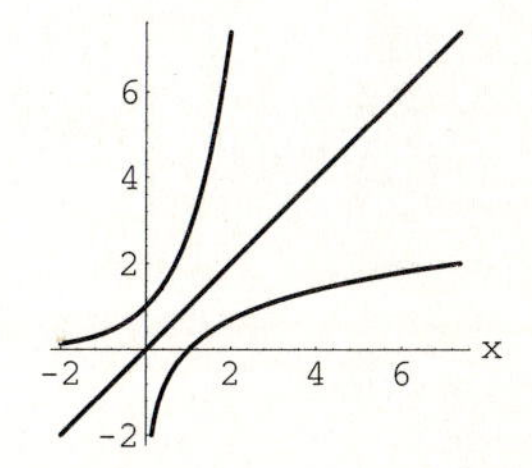

G.9.a) Describe what you see including any eye-catching patterns.

G.9.b) Take the same plots and throw in some points $\{x, e^x\}$ for a selection of x's running from -2 to 2 in x-increments of 0.5:

In[40]:=
```
points = Table[{x,E^x},{x,-2,2,0.5}];
pointplot =  ListPlot[points,
PlotStyle->PointSize[0.04],
DisplayFunction->Identity];
Show[plot1,plot2,plot3,pointplot,
AspectRatio->Automatic,
PlotRange->All, AxesLabel->{"x",""},
 DisplayFunction->$DisplayFunction];
```

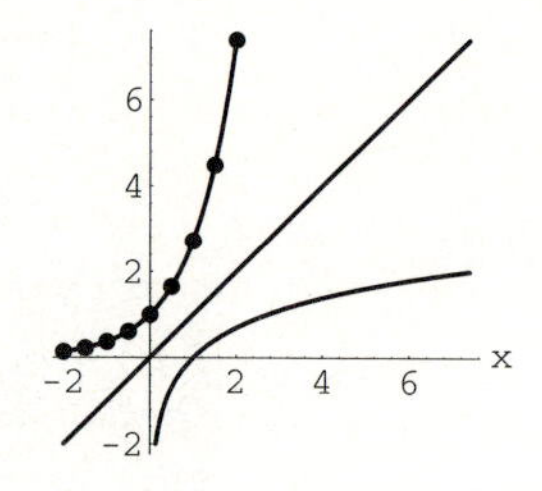

Nothing surprising here. Now take the same points but reverse the order from $\{x, e^x\}$ to $\{e^x, x\}$ and plot:

In[41]:=
```
reversedpoints = Table[{E^x,x},{x,-2,2,0.5}];
reversedpointplot =
ListPlot[reversedpoints,
PlotStyle->PointSize[0.04],
DisplayFunction->Identity];
Show[plot1,plot2,plot3,reversedpointplot,
AspectRatio->Automatic,
PlotRange->All, AxesLabel->{"x",""},
DisplayFunction->$DisplayFunction];
```

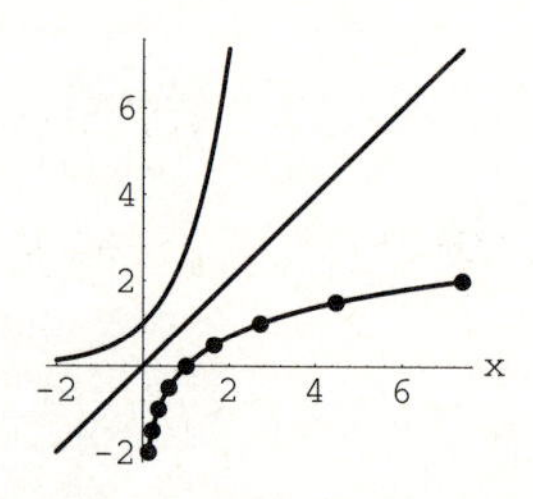

Why do you think the reversed points moved from the e^x curve to the $\log[x]$ curve?

LESSON 1.03

Instantaneous Growth

Basics

B.1) Instantaneous growth rates

Here is a friendly function

$$f[x] = 1 + 2\,x^3 - x^4$$

and a plot:

In[1]:=

```
Clear[f,x]
f[x_] = 1 + 2 x^3 - x^4;
fplot = Plot[f[x],{x,-1,2},
PlotStyle->{{Thickness[0.01],Blue}},
AxesLabel->{"x",""}];
```

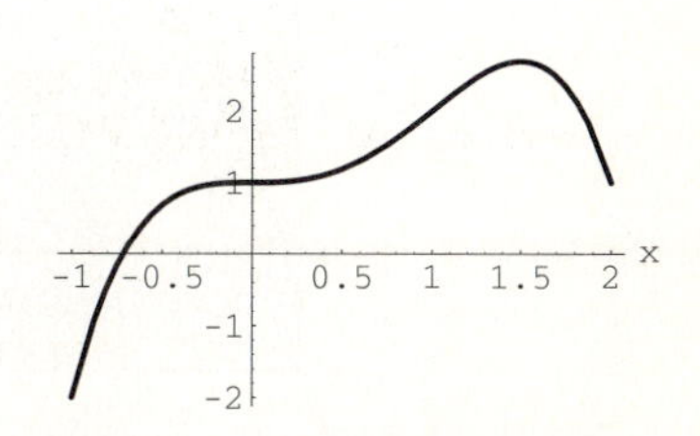

B.1.a.i) Measure the net growth of

$$f[x] = 1 + 2\,x^3 - x^4$$

over the interval $[-1, 2]$. Then measure the average growth rate of $f[x]$ over the interval $[-1, 2]$.

Answer: Here you go: Over the interval $[-1, 2]$, the function starts out at:

In[2]:=

```
f[-1]
```

Out[2]=
```
-2
```

And it ends up at:

In[3]:=
```
f[2]
```

Out[3]=
```
1
```

Its net growth is:

In[4]:=
```
fgrowth = f[2] - f[-1]
```

Out[4]=
```
3
```

Its average growth rate in units on the y-axis per unit on the x-axis over the interval $[-1, 2]$ is:

In[5]:=
```
xgrowth = (2 -(-1))
```

Out[5]=
```
3
```

In[6]:=
```
fgrowth/xgrowth
```

Out[6]=
```
1
```

As x grows from -1 to 2, on the average, $f[x]$ grows at a rate of one unit every time x grows by one unit. The average growth rate of $f[x]$ over the interval $[-1, 2]$ is 1.

B.1.a.ii)

> Measure the average growth rate of
>
> $$f[x] = 1 + 2\,x^3 - x^4$$
>
> over the interval $[x, x + 0.5]$. Interpret the result.

Answer:

In[7]:=
```
Clear[f,x]; f[x_] = 1 + 2 x^3 - x^4;
```

Over the interval $[x, x + 0.5]$, the function starts out at:

In[8]:=
```
f[x]
```

Out[8]=
```
           3    4
1 + 2 x  - x
```

and it ends up at:

In[9]:=
```
f[x + 0.5]
```
Out[9]=
```
                3            4
1 + 2 (0.5 + x)  - (0.5 + x)
```

Its net growth over the interval $[x, x + 0.5]$ is:

In[10]:=
```
fgrowth = Expand[f[x + 0.5] - f[x]]
```
Out[10]=
```
                      2       3
0.1875 + 1. x + 1.5 x  - 2. x
```

Its average growth rate in units on the y-axis per unit on the x-axis over the interval $[x, x + 0.5]$ is:

In[11]:=
```
xgrowth = 0.5; Clear[fAverageGrowthRate];
fAverageGrowthRate[x_] = Expand[fgrowth/xgrowth]
```
Out[11]=
```
                   2       3
0.375 + 2. x + 3. x  - 4. x
```

On the interval $[x, x + 0.5]$, the average growth rate of $f[x]$ is

$$0.375 + 2\,x + 3\,x^2 - 4\,x^3.$$

For instance, when you look at:

In[12]:=
```
fAverageGrowthRate[0]
```
Out[12]=
```
0.375
```

Then you see that, on the average, $f[x]$ grows 0.375 times as fast as x grows as x advances from 0 to $0 + 0.5 = 0.5$. But when you look at:

In[13]:=
```
fAverageGrowthRate[1.5]
```
Out[13]=
```
-3.375
```

then you see that, on the average, f goes down 3.375 times as fast as x grows as x advances from 1.5 to $1.5 + 0.5 = 2$.

B.1.a.iii) Given a positive number h, measure the average growth rate of

$$f[x] = 1 + 2\,x^3 - x^4$$

over the interval $[x, x + h]$. Interpret the result.

Answer:

In[14]:=

```
Clear[f,x]; f[x_] = 1 + 2 x^3 - x^4;
```

Over the interval $[x, x+h]$, the function starts out at:

In[15]:=

```
f[x]
```

Out[15]=

```
         3    4
1 + 2 x  - x
```

and it ends up at:

In[16]:=

```
Clear[h]; Expand[f[x + h]]
```

Out[16]=

```
         3    4      2        3            2      2  2
1 + 2 h  - h  + 6 h  x - 4 h  x + 6 h x  - 6 h  x  +

      3          3    4
  2 x  - 4 h x  - x
```

Its net growth over the interval $[x, x+h]$ is:

In[17]:=

```
fgrowth = Expand[f[x + h] - f[x]]
```

Out[17]=

```
   3    4      2        3            2      2  2          3
2 h  - h  + 6 h  x - 4 h  x + 6 h x  - 6 h  x  - 4 h x
```

Its average growth rate in units on the y-axis per unit on the x-axis over the interval $[x, x+h]$ is:

In[18]:=

```
xgrowth = h; Clear[fAverageGrowthRate];
fAverageGrowthRate[x_,h_] = Expand[(f[x + h] - f[x])/h]
```

Out[18]=

```
   2    3                2          2          2      3
2 h  - h  + 6 h x - 4 h  x + 6 x  - 6 h x  - 4 x
```

This gives you the measurement of the average growth rate of $f[x]$ on the interval $[x, x+h]$. For instance, when you look at:

In[19]:=

```
fAverageGrowthRate[0,h]
```

Out[19]=

```
   2    3
2 h  - h
```

Then you see that, on the average, $f[x]$ grows

$$2h^2 - h^3$$

times as fast as x grows as x advances from 0 to $0+h$. But when you look at:

In[20]:=

```
fAverageGrowthRate[1,h]
```

Out[20]=

```
         2    3
2 - 2 h  - h
```

Then you see that, on the average, f grows

$$2 - 2h^2 - h^3$$

times as fast as x grows as x advances from 1 to $1 + h$.

B.1.a.iv) Measure the instantaneous growth rate $f'[x]$ of

$$f[x] = 1 + 2x^3 - x^4$$

at a point x.

Answer:

In[21]:=

```
Clear[f,x]; f[x_] = 1 + 2 x^3 - x^4;
```

Its average growth rate in units on the y-axis per unit on the x-axis over the interval $[x, x+h]$ is:

In[22]:=

```
xgrowth = h; Clear[fAverageGrowthRate];
fAverageGrowthRate[x_,h_] = Expand[(f[x + h] - f[x])/h]
```

Out[22]=

```
   2    3                 2         2             2       3
2 h  - h  + 6 h x - 4 h  x + 6 x  - 6 h x  - 4 x
```

The instantaneous rate of change at x is the limiting case of the above average rate as h closes in on 0. Evidently this is

$$0 - 0 + 0 - 0 + 6x^2 - 0 - 4x^3 = 6x^2 - 4x^3.$$

So the instantaneous growth rate of $f[x] = 1 + 2x^3 - x^4$ at a point x in units on the y-axis per unit on the x-axis is given by

$$f'[x] = 6x^2 - 4x^3.$$

Mathematica knows how to calculate the instantaneous growth rate $f'[x]$. You just have to ask for it:

In[23]:=

```
f'[x]
```

Out[23]=

```
   2      3
6 x  - 4 x
```

Nice work, *Mathematica*. For instance, when you look at:

In[24]:=
```
f'[1.2]
```
Out[24]=
```
1.728
```

then you see that, as x advances through 1.2, $f[x]$ grows 1.728 times as fast as x grows. This tells you that $f[x]$ goes up as x advances through 1.2. But when you look at:

In[25]:=
```
f'[1.6]
```
Out[25]=
```
-1.024
```

then you see that, as x advances through 1.6, $f[x]$ grows -1.024 times as fast as x grows. This tells you that $f[x]$ goes down as x advances through 1.6.

B.1.b.i) Here is a plot of

$$f[x] = 1 + 2\,x^3 - x^4$$

together with a plot of the function $f'[x]$ that measures the instantaneous growth rate of $f[x]$ at a point x.

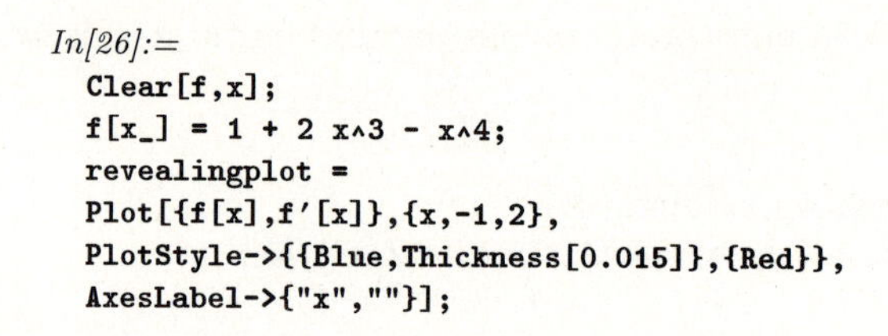

In[26]:=
```
Clear[f,x];
f[x_] = 1 + 2 x^3 - x^4;
revealingplot =
Plot[{f[x],f'[x]},{x,-1,2},
PlotStyle->{{Blue,Thickness[0.015]},{Red}},
AxesLabel->{"x",""}];
```

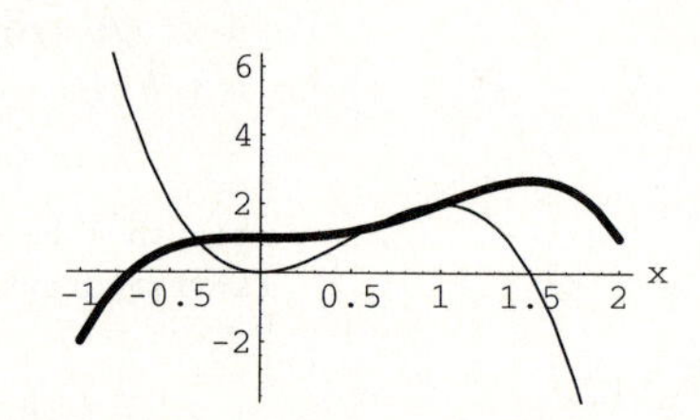

The plot of the instantaneous growth rate $f'[x]$ is the thinner of the two plots.

> Interpret the relationship between the two plots.

Answer: Here are the things that should grab your attention:

→ When $f'[x]$ is positive, then $f[x]$ is going up.

→ When $f'[x]$ is negative, then $f[x]$ is going down.

These are both natural because $f'[x]$ measures the instantaneous growth rate of $f[x]$ at x.

B.1.b.ii) Here is a plot of the function $f'[x]$ that measures the instantaneous growth rate of

$$f[x] = 1 + 2\,x^3 - x^4$$

at a point x.

In[27]:=
```
Clear[f,x]
f[x_] = 1 + 2 x^3 - x^4;
growthrateplot =
Plot[f'[x],{x,-1,2},
PlotStyle->{{Red}},PlotRange->All,
AxesLabel->{"x","f'[x]"}];
```

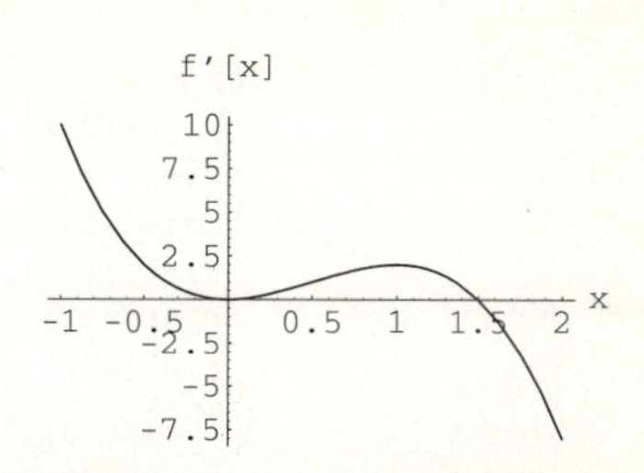

> Use this plot of the instantaneous growth rate $f'[x]$ to read off the answers to the following three questions:
>
> At what point in $[-1, 2]$ is $f[x]$ going up most rapidly?
>
> At what point in $[-1, 2]$ is $f[x]$ going down most rapidly?
>
> At what points in $[-1, 2]$ is $f[x]$ changing least rapidly?

Answer: Look again at the plot of $f'[x]$ for $-1 \leq x \leq 2$. This plot tells you that:

→ The instantaneous growth rate $f'[x]$ is highest at the left endpoint $x = -1$, so the function $f[x]$ is going up most rapidly at $x = -1$.

→ The instantaneous growth rate $f'[x]$ is lowest at the right endpoint $x = 2$, so the function $f[x]$ is going down most rapidly at $x = 2$.

→ The instantaneous growth rate $f'[x]$ is 0 at $x = 0$ and $x = 1.5$, so the function $f[x]$ is changing the least rapidly at $x = 0$ and $x = 1.5$.

Take a look at the plot of $f[x]$ to make sure that this makes sense:

In[28]:=
```
fplot = Plot[f[x],{x,-1,2},
PlotStyle->{{Blue,Thickness[0.015]}},
AxesLabel->{"x","f[x]"}];
```

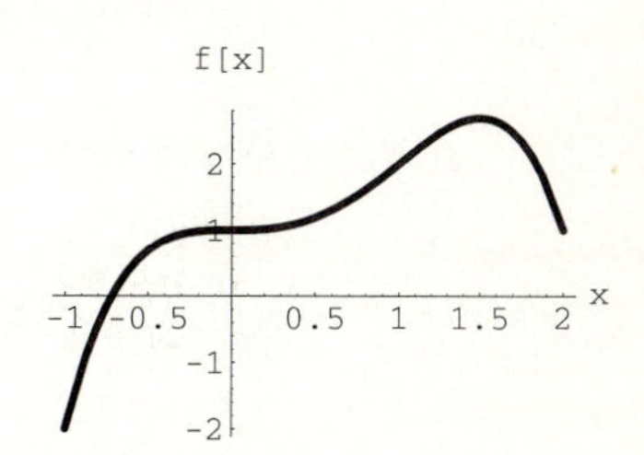

Sure enough: Rapid increase at the left endpoint, rapid decrease at the right endpoint, and lazy growth at $x = 0$ and $x = 1.5$.

B.1.c.i) Here is a plot on $[0, 12]$ of

$$f[x] = x^2 e^{-x}$$

together with a plot of the function $f'[x]$ that measures the instantaneous growth rate of $f[x]$ at a point x.

In[29]:=
```
Clear[f,x]
f[x_] = x^2 E^(-x);
revealingplot =
Plot[{f[x],f'[x]},{x,0,12},
PlotStyle->{{Blue,Thickness[0.015]},{Red}},
AxesLabel->{"x",""}];
```

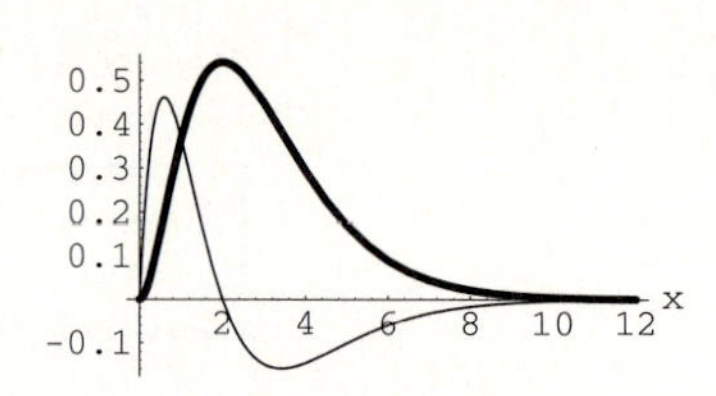

The plot of the instantaneous growth rate $f'[x]$ is the thinner of the two plots.

> Interpret the relationship between the two plots.

Answer: Again:

→ When $f'[x]$ is positive, then $f[x]$ is going up.

→ When $f'[x]$ is negative, then $f[x]$ is going down.

These are both natural because $f'[x]$ measures the instantaneous growth rate of $f[x]$ at x.

B.1.c.ii) This problem appears only in the electronic version.

■ B.2) The instantaneous growth rate of x^k is measured by $k\,x^{k-1}$

B.2.a)

> Measure the instantaneous growth rate $f'[x]$ of $f[x] = x$ at any point x.

Answer:

In[30]:=
```
Clear[x,f]; f[x_] = x
```

Out[30]=
```
x
```

Look at the average growth rate on the interval $[x, x + h]$ in units on the y-axis per unit on the x-axis:

In[31]:=
```
Clear[h]; Expand[(f[x + h] - f[x])/h]
```

Out[31]=
```
1
```

The instantaneous rate of change at x is the limiting case of the above average rates as h closes in on 0. Evidently this is 1. So the instantaneous growth rate of $f[x] = x$ at a point x in units on the y-axis per unit on the x-axis is

$$f'[x] = 1.$$

This is not much of a surprise because $f[x] = x$ goes up h units every time x goes up h units. *Mathematica* knows how to calculate $f'[x]$:

In[32]:=
```
f'[x]
```
Out[32]=
```
1
```

Good.

B.2.b.i) Measure the instantaneous growth rate $f'[x]$ of $f[x] = x^2$ at any point x.

Answer:

In[33]:=
```
Clear[f,x]; f[x_] = x^2;
```

Its average growth rate in units on the y-axis per unit on the x-axis over the interval $[x, x+h]$ is:

In[34]:=
```
Expand[(f[x + h] - f[x])/h]
```
Out[34]=
```
h + 2 x
```

The instantaneous rate of change at x is the limiting case of the above average growth rate as h closes in on 0. Evidently this is

$$0 + 2\,x.$$

So the instantaneous growth rate of $f[x] = x^2$ at a point x in units on the y-axis per unit on the x-axis is given by

$$f'[x] = 2\,x.$$

Mathematica knows how to calculate the instantaneous growth rate $f'[x]$:

In[35]:=
```
f'[x]
```
Out[35]=
```
2 x
```

B.2.b.ii) Continue to go with $f[x] = x^2$ and illustrate how the

$$\frac{f[x+h] - f[x]}{h}$$

curves crawl onto the $f'[x]$ curve as h closes in on 0.

Answer: Here are the $(f[x+h] - f[x])/h$ and $f'[x]$ curves plotted for

$h = 0.5$:

In[36]:=

```
Clear[f,x]
f[x_] = x^2;
h = 0.5;
Plot[{f'[x],(f[x + h] - f[x])/h},{x,-2,4},
PlotStyle->
{{GrayLevel[0.5],Thickness[0.01]},{Red}},
AxesLabel->{"x",""},PlotRange->{-4,8},
PlotLabel->h"= h"];
```

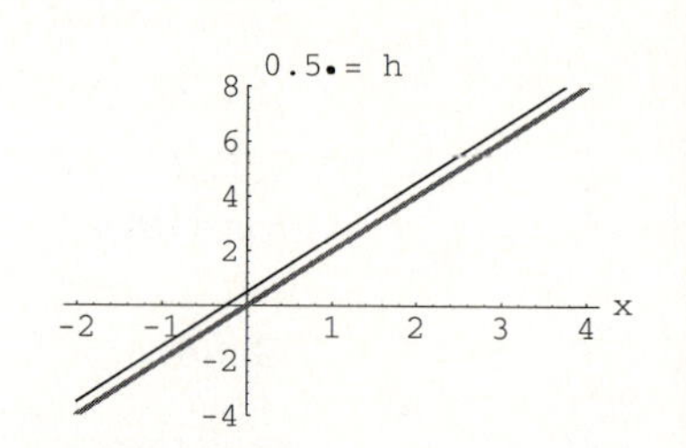

The $f'[x]$ curve is the darker curve.

$h = 0.1$:

In[37]:=

```
h = 0.1;
Plot[{f'[x],(f[x + h] - f[x])/h},{x,-2,4},
PlotStyle->
{{GrayLevel[0.5],Thickness[0.01]},{Red}},
AxesLabel->{"x",""},PlotRange->{-4,8},
PlotLabel->h"= h"];
```

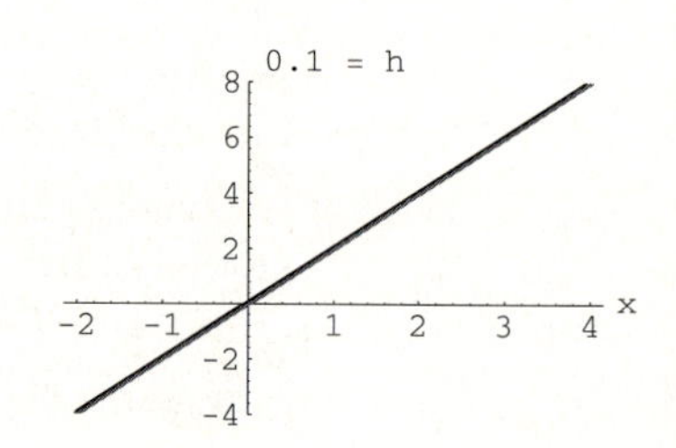

$h = 0.01$:

In[38]:=

```
h = 0.01;
Plot[{f'[x],(f[x + h] - f[x])/h},{x,-2,4},
PlotStyle->
{{GrayLevel[0.5],Thickness[0.01]},{Red}},
AxesLabel->{"x",""},PlotRange->{-4,8},
PlotLabel->h"= h"];
```

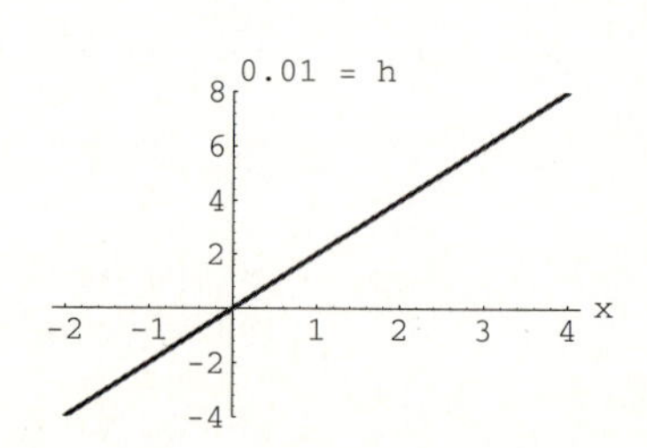

As h closes in on 0, the $(f[x+h] - f[x])/h$ curves crawl onto the $f'[x]$ curve.

B.2.c.i) Measure the instantaneous growth rate $f'[x]$ of $f[x] = x^3$ at any point x.

Answer: Its average growth rate in units on the y-axis per unit on the x-axis over the interval $[x, x+h]$ is:

In[39]:=

```
Clear[f,h,x]; f[x_] = x^3;
Expand[(f[x + h] - f[x])/h]
```

Out[39]=

```
 2              2
h  + 3 h x + 3 x
```

The instantaneous rate of change at x is the limiting case of the above average growth rate as h closes in on 0. Evidently this is

$$0 + 0 + 3x^2.$$

So the instantaneous growth rate of $f[x] = x^3$ at a point x in units on the y-axis per unit on the x-axis is given by:

$$f'[x] = 3x^2.$$

Mathematica knows how to calculate the instantaneous growth rate $f'[x]$:

In[40]:=
```
f'[x]
```
Out[40]=
```
   2
3 x
```

B.2.c.ii) Continue to go with

$$f[x] = x^3$$

and illustrate how the $(f[x+h] - f[x])/h$ curves crawl onto the $f'[x]$ curve as h closes in on 0.

Answer: Here are the $(f[x+h] - f[x])/h$ and $f'[x]$ curves plotted for

$h = 0.5$:

In[41]:=
```
Clear[f,x]
f[x_] = x^3;
h = 0.5;
Plot[{f'[x],(f[x + h] - f[x])/h},{x,-2,2},
PlotStyle->
{{GrayLevel[0.5],Thickness[0.01]},{Red}},
AxesLabel->{"x",""},PlotRange->{-1,12},
PlotLabel->h"= h"];
```

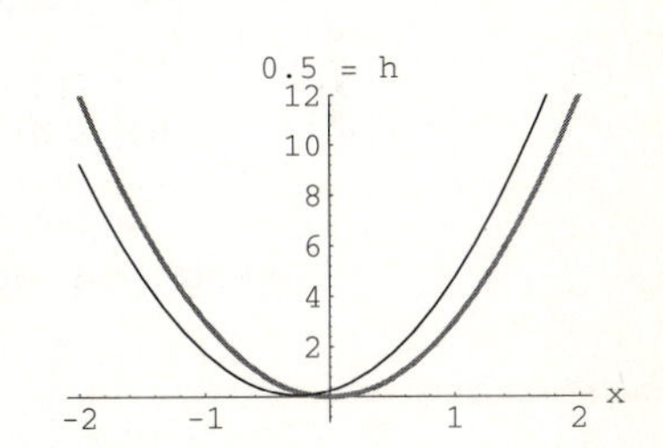

$h = 0.1$:

In[42]:=
```
h = 0.1;
Plot[{f'[x],(f[x + h] - f[x])/h},{x,-2,2},
PlotStyle->
{{GrayLevel[0.5],Thickness[0.01]},{Red}},
AxesLabel->{"x",""},PlotRange->{-1,12},
PlotLabel->h"= h"];
```

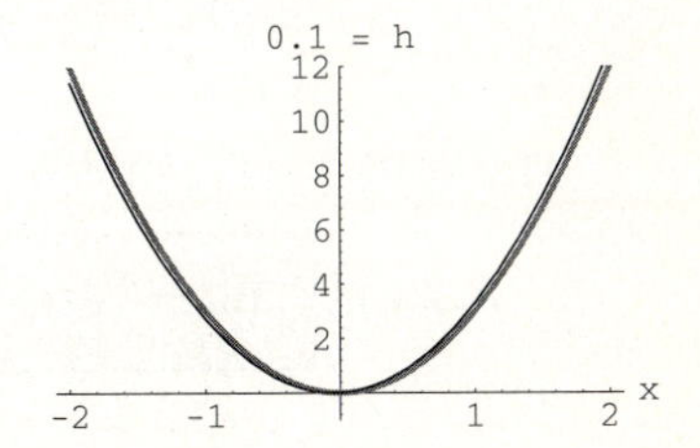

$h = 0.001$:

In[43]:=
```
h = 0.001;
Plot[{f'[x],(f[x + h] - f[x])/h},{x,-2,2},
PlotStyle->
{{GrayLevel[0.5],Thickness[0.01]},{Red}},
AxesLabel->{"x",""},PlotRange->{-1,12},
PlotLabel->h"= h"];
```

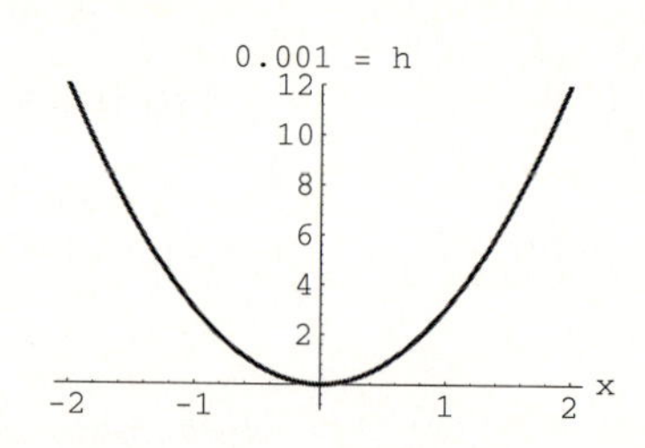

The $f'[x]$ curve is the thicker curve. As h closes in on 0, the $(f[x+h] - f[x])/h$ curves crawl onto the $f'[x]$ curve.

B.2.d.i) Measure the instantaneous growth rate $f'[x]$ of $f[x] = x^6$ at any point x.

Answer: Its average growth rate in units on the y-axis per unit on the x-axis over the interval $[x, x+h]$ is:

In[44]:=
```
Clear[f,h,x]; f[x_] = x^6;
Expand[(f[x + h] - f[x])/h]
```

Out[44]=
```
 5      4         3  2       2  3         4      5
h  + 6 h  x + 15 h  x  + 20 h  x  + 15 h x  + 6 x
```

The instantaneous rate of change at x is the limiting case of the above average growth rate as h closes in on 0. Evidently this is

$$0 + 0 + 0 + 0 + 0 + 6\,x^5.$$

So the instantaneous growth rate of $f[x] = x^6$ at a point x in units on the y-axis per unit on the x-axis is given by:

$$f'[x] = 6\,x^5.$$

Mathematica knows how to calculate the instantaneous growth rate $f'[x]$:

In[45]:=
```
f'[x]
```

Out[45]=
```
   5
6 x
```

B.2.d.ii) This problem appears only in the electronic version.

B.2.e) What is the instantaneous growth rate $f'[x]$ of $f[x] = x^k$ at any point x?

Answer: It would be silly to bet against

$$f'[x] = k\,x^{k-1}.$$

Try it:

In[46]:=
```
Clear[f,k,x]; f[x_] = x^k; f'[x]
```

Out[46]=
```
     -1 + k
k x
```

There it is: $f'[x] = k\,x^{k-1}$.

■ B.3) The instantaneous growth rate of $\sin[x]$ is measured by $\cos[x]$. The instantaneous growth rate of $\cos[x]$ is measured by $-\sin[x]$.

B.3.a) Put $f[x] = \sin[x]$ and look at the following plots of the average growth rates

$$\frac{f[x+h]-f[x]}{h}$$

for $h = 1.0$, $h = 0.5$, $h = 0.1$, and $h = 0.001$:

In[47]:=
```
Clear[f,x]
f[x_] = Sin[x];
h = 1.0;
plot1 = Plot[(f[x + h] - f[x])/h,{x,0,2 Pi},
PlotStyle->{{Thickness[0.01],Red}},
AxesLabel->{"x",""},
PlotLabel->h"= h"];
```

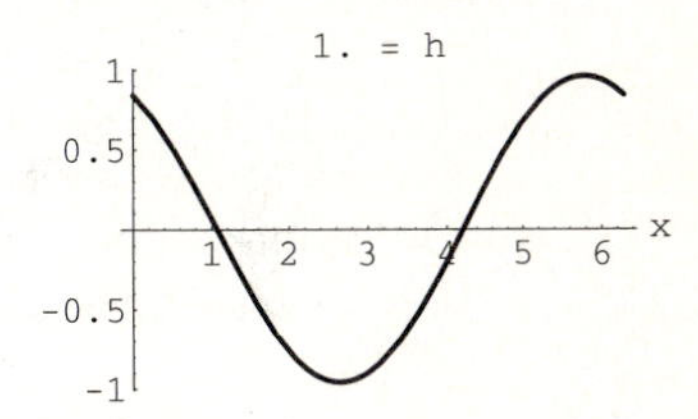

In[48]:=

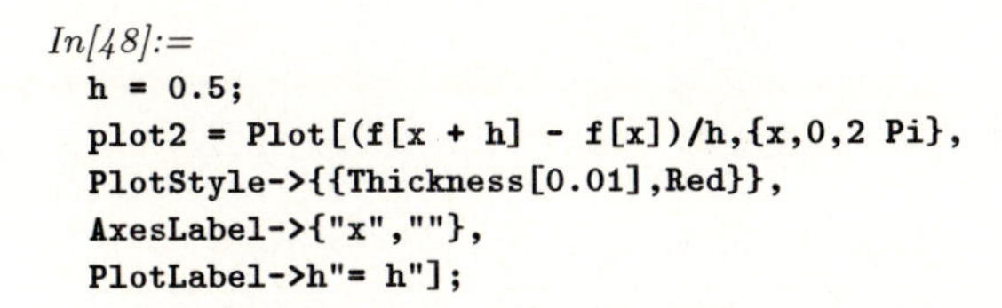

```
h = 0.5;
plot2 = Plot[(f[x + h] - f[x])/h,{x,0,2 Pi},
PlotStyle->{{Thickness[0.01],Red}},
AxesLabel->{"x",""},
PlotLabel->h"= h"];
```

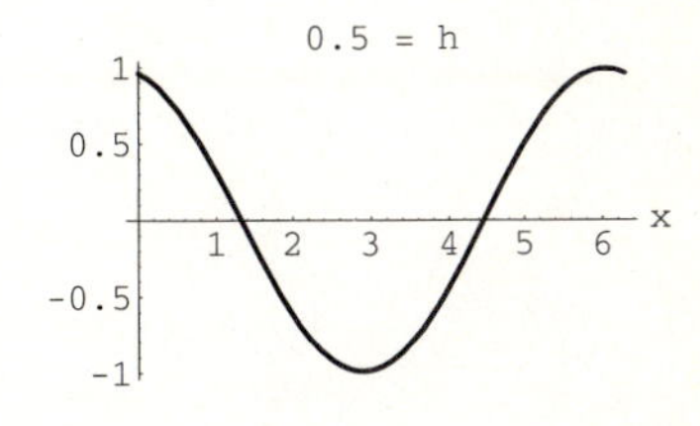

In[49]:=

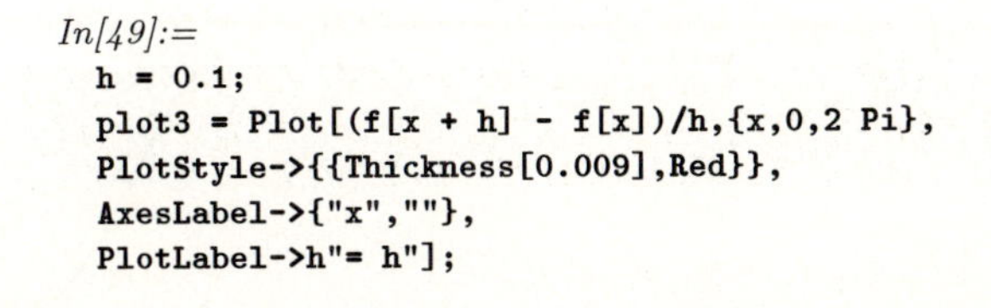

```
h = 0.1;
plot3 = Plot[(f[x + h] - f[x])/h,{x,0,2 Pi},
PlotStyle->{{Thickness[0.009],Red}},
AxesLabel->{"x",""},
PlotLabel->h"= h"];
```

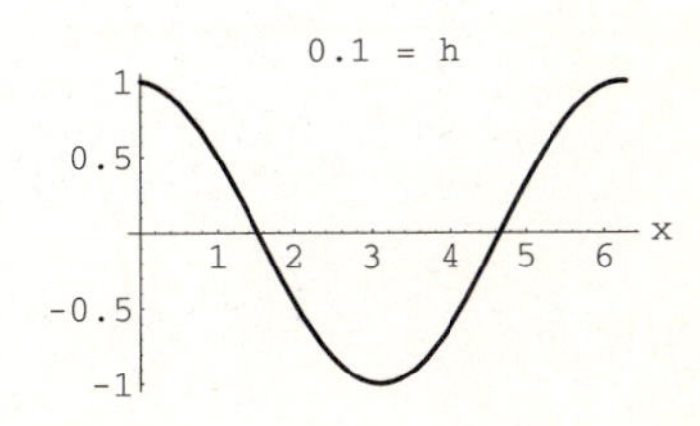

In[50]:=

```
h = 0.001;
plot4 = Plot[(f[x + h] - f[x])/h,{x,0,2 Pi},
PlotStyle->{{Thickness[0.009],Red}},
AxesLabel->{"x",""},
PlotLabel->h"= h"];
```

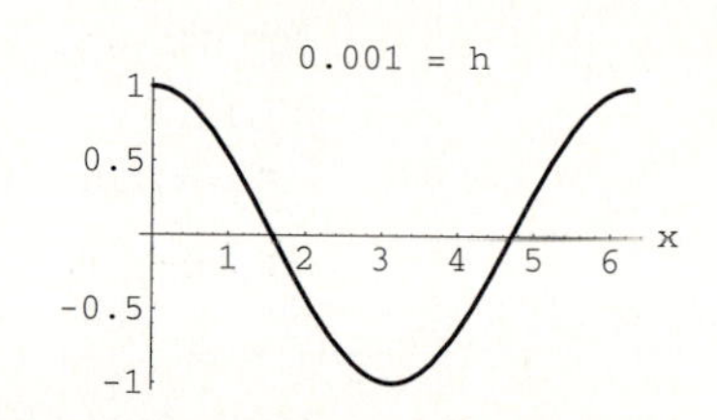

> What do these plots tell you about the instantaneous growth rate $f'[x]$ of $f[x] = \sin[x]$?

Answer: Damned if these curves don't look like cosine curves! Check this out by superimposing a plot of $\cos[x]$ on each of the plots above. Here is a plot of $\cos[x]$:

In[51]:=

```
cosineplot = Plot[Cos[x],{x,0,2 Pi},
PlotStyle->GrayLevel[0.2]];
```

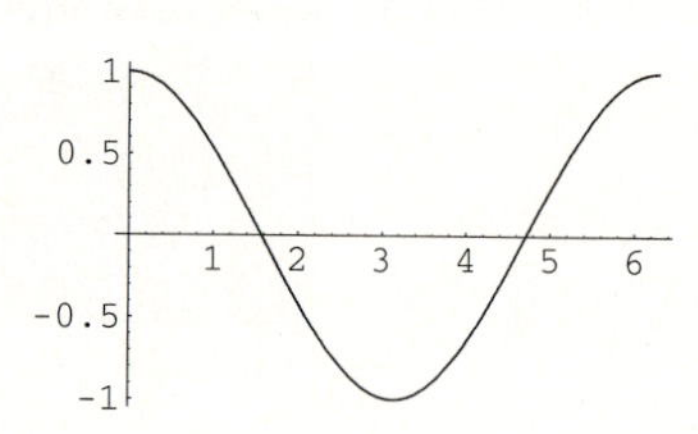

Here is the plot of $(f[x+h] - f[x])/h$ and $\cos[x]$ on the same axes for $h = 1.0$:

In[52]:=

```
Show[plot1,cosineplot];
```

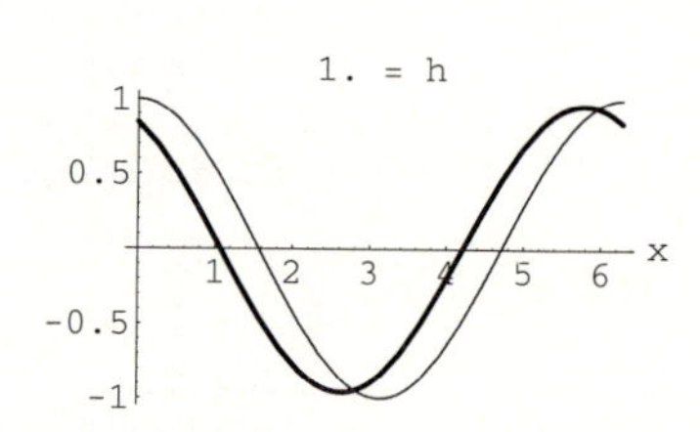

The plot of $\cos[x]$ is the thinner plot.

Kissing cousins. Here is the plot of $(f[x+h] - f[x])/h$ and $\cos[x]$ on the same axes for $h = 0.5$:

In[53]:=

```
Show[plot2,cosineplot];
```

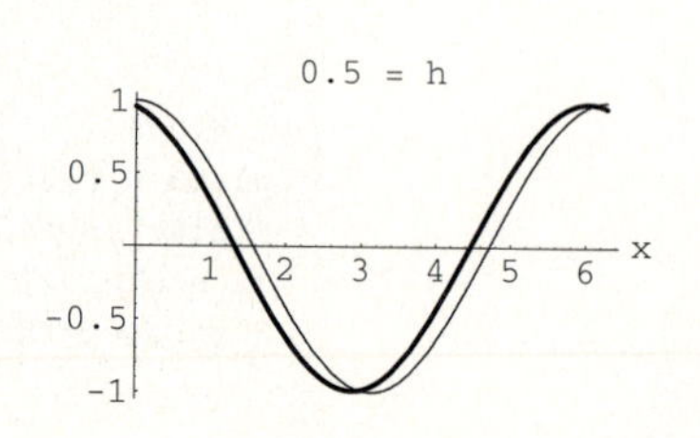

Getting closer. Here is the plot of $(f[x+h]-f[x])/h$ and $\cos[x]$ on the same axes for $h = 0.1$:

In[54]:=
```
Show[plot3,cosineplot];
```

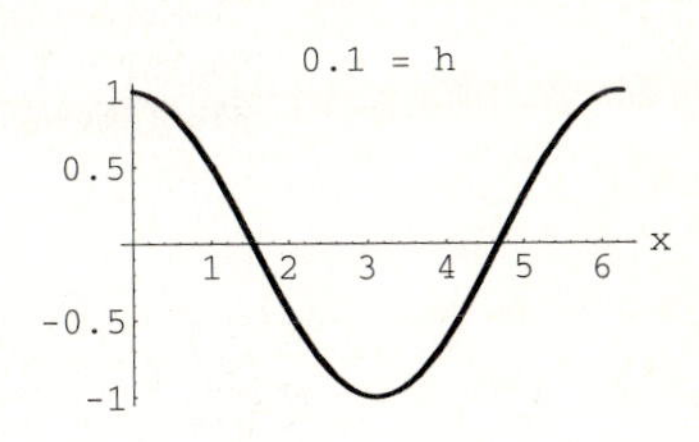

Getting really close. Here is the plot of $(f[x+h]-f[x])/h$ and $\cos[x]$ on the same axes for $h = 0.001$:

In[55]:=
```
Show[plot4,cosineplot];
```

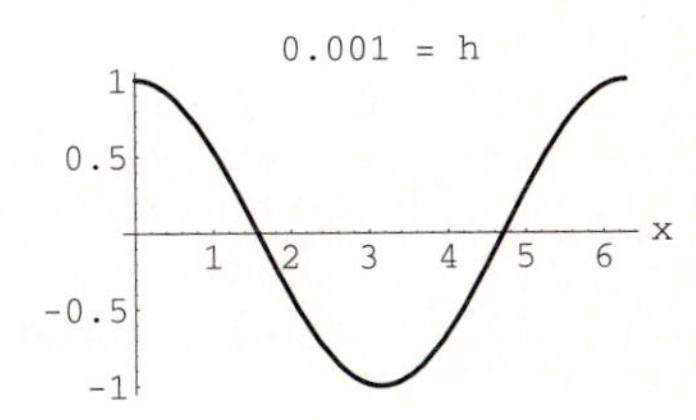

Sharing the same ink. When you go with $f[x] = \sin[x]$, then as h closes in on 0, the plots of $(f[x+h]-f[x])/h$ crawl right onto the $\cos[x]$ plot.

The evidence is that

$$f'[x] = \cos[x].$$

Check this out with *Mathematica*:

In[56]:=
```
f'[x]
```

Out[56]=
```
Cos[x]
```

How sweet it is. Math happens.

B.3.b) How well does the average growth rate

$$\frac{\sin[x+0.00001]-\sin[x]}{0.00001}$$

approximate the instantaneous growth rate

$$f'[x] = \cos[x]$$

of $f[x] = \sin[x]$ at a point x for $0 \leq x \leq 2\pi$?

Answer: Plot their difference for $0 \leq x \leq 2\pi$:

In[57]:=
```
Clear[x,difference]; difference[x_] =
Cos[x] - ((Sin[x + 0.00001] - Sin[x])/0.00001);
Plot[difference[x],{x,0,2 Pi}];
```

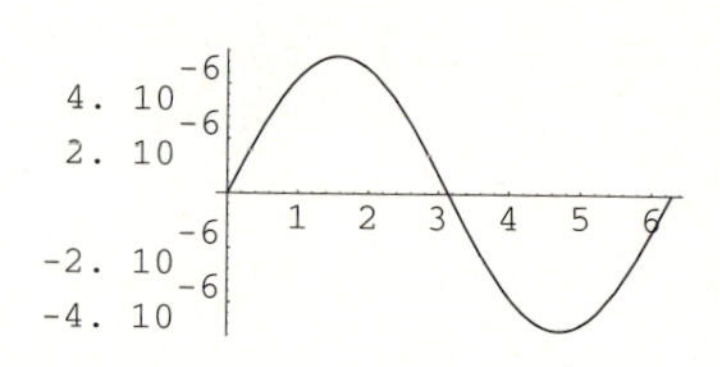

The plot shows $(\sin[x + 0.00001] - \sin[x])/0.00001$ is $f'[x] = \cos[x]$ within 5.0×10^{-6}. This means

$$\frac{\sin[x + 0.00001] - \sin[x]}{0.00001}$$

is

$$f'[x] = \cos[x]$$

to at least five accurate decimals. For most practical purposes, there is no difference between the average growth rate $(\sin[x + 0.00001] - \sin[x])/0.00001$ and the instantaneous growth rate $f'[x] = \cos[x]$.

B.3.c) This time go with

$$f[x] = \cos[x]$$

and say what function measures the instantaneous growth rate $f'[x]$, of $f[x]$.

Answer: If you don't know, then ask *Mathematica*:

In[58]:=
```
Clear[f,x]; f[x_] = Cos[x]; f'[x]
```

Out[58]=
```
-Sin[x]
```

To check that this is plausible, look at a plot of $(f[x+h] - f[x])/h$ together with a plot of $-\sin[x]$ for a pretty small h:

In[59]:=
```
h = 0.01;
Plot[{(f[x + h] - f[x])/h,-Sin[x]},
{x,0,2 Pi},
PlotStyle->
{{Thickness[0.009],Red},{GrayLevel[0.2]}},
AxesLabel->{"x",""}, PlotLabel->h"= h"];
```

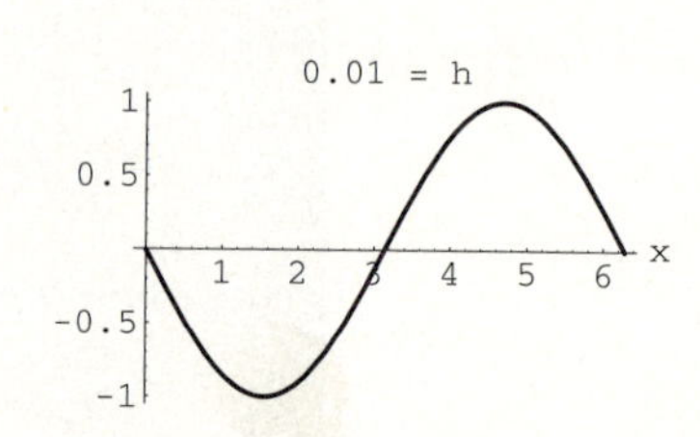

Sharing lots of ink; the plot confirms that when you go with $f[x] = \cos[x]$, then the instantaneous growth rate $f'[x]$ of $f[x]$ is measured by $f'[x] = -\sin[x]$.

■ B.4) The instantaneous growth rate of $\log[x]$ is measured by $1/x$. The instantaneous growth rate of e^x is measured by e^x.

B.4.a.i) Look at *Mathematica*'s calculation of the instantaneous growth rate of $f[x] = \log[x]$.

In[60]:=
```
Clear[f,x]; f[x_] = Log[x]; f'[x]
```

Out[60]=
```
1
-
x
```

> *Mathematica* says that the instantaneous growth rate of $f[x] = \log[x]$ is $f'[x] = 1/x$. Confirm with a plot.

Answer: To check that this is plausible, go with $f[x] = \log[x]$ and look at a plot of $(f[x+h] - f[x])/h$ together with a plot of $1/x$ for a pretty small h:

In[61]:=
```
Clear[f,x]
f[x_] = Log[x];
h = 0.01;
Plot[{(f[x + h] - f[x])/h,1/x},{x,0.5,8},
PlotStyle->
{{Thickness[0.009],Red},{GrayLevel[0.2]}},
AxesLabel->{"x",""},
PlotLabel->h"= h"];
```

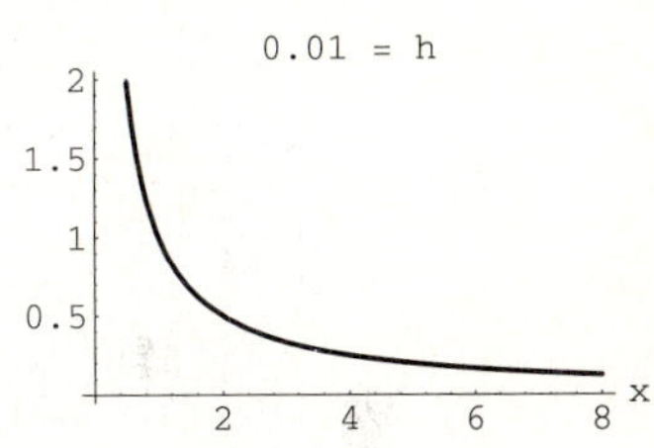

Sharing lots of ink; the plot confirms that when you go with $f[x] = \log[x]$, then the instantaneous growth rate $f'[x]$ of $f[x]$ is measured by $f'[x] = 1/x$.

B.4.a.ii)

> Recall that $\log[x]$ is the logarithm whose base is e. Why do nearly all folks like to call $\log[x]$ by the name "natural logarithm"?

Answer: Look at *Mathematica*'s calculation of the instantaneous growth rate of the base 10 logarithm:

In[62]:=
```
Clear[f,x]; f[x_] = Log[10,x]; N[f'[x]]
```

Out[62]=
```
0.434294
--------
   x
```

A weird rounded number in the numerator. That's bad for precise calculation. Look at *Mathematica*'s calculation of the instantaneous growth rate of the base 2 logarithm:

In[63]:=
```
Clear[f,x]; f[x_] = Log[2,x]; N[f'[x]]
```
Out[63]=
```
1.4427
------
  x
```
Another weird rounded number in the numerator. That's still bad for precise calculation. Look at some more:

In[64]:=
```
Clear[f,x]; f[x_] = Log[3,x]; N[f'[x]]
```
Out[64]=
```
0.910239
--------
   x
```
In[65]:=
```
Clear[f,x]; f[x_] = Log[Pi,x]; N[f'[x]]
```
Out[65]=
```
0.873569
--------
   x
```
But when you go with base e, then you get:

In[66]:=
```
Clear[f,x]; f[x_] = Log[E,x]; N[f'[x]]
```
Out[66]=
```
1
-
x
```
$\log[e, x]$ and $\log[x]$ are the very same thing:

In[67]:=
```
Clear[f,x]; f[x_] = Log[x]; N[f'[x]]
```
Out[67]=
```
1
-
x
```
No weird rounded numbers here; just a clean 1 in the numerator. This instantaneous growth rate is beautiful for precise calculation. This is why folks call $\log[x]$, the logarithm whose base is e, by the name "natural log."

B.4.b.i) Look at *Mathematica*'s calculation of the instantaneous growth rate of $f[x] = e^x$.

In[68]:=
```
Clear[f,x]; f[x_] = E^x; f'[x]
```
Out[68]=
```
 x
E
```
> *Mathematica* says that the instantaneous growth rate of $f[x] = e^x$ is $f'[x] = e^x$. Confirm with a plot.

Answer: To check that this is plausible, go with $f[x] = e^x$ and look at a plot of $(f[x+h] - f[x])/h$ together with a plot of e^x for a pretty small h:

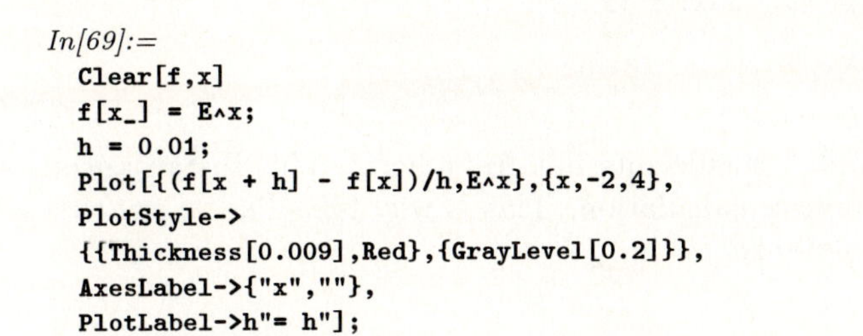

```
In[69]:=
  Clear[f,x]
  f[x_] = E^x;
  h = 0.01;
  Plot[{(f[x + h] - f[x])/h,E^x},{x,-2,4},
  PlotStyle->
  {{Thickness[0.009],Red},{GrayLevel[0.2]}},
  AxesLabel->{"x",""},
  PlotLabel->h"= h"];
```

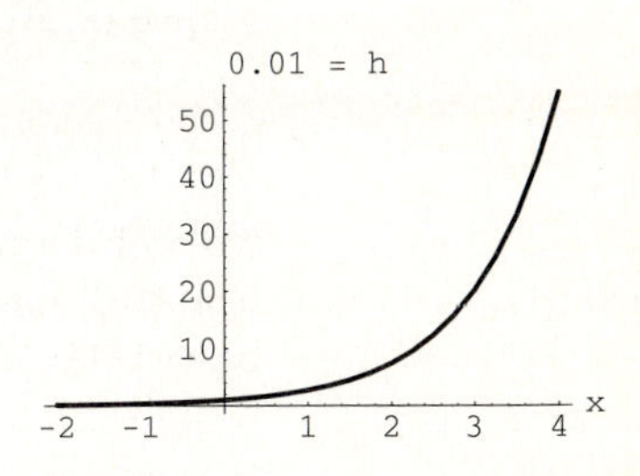

Sharing lots of ink; the plot confirms that when you go with $f[x] = e^x$, then the instantaneous growth rate $f'[x]$ of $f[x]$ is measured by $f'[x] = e^x$.

B.4.b.ii) Why do nearly all folks like to say that e is the natural base for exponentials?

Answer: Well, for one thing, e is the base of the natural logarithm. But that's selling e short. Look at *Mathematica*'s calculation of the instantaneous growth rate of 10^x:

```
In[70]:=
  Clear[f,x]; f[x_] = 10^x; N[f'[x]]

Out[70]=
              x
  2.30259 10.
```

A weird rounded number out front. That's bad for precise calculation.

Look at *Mathematica*'s calculation of the instantaneous growth rate of 2^x:

```
In[71]:=
  Clear[f,x]; f[x_] = 2^x; N[f'[x]]

Out[71]=
              x
  0.693147 2.
```

Another weird rounded number out front. Look at some more:

```
In[72]:=
  Clear[f,x]; f[x_] = 3^x; N[f'[x]]

Out[72]=
             x
  1.09861 3.

In[73]:=
  Clear[f,x]; f[x_] = Pi^x; N[f'[x]]

Out[73]=
                   x
  1.14473 3.14159
```

But when you go with base e, then you get:

In[74]:=

```
Clear[f,x]; f[x_] = E^x; f'[x]
```

Out[74]=

```
 x
E
```

No weird rounded coefficients out front here. This instantaneous growth rate is beautiful for precise calculation. This is why folks like to say that e is the natural base for exponentials.

Tutorials

■ T.1) Average growth rate versus instantaneous growth rate

Take $f[x] = -9x^4 + 10x^3 + 20x^2$:

In[1]:=

```
Clear[f,x]; f[x_] =  -9 x^4 + 10 x^3 + 20 x^2
```

Out[1]=

```
     2       3      4
20 x  + 10 x  - 9 x
```

Here is a plot of the average growth rate $(f[x+h] - f[x])/h$ of $f[x]$ on the interval $[x, x+h]$ for $h = 1$:

In[2]:=

```
h = 1;
Plot[(f[x + h] - f[x])/h,{x,-1,1},
PlotStyle->{Red},
AxesLabel->{"x",""}];
```

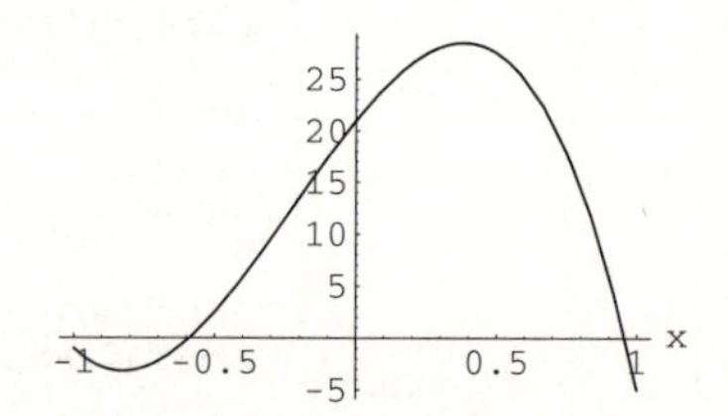

T.1.a.i) Use the plot to estimate the x_0 with $-1 \leq x_0 \leq 1$ at which $f[x]$ is showing the greatest average rate of increase on the interval $[x_0, x_0 + 1]$.

Answer: Take another look at the plot to see that a reasonable estimate of the x_0 for which $f[x]$ shows the greatest average rate of increase on the interval $[x_0, x_0 + 1]$ is $x_0 = 0.4$.

T.1.a.ii) Continue to go with the same $f[x]$. Here is a plot of the average growth rate $(f[x+h] - f[x])/h$ of $f[x]$ on the interval $[x, x+h]$ for $h = 0.1$:

In[3]:=
```
h = 0.1;
Plot[(f[x + h] - f[x])/h,{x,-1,1},
PlotStyle->{RGBColor[1,0,0]},
AxesLabel->{"x",""}];
```

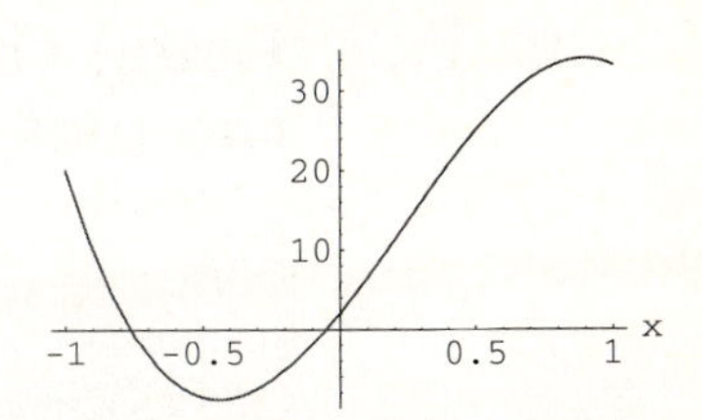

> Use the plot to estimate the x_0 with $-1 \leq x_0 \leq 1$ for which $f[x]$ is showing the greatest average rate of increase on the interval $[x_0, x_0 + 0.1]$.

Answer: A reasonable estimate is $x_0 = 0.9$.

T.1.a.iii) Continue to go with the same $f[x]$. Here is a plot of the instantaneous growth rate $f'[x]$:

In[4]:=
```
h = 0.1;
Plot[f'[x],{x,-1,1}, PlotStyle->{Red},
AxesLabel->{"x",""}];
```

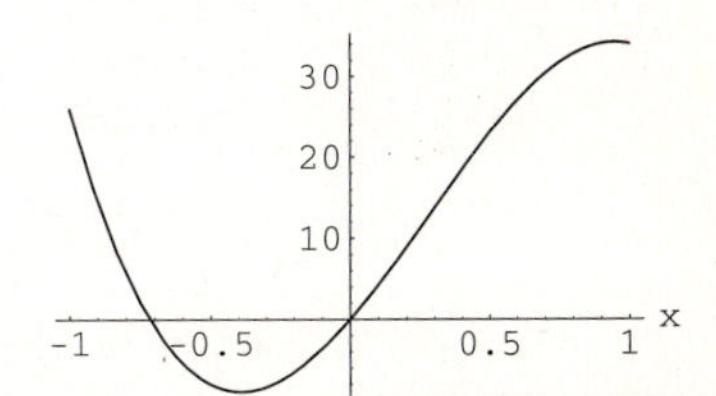

> Use the plot to estimate the x_0 with $-1 \leq x_0 \leq 1$ at which $f[x]$ is showing the greatest instantaneous rate of increase.

Answer: A reasonable estimate of the x_0 at which $f[x]$ shows the greatest instantaneous growth rate is $x_0 = 0.95$.

T.1.a.iv)

> Was it an accident that the answers to parts T.1.a.ii) and T.1.a.iii) above were nearly the same?

Answer: Get off it! In mathematics there are no accidents! Remember that $f'[x]$ is the limiting case of $(f[x+h] - f[x])/h$ as h closes in on 0. As a result, when you take $h = 0.1$, you can expect $f'[x]$ and $(f[x+h] - f[x])/h$ to be nearly the same functions. Check it out:

In[5]:=
```
h = 0.1;
Plot[{(f[x + h] - f[x])/h,f'[x]},
{x,-1,1},
PlotStyle->{{Red},
{Thickness[0.009],Blue}},
AxesLabel->{"x",""}];
```

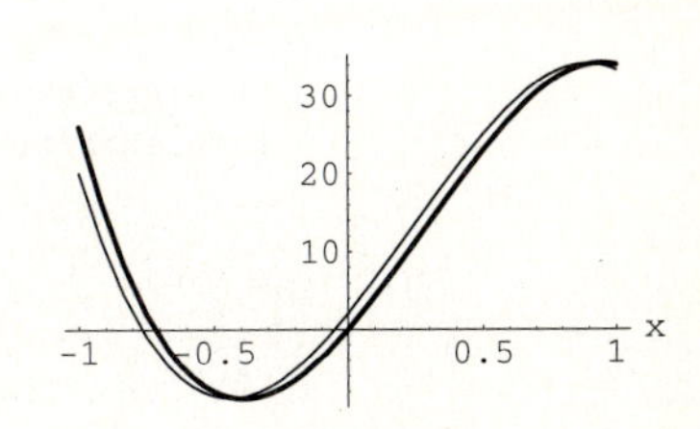

■ T.2) Using the instantaneous growth rate $f'[x]$ to predict the plot of $f[x]$

T.2.a)

> Take the function
>
> $$f[x] = x^3 e^{-x^2}$$
>
> and plot the instantaneous growth rate $f'[x]$ on $[-3, 3]$. Use it to predict how the plot of $f[x]$ on $[-3, 3]$ looks. Finally, test your prediction by plotting $f[x]$.

Answer: Here is a plot of the instantaneous growth rate $f'[x]$.

In[6]:=
```
Clear[x,f]
f[x_] = x^3 E^(-x^2);
growthplot =
Plot[f'[x],{x,-3,3},
PlotStyle->Red,
AxesLabel->{"x","f[x]"}];
```

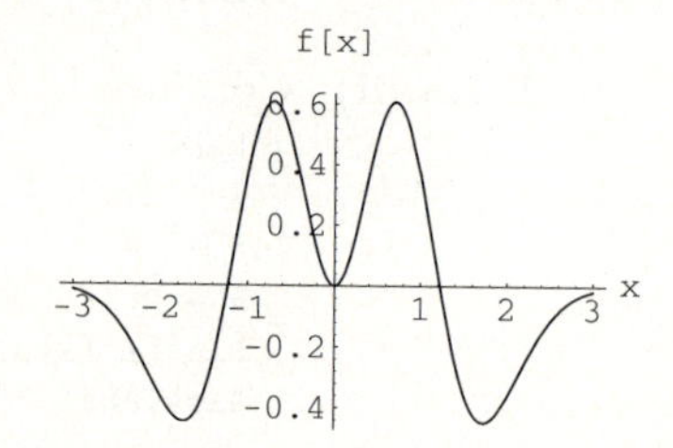

When the instantaneous growth rate $f'[x]$ is positive, then $f[x]$ is going up. Consequently, the smart money predicts that the graph of the function $f[x]$ will go up as x runs from -1.2 to 1.2.

When the instantaneous growth rate $f'[x]$ is negative, then $f[x]$ is going down. Consequently, the smart money predicts that the graph of the function $f[x]$ will go down on the intervals $[-3, 1.2]$ and $[1.2, 3]$.

Test the prediction:

In[7]:=
```
functplot = Plot[f[x],{x,-3,3},
PlotStyle->{{Blue,Thickness[0.01]}},
AxesLabel->{"x","f[x]"}];
```

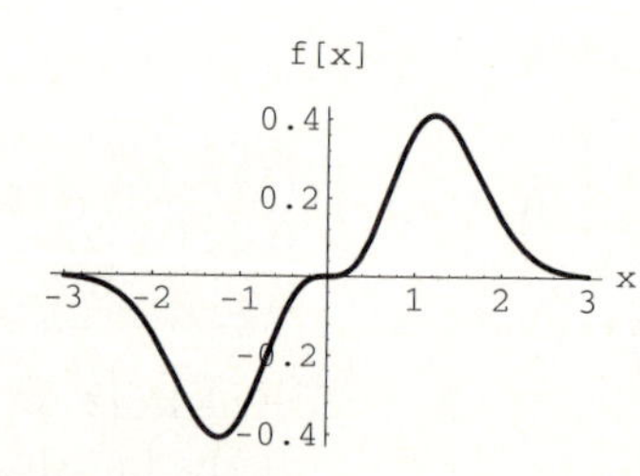

The smart money wins. Here is a plot of $f[x]$ and $f'[x]$ on the same axes for your enjoyment and good use:

In[8]:=
```
Show[growthplot,functplot,
AxesLabel->{"x",""}];
```

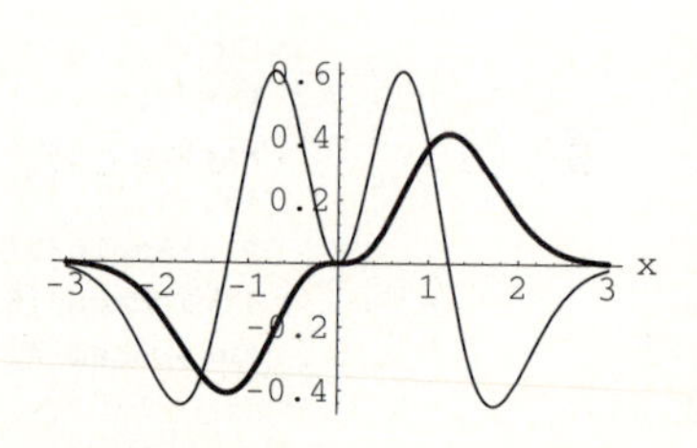

There you go:

→ When $f'[x]$ is positive, then $f[x]$ is going up.

→ When $f'[x]$ is negative, then $f[x]$ is going down.

■ T.3) Spread of disease

This problem appears only in the electronic version.

■ T.4) Instantaneous growth rates in context

The idea of instantaneous growth rate is pervasive in science, mathematics, and economics. Here are just a few of its manifestations. Browse the list, but don't try to memorize it.

→ *Velocity* is the instantaneous growth rate of distance as a function of time.

→ *Acceleration* is the instantaneous growth rate of velocity as a function of time.

→ *Jerk* is the instantaneous growth rate of acceleration as a function of time.

→ *Slope* is the instantaneous growth rate of $f[x]$ as a function of x. Thus $f'[x]$ measures the slope of the graph of $y = f[x]$ at the point $\{x, f[x]\}$.

→ *Density* is the instantaneous growth rate of weight as a function of volume.

→ *Force* is the instantaneous growth rate of momentum as a function of time.

→ *Marginal revenue* is the instantaneous growth rate of revenue as a function of level of production.

→ *Current* is the instantaneous growth rate of electrical charge as a function of time.

Give It a Try

Experience with the starred ($\star$) problems will be especially beneficial for understanding later lessons.

■ G.1) Relating $f[x]$ and $f'[x]^\star$

G.1.a) Go with $f[x] = 2 + x^2$ and look at a plot of $f[x]$ and its instantaneous growth rate $f'[x]$ on the same axes:

In[1]:=
```
Clear[x,f]
f[x_] = 2 + x^2;
Plot[{f'[x],f[x]},{x,-4,4},
PlotStyle->{Red,
{Thickness[0.01],Blue}},
AxesLabel->{"x",""},
PlotLabel->"f[x] and f'[x]"];
```

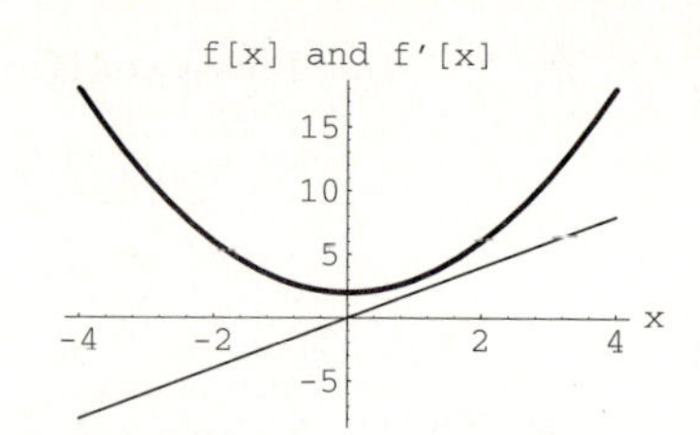

The plot of $f[x]$ is the thicker of the two.

> Discuss what you see, paying special attention to what $f[x]$ is doing when $f'[x]$ is positive and to what $f[x]$ is doing when $f'[x]$ is negative. How is $f[x]$ behaving when $f'[x]$ is near 0? Comment on the statement: The bigger $|f'[x]|$ is, the faster $f[x]$ changes.

G.1.b) Go with $f[x] = x^4e^{-x}$ and look at a plot of $f[x]$ and its instantaneous growth rate $f'[x]$ on the same axes for $-1 \leq x \leq 15$:

In[2]:=
```
Clear[x,f]
f[x_] = x^4 E^(-x);
Plot[{f'[x],f[x]},{x,-1,15},
PlotStyle->{Red,
{Thickness[0.01],Blue}},
AxesLabel->{"x",""},
PlotLabel->"f[x] and f'[x]"];
```

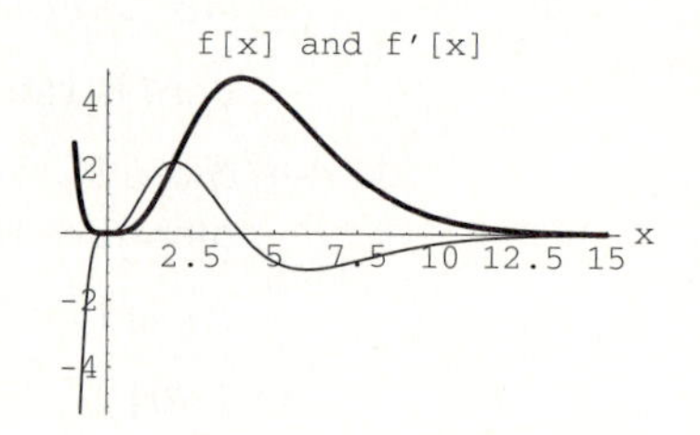

The plot of $f[x]$ is the thicker of the two.

> Discuss what you see, paying special attention to what $f[x]$ is doing when $f'[x]$ is positive and to what $f[x]$ is doing when $f'[x]$ is negative. How is $f[x]$ behaving when $f'[x]$ is near 0? Comment on the statement: The bigger $|f'[x]|$ is, the faster $f[x]$ changes.

G.1.c)

> Go with a function $f[x]$ of your own choice.
>
> Give a plot of $f[x]$ and its instantaneous growth rate $f'[x]$ on the same axes.
>
> Then discuss what you see, paying special attention to what $f[x]$ is doing when $f'[x]$ is positive and to what $f[x]$ is doing when $f'[x]$ is negative. How is $f[x]$ behaving when $f'[x]$ is near 0? Comment on the statement: The bigger $|f'[x]|$ is, the faster $f[x]$ changes.

■ G.2) Explaining *Mathematica* output★

You can explain the *Mathematica* output from:

In[3]:=
```
Clear[f,x]; f[x_] = x^5; f'[x]
```

as follows:

In[4]:=
```
Clear[h]; Expand[(f[x + h] - f[x])/h]
```

Out[4]=
```
 4      3           2  2          3      4
h  + 5 h  x + 10 h  x  + 10 h x  + 5 x
```

As h closes in on 0, this closes in on the instantaneous growth rate

$$f'[x] = 0 + 0 + 0 + 0 + 5\,x^4 = 5\,x^4.$$

For each of the following, explain the *Mathematica* output from:

G.2.a.i) *In[5]:=*
```
Clear[f,x]; f[x_] = x^5
```

In[6]:=
```
f'[x]
```

G.2.a.ii) *In[7]:=*
```
Clear[f,x]; f[x_] = 5 x^4 - 3 x^2 + 8
```

In[8]:=
```
f'[x]
```

G.2.b.i) *In[9]:=*
```
Clear[f,x]; f[x_] = Sin[2 x]
```

Out[9]=
```
Sin[2 x]
```

In[10]:=
```
f'[x]
```

Out[10]=
```
2 Cos[2 x]
```

When you try the same thing for this one, you get:

In[11]:=
```
Clear[h]; Expand[(f[x + h] - f[x])/h]
```

Out[11]=
```
   Sin[2 x]     Sin[2 (h + x)]
-(--------) + --------------
      h              h
```

It's hard to read off what happens as h closes in on 0; so you can plot this and $2\cos[2\,x]$ to try to explain the output:

In[12]:=
```
h = 0.1;
Plot[{(f[x + h] - f[x])/h,2 Cos[2 x]},{x,0,2 Pi},
PlotStyle->
{{Thickness[0.009],Red},{GrayLevel[0.2]}},
AxesLabel->{"x",""},
PlotLabel-> h "= h"];
```

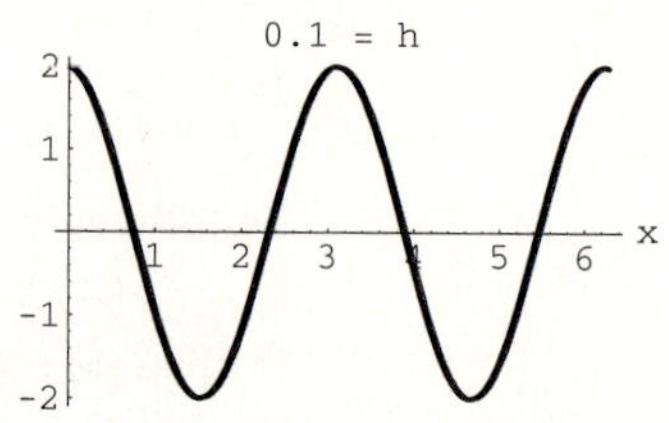

Looking good; it will look better if you go with a smaller h:

In[13]:=
```
h = 0.0001;
Plot[{(f[x + h] - f[x])/h,2 Cos[2 x]},{x,0,2 Pi},
PlotStyle->
{{Thickness[0.009],Red},{GrayLevel[0.2]}},
AxesLabel->{"x",""},
PlotLabel-> h "= h"];
```

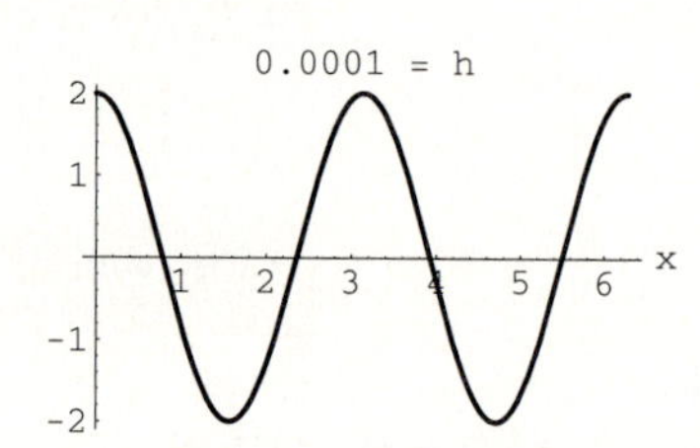

Sharing ink all the way. This is a superb visual explanation of the *Mathematica* output:

In[14]:=
```
f'[x]
```

Out[14]=
```
2 Cos[2 x]
```

Now go ahead and explain the output:

In[15]:=
```
Clear[f,x]; f[x_] = E^(2 x)
```

Out[15]=
```
 2 x
E
```

In[16]:=
```
f'[x]
```

Out[16]=
```
   2 x
2 E
```

G.2.b.ii) Explain the output:

In[17]:=
```
Clear[f,x]; f[x_] = Cos[2 x]
```
Out[17]=
```
Cos[2 x]
```
In[18]:=
```
f'[x]
```
Out[18]=
```
-2 Sin[2 x]
```

G.2.b.iii) Explain the output:

In[19]:=
```
Clear[f,x]; f[x_] = x E^(-x^2)
```
Out[19]=
```
 x
---
  2
 x
E
```
In[20]:=
```
f'[x]
```
Out[20]=
```
   2      2
 -x    2 x
E    - ----
          2
         x
        E
```
This is the same as:

In[21]:=
```
Together[f'[x]]
```
Out[21]=
```
       2
1 - 2 x
--------
    2
   x
  E
```

■ G.3) Approximation of the instantaneous growth rate $f'[x]$ by average growth rates $(f[x+h] - f[x])/h^\star$

G.3.a.i) Go with $f[x] = \log[x]$ so that $f'[x] = 1/x$. You can get definitive information on how well the average growth rate $(f[x+h] - f[x])/h$ approximates the true instantaneous growth rate $f'[x]$ on the interval $[2, 10]$ for $h = 0.01$ by plotting the difference $f'[x] - (f[x+h] - f[x])/h$:

In[22]:=
```
Clear[f,x]
f[x_] = Log[x];
h = 0.01;
Plot[f'[x] - (f[x + h] - f[x])/h ,{x,2,10},
PlotStyle->{{Red,Thickness[0.01]}},
PlotRange->All,AxesLabel->{"x",""},
PlotLabel->h "= h"];
```

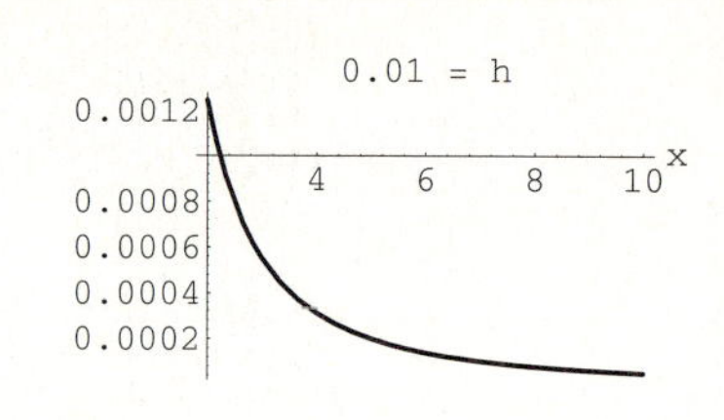

This tells you that when you go with $h = 0.01$, then as x runs from 2 to 10,

$$\frac{f[x+h] - f[x]}{h}$$

always runs within 0.0012 of $f'[x]$. Consequently, for $h = 0.01$, $(f[x+h] - f[x])/h$ calculates $f'[x]$ to at least two accurate decimals for all the x's with $2 \leq x \leq 10$.

Come up with a smaller h so that $(f[x+h] - f[x])/h$ calculates $f'[x]$ to at least four accurate decimals for all the x's with $2 \leq x \leq 10$.

G.3.a.ii) Continue to go with $f[x] = \log[x]$ so that $f'[x] = 1/x$. But this time come up with a small h so that $(f[x+h] - f[x])/h$ calculates $f'[x]$ to at least four accurate decimals for all the x's with $0.25 \leq x \leq 2.5$.

G.3.b.i) This time go with $f[x] = \sin[x]$ so that $f'[x] = \cos[x]$. Come up with a small h so that $(f[x+h] - f[x])/h$ calculates $f'[x]$ to at least six accurate decimals for all the x's with $0 \leq x \leq 2\pi$.

■ G.4) Using the instantaneous growth rate $f'[x]$ to predict the plot of $f[x]^\star$

G.4.a) Take the function

$$f[x] = x^2 e^{-x^2};$$

plot the instantaneous growth rate $f'[x]$ on $[-3, 3]$ and use it to predict how the plot of $f[x]$ on $[-3, 3]$ looks. Finally, test your prediction by plotting $f[x]$.

G.4.b) Take the function

$$f[x] = x + \sin[2\,x];$$

plot the instantaneous growth rate $f'[x]$ on $[0, 9]$ and use it to predict how the plot of $f[x]$ on $[0, 9]$ looks. Finally, test your prediction by plotting $f[x]$.

G.4.c) Take the function

$$f[x] = x \log[x];$$

plot the instantaneous growth rate $f'[x]$ on $[0.1, 2.1]$ and use it to predict how the plot of $f[x]$ on $[0.1, 2.1]$ looks. Finally, test your prediction by plotting $f[x]$.

■ G.5) Graphics action

Go with:

In[23]:=
```
Clear[x,f]; f[x_] = 5 x^4 - 20 x^3 + 21 x^2
```

Out[23]=
```
     2       3      4
21 x  - 20 x  + 5 x
```

Execute all the cells below. The plot of $f'[x]$ is the thicker of the two.

In[24]:=
```
h = 0.5;
plot1 =
Plot[{f'[x],(f[x + h] - f[x])/h},{x,0,2},
PlotStyle->{{Blue,Thickness[0.01]},{Red}},
PlotRange->{-5,10},
PlotLabel-> h "= h"];
```

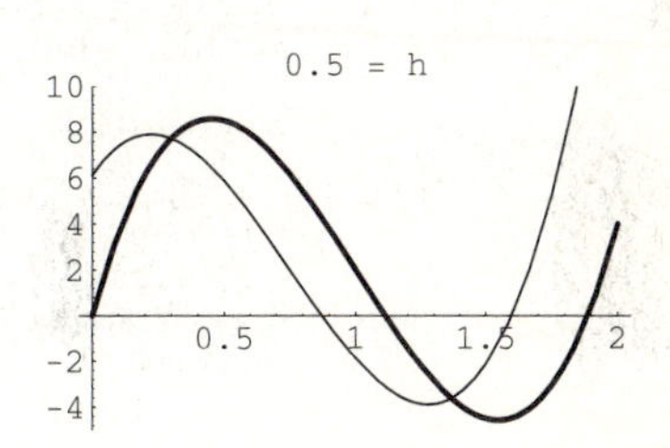

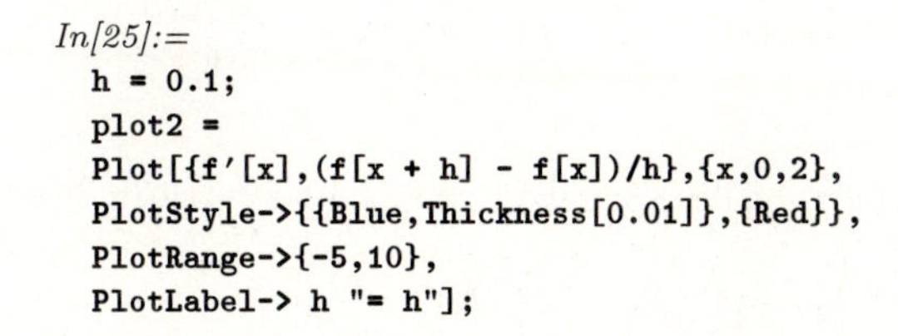

In[25]:=
```
h = 0.1;
plot2 =
Plot[{f'[x],(f[x + h] - f[x])/h},{x,0,2},
PlotStyle->{{Blue,Thickness[0.01]},{Red}},
PlotRange->{-5,10},
PlotLabel-> h "= h"];
```

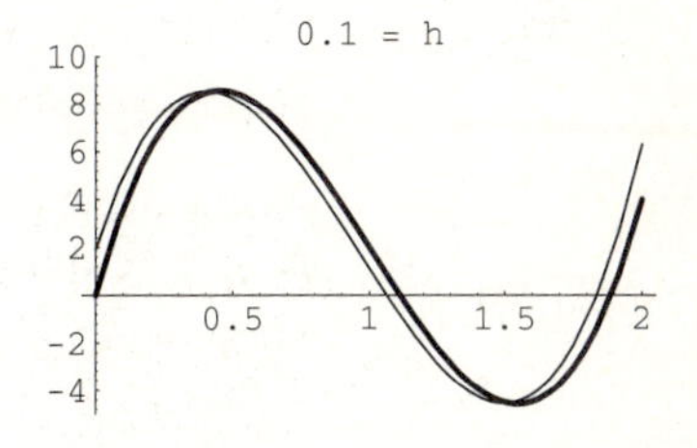

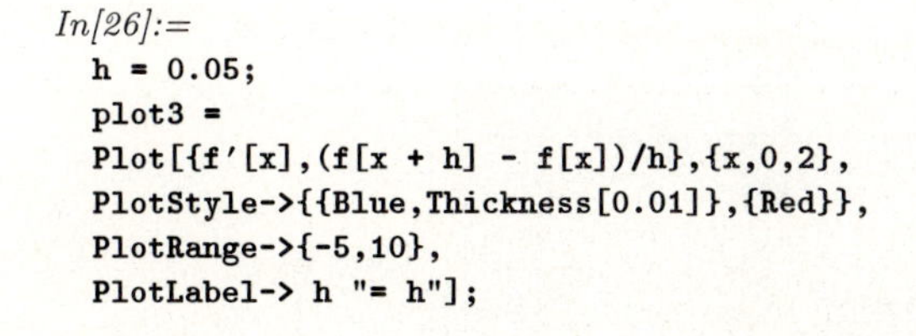

In[26]:=
```
h = 0.05;
plot3 =
Plot[{f'[x],(f[x + h] - f[x])/h},{x,0,2},
PlotStyle->{{Blue,Thickness[0.01]},{Red}},
PlotRange->{-5,10},
PlotLabel-> h "= h"];
```

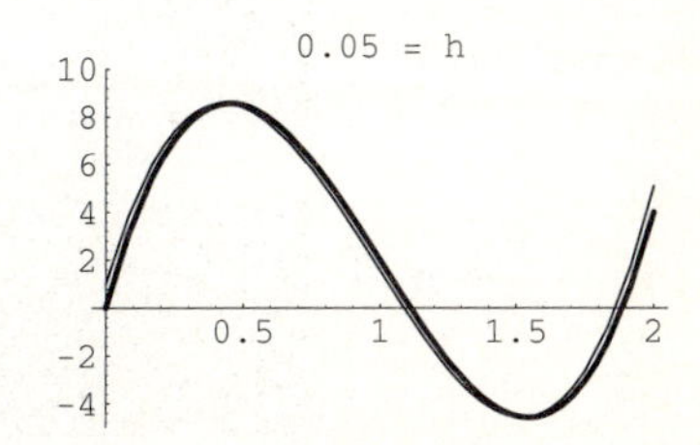

In[27]:=
```
h = 0.01;
plot4 =
Plot[{f'[x],(f[x + h] - f[x])/h},{x,0,2},
PlotStyle->{{Blue,Thickness[0.01]},{Red}},
PlotRange->{-5,10},
PlotLabel-> h "= h"];
```

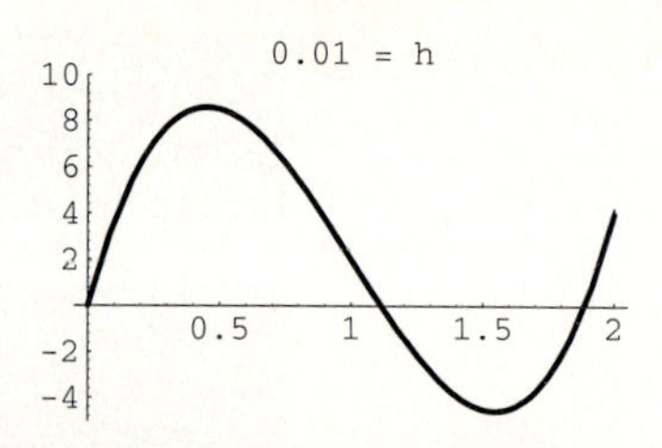

Discuss what you think is being depicted.

■ G.6) Up and down, maximum and minimum★

G.6.a) You can tell what happens to

$$f[x] = x^3 - 3\,x^2$$

as x leaves $x = 2.6$ and advances a little bit by looking at $f'[2.6]$:

In[28]:=
```
Clear[f,x]; f[x_] = x^3 - 3 x^2;; f'[2.6]
```

Out[28]=
```
4.68
```

Positive. This means $f[x]$ increases as x leaves 2.6 and advances a little bit. Check with a plot:

In[29]:=
```
Plot[f[x],{x,2.6,2.8},
AxesLabel->{"x","f[x]"}];
```

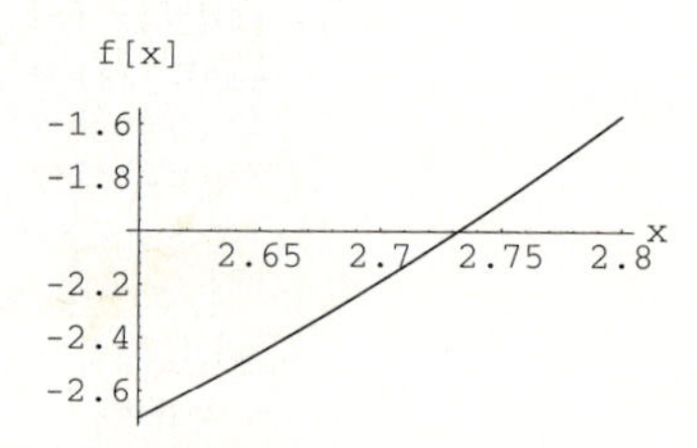

Yep. As x leaves 2.6 and advances a little bit, $f[x]$ goes up. Stay with $f[x] = x^3 - 3\,x^2$ and look at:

In[30]:=
```
f'[1.7]
```

Out[30]=
```
-1.53
```

As x leaves $x = 1.7$ and advances a little bit, does

$$f[x] = x^3 - 3\,x^2$$

go up or down? Confirm with a plot.

G.6.b) This time go with

$$f[x] = x^4 - 4\,x^2$$

and look at:

In[31]:=
```
Clear[f,x]; f[x_] = x^4 - 4 x^2; f'[1.3]
```

Out[31]=
```
-1.612
```

In[32]:=
```
f'[2.6]
```

What happens to $f[x]$ as x leaves 1.3 and increases a little bit?

What happens to $f[x]$ as x leaves 1.3 and decreases a little bit?

What happens to $f[x]$ as x leaves 2.6 and increases a little bit?

What happens to $f[x]$ as x leaves 2.6 and decreases a little bit?

G.6.c.i) You've got a function $f[x]$ and a point $x = a$. If $f'[a] > 0$, is it possible that $f[a] \leq f[x]$ for all other x's? Why?

G.6.c.ii) You've got a function $f[x]$ and a point $x = a$. If $f'[a] < 0$, is it possible that $f[a] \geq f[x]$ for all other x's? Why?

G.6.c.iii) You've got a function $f[x]$ and a point $x = a$. If $f'[a] > 0$, is it possible that $f[a] \geq f[x]$ for all other x's? Why?

G.6.c.iv) You've got a function $f[x]$ and a point $x = a$. If $f'[a] > 0$, is it possible that $f[a] \geq f[x]$ for all other x's? Why?

G.6.d) Here is a plot of $f[x] = x\,e^{-x^2}$ for $-3 \leq x \leq 3$:

In[33]:=
```
Clear[f,x]
f[x_] = x E^(-x^2);
fplot = Plot[f[x],{x,-3,3},
PlotStyle->{{VioletRed,Thickness[0.01]}},
AxesLabel->{"x","f[x]"}];
```

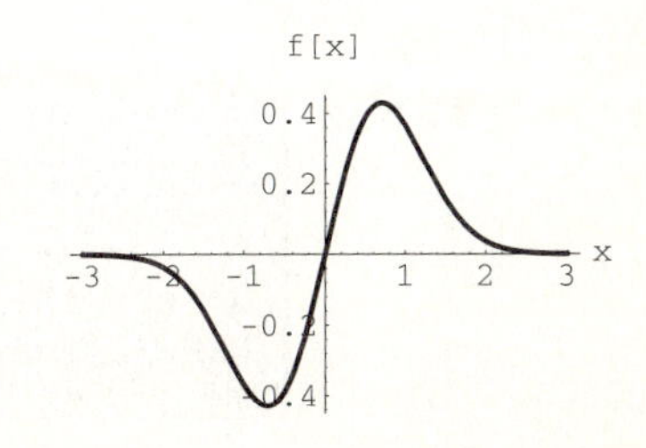

Note the dip on the left and the crest on the right. Now look at this:

In[34]:=
```
FindRoot[f'[x] == 0,{x,1}]
```
Out[34]=
```
{x -> 0.707107}
```

Now display the points $\{a, f[a]\}$ and $\{-a, f[-a]\}$ for $a = 0.707107$ right on the plot of $f[x]$.

In[35]:=
```
a = 0.707107;
crest = Graphics[
{PointSize[0.04],Point[{a,f[a]}]}];
dip = Graphics[
{PointSize[0.04],Point[{-a,f[-a]}]}];

Show[fplot,crest,dip];
```

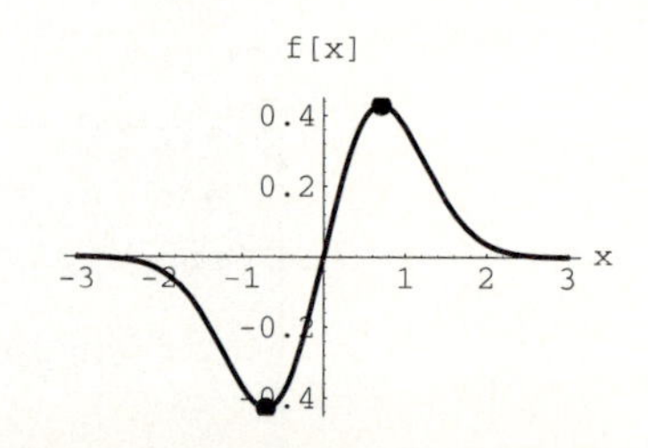

> Explain why you solve $f'[x] = 0$, as above, to hit the top of the crest and the bottom of the dip.

G.6.e) Here is a plot of $f[x] = 2\,x/(1 + 3\,x^2)$ for $-6 \leq x \leq 6$:

In[36]:=
```
Clear[f,x]
f[x_] = 2 x/(1 + 3 x^2);
fplot = Plot[f[x],{x,-6,6},
PlotStyle->{{VioletRed,Thickness[0.01]}},
AxesLabel->{"x","f[x]"}];
```

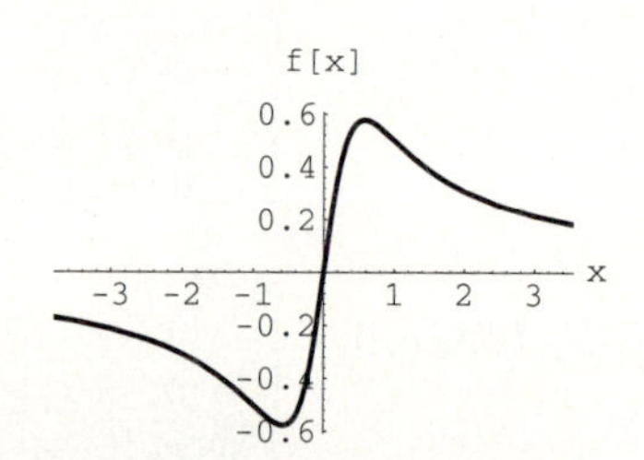

> Use $f'[x]$ to nail down the precise locations of the top of the crest and the bottom of the dip and show them in the plot of $f[x]$.

■ G.7) Spread of disease

G.7.a)

> You find that you are dealing with a disease with
>
> $$\text{infect}'[x] = 0.00018\,\text{suscept}[x]\,\,\text{infect}[x] - 0.023\,\text{infect}[x].$$
>
> Discuss the outlook on day 13 ($x = 13$) if you know suscept[13] = 5028 and infect[13] = 2012.

G.7.b) You find that you are dealing with a disease with

$$\text{infect}'[x] = 0.000004 \text{ suscept}[x] \text{ infect}[x] - 0.047 \text{ infect}[x].$$

Discuss the outlook on day 13 ($x = 13$) if you know suscept[13] = 5028 and infect[13] = 4012.

G.8) Average growth rate versus instantaneous growth rate

Take $f[x] = \cos[x^2]$:

In[37]:=
```
Clear[f,x]; f[x_] = Cos[x^2]
```
Out[37]=
```
      2
Cos[x ]
```

Here is a plot of the average growth rate

$$\frac{f[x+h] - f[x]}{h}$$

of $f[x]$ on the interval $[x, x+h]$ for $h = 1$:

In[38]:=
```
h = 1;
Plot[(f[x + h] - f[x])/h,{x,0,4},
PlotStyle->{Red},
PlotRange->All,
AxesLabel->{"x",""}];
```

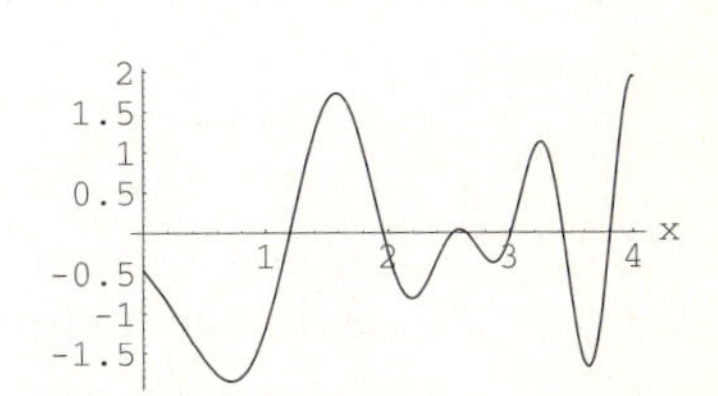

G.8.a.i) Use the plot to give a reasonable estimate of that x_0 with $0 \leq x_0 \leq 4$ at which $f[x]$ is showing the greatest average rate of increase on the interval $[x_0, x_0 + 1]$.

G.8.a.ii) Continue to go with the same $f[x]$. Here is a plot of the average growth rate $(f[x+h] - f[x])/h$ of $f[x]$ on the interval $[x, x+h]$ for $h = 0.1$:

In[39]:=
```
h = 0.1;
Plot[(f[x + h] - f[x])/h,{x,0,4},
PlotStyle->{Red},
AxesLabel->{"x",""}];
```

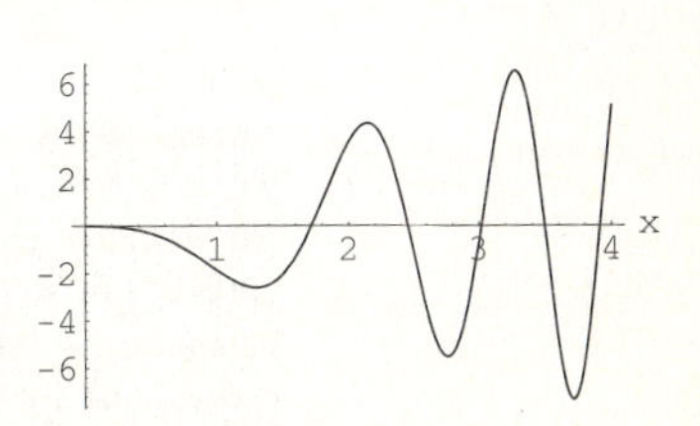

Use the plot to estimate the x_0 with $0 \leq x_0 \leq 4$ for which $f[x]$ is showing the greatest average rate of increase on the interval $[x_0, x_0 + 0.1]$.

G.8.a.iii) Continue to go with the same $f[x]$. Use the plot to estimate the x_0 with $0 \leq x_0 \leq 4$ at which $f[x]$ is showing the greatest instantaneous rate of increase.

G.8.a.iv) Was it an accident that the answers to parts G.8.a.ii) and G.8.a.iii) above were nearly the same?

■ G.9) Why folks study the instantaneous growth rate instead of instantaneous growth

Once in a while in mathematical research, you get a good idea and it pans out. But other times you get a good idea and it doesn't pan out. Here is what might be a good idea:

Instead of studying the instantaneous growth rate $f'[x]$ which is given by the limiting case of

$$\frac{f[x+h] - f[x]}{h}$$

as h closes in on 0, why not study the instantaneous growth $f^*[x]$ which is given by the limiting case of

$$f[x+h] - f[x]$$

as h closes in on 0?

Good idea? Maybe.

G.9.a) Run the following cells to see if you can spot what the limiting case of $f[x+h] - f[x]$ is as h closes in 0 for $f[x] = \sin[x]$:

In[40]:=
```
Clear[f,x]
f[x_] = Sin[x];
h = 0.5;
Plot[f[x + h] - f[x],{x,0,2 Pi},
PlotStyle->{{Navy,Thickness[0.01]}},
AxesLabel->{"x",""},
PlotRange->{-1,1},PlotLabel->h"= h"];
```

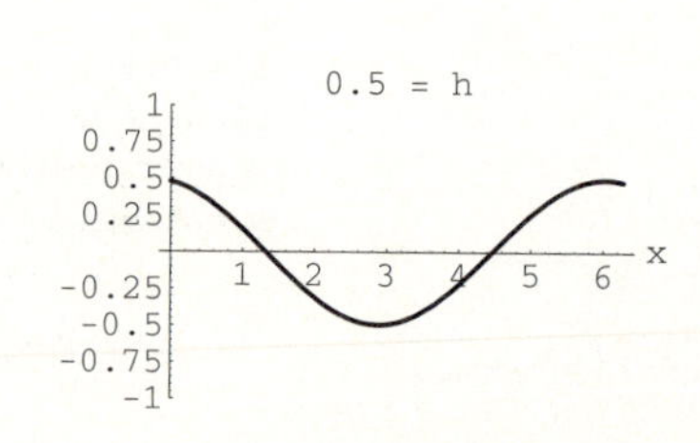

In[41]:=
```
h = 0.1;
Plot[f[x + h] - f[x],{x,0,2 Pi},
PlotStyle->{{Navy,Thickness[0.01]}},
AxesLabel->{"x",""},
PlotRange->{-1,1},PlotLabel->h"= h"];
```

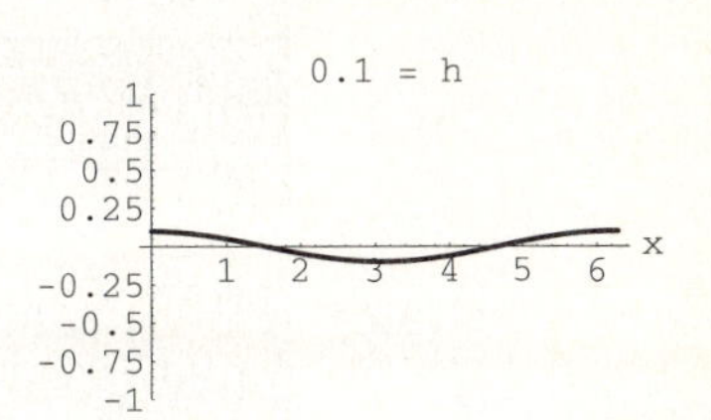

What in the . . . ?

In[42]:=
```
h = 0.00001;
Plot[f[x + h] - f[x],{x,0,2 Pi},
PlotStyle->{{Navy,Thickness[0.01]}},
AxesLabel->{"x",""},
PlotRange->{-1,1},PlotLabel->h"= h"];
```

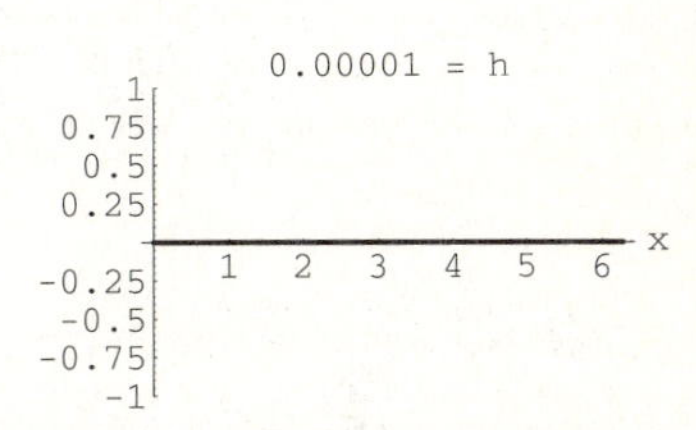

What's going on here?

G.9.b) Just to see whether what happened above is just a special case, go with $f[x] = x^2$:

In[43]:=
```
Clear[f,x]
f[x_] = x^2;
h = 0.5;
Plot[f[x + h] - f[x],{x,0,1},
PlotStyle->{{Navy,Thickness[0.01]}},
AxesLabel->{"x",""},
PlotRange->{-1,2},PlotLabel->h"= h"];
```

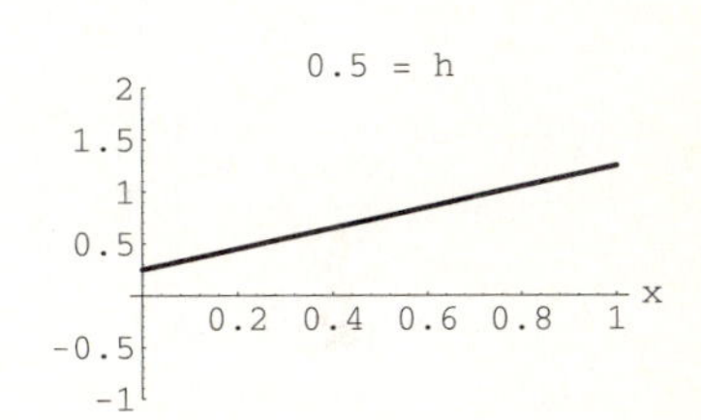

In[44]:=
```
h = 0.1;
Plot[f[x + h] - f[x],{x,0,1},
PlotStyle->{{Navy,Thickness[0.01]}},
AxesLabel->{"x",""},
PlotRange->{-1,2},PlotLabel->h"= h"];
```

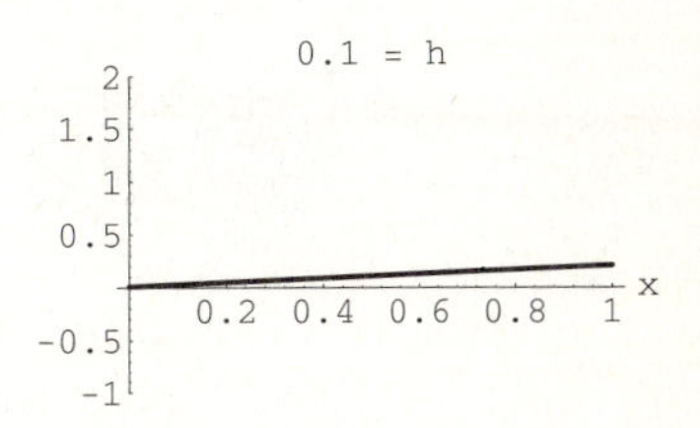

In[45]:=
```
h = 0.00001;
Plot[f[x + h] - f[x],{x,0,1},
PlotStyle->{{Navy,Thickness[0.01]}},
AxesLabel->{"x",""},
PlotRange->{-1,2},PlotLabel->h"= h"];
```

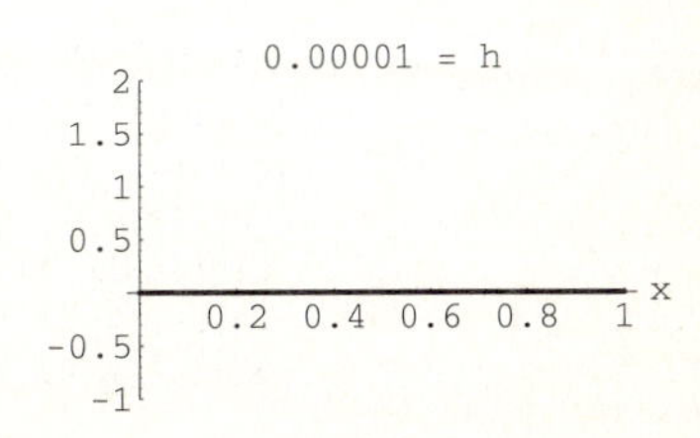

What's going on here?

G.9.c) Given a function $f[x]$, what do you expect the limiting case of $f[x+h] - f[x]$ as h closes in on 0 to be? Why?

LESSON 1.04

Rules of the Derivative

Basics

■ B.1) Derivatives

B.1.a) What is the derivative of a function $f[x]$? How do you use *Mathematica* to calculate derivatives?

Answer: The derivative $f'[x]$ of $f[x]$ is another function that measures the instantaneous growth rate of $f[x]$ at each point x. Thus "derivative" is just a short-hand way of saying "instantaneous growth rate." Mathematicians have programmed *Mathematica* to find derivatives of most common functions. Let's see how *Mathematica* handles some simple functions. To get the derivative of $f[x] = a\,x^2 + b\,x + c$, you have several options. Go with:

In[1]:=
```
Clear[f,x,a,b,c]
f[x_] = a x^2 + b x + c
```
Out[1]=
```
             2
c + b x + a x
```

One way to evaluate the derivative is to look at:

In[2]:=
```
Factor[(f[x + h] - f[x])/h]
```
Out[2]=
```
b + a h + 2 a x
```

As h closes in on 0, you see that $(f[x+h]-f[x])/h$ closes in on $f'[x]=2\,a\,x+b$. To get the same result from *Mathematica*, simply type:

In[3]:=
```
f'[x]
```
Out[3]=
```
b + 2 a x
```

It works! Or you can type:

In[4]:=
```
D[f[x],x]
```
Out[4]=
```
b + 2 a x
```

You can also eliminate the need to define $f[x]$:

In[5]:=
```
D[a x^2 + b x + c,x]
```
Out[5]=
```
b + 2 a x
```

Another way of doing this is to type:

In[6]:=
```
y = a x^2 + b x + c
```
Out[6]=
```
                2
c + b x + a x
```

The command:

In[7]:=
```
y'
```
Out[7]=
```
                 2
(c + b x + a x )'
```

is not successful. But you can use:

In[8]:=
```
D[y,x]
```
Out[8]=
```
b + 2 a x
```

This is reminiscent of the Leibniz notation $\frac{dy}{dx}$ that many folks like to use for the derivative of y with respect to x.

B.1.b) The $D[\ \ ,x]$ notation is particularly useful when the expression you want to differentiate contains parameters or other independent variables. For example, if you are given:

In[9]:=
```
Clear[a,f,x,y,z]; f[x_,z_] = a x^2 + 5 x Sin[z] - 3
```

Out[9]=

```
              2
 -3 + a x  + 5 x Sin[z]
```

In[10]:=

```
Factor[(f[x + h,z] - f[x,z])/h]
```

Out[10]=

```
a h + 2 a x + 5 Sin[z]
```

as h closes in on 0, you see that $(f[x+h,z]-f[x,z])\,/h$ closes in on the derivative with respect to x

$$a\,0+2\,a\,x+5\sin[z]=2\,a\,x+5\sin[z].$$

You can use *Mathematica* to get a formula for this derivative with respect to x:

In[11]:=

```
D[f[x,z],x]
```

Out[11]=

```
2 a x + 5 Sin[z]
```

Here you think of z as a second variable and think of a as a parameter. This distinction is not God given. There is nothing but custom preventing you from thinking of a as another variable and z as a parameter. In fact, you could think of both as parameters or you could think of both as new independent variables. Here are two calculations for specific values of z:

In[12]:=

```
D[f[x,Pi/8],x]
```

Out[12]=

```
                Pi
2 a x + 5 Sin[--]
                8
```

In[13]:=

```
D[f[x,Pi/4],x]
```

Out[13]=

```
   5
------- + 2 a x
Sqrt[2]
```

You can calculate the derivative at a particular x:

In[14]:=

```
D[f[x,Pi/4],x]/.x->8
```

Out[14]=

```
   5
------- + 16 a
Sqrt[2]
```

This is not the same as:

In[15]:=

```
D[f[8,Pi/4],x]
```

Out[15]=

```
0
```

Why?

Answer: Because

In[16]:=
```
f[8,Pi/4]
```

Out[16]=
```
            5/2
-3 + 5 2        + 64 a
```

is a constant. It does not vary as x varies. Thus its instantaneous growth rate is 0.

B.1.c) Use *Mathematica* to calculate the derivative of

$$a\,x^2 + b\,x\,y + c\,y^2 + d\,x + g\,y + h$$

with respect to x. Use *Mathematica* to calculate the derivative of

$$a\,x^2 + b\,x\,y + c\,y^2 + d\,x + g\,y + h$$

with respect to y. Use *Mathematica* to calculate the derivative of

$$a\,x^2 + b\,x\,y + c\,y^2 + d\,x + g\,y + h$$

with respect to a.

Answer: The derivative with respect to x is:

In[17]:=
```
Clear[x,y,a,b,c,d,g,h]
D[a x^2 + b x y + c y^2 + d x + g y + h,x]
```

Out[17]=
```
d + 2 a x + b y
```

The derivative with respect to y is:

In[18]:=
```
D[a x^2 + b x y + c y^2 + d x + g y + h,y]
```

Out[18]=
```
g + b x + 2 c y
```

The derivative with respect to a is:

In[19]:=
```
D[a x^2 + b x y + c y^2 + d x + g y + h,a]
```

Out[19]=
```
 2
x
```

■ B.2) The chain rule

Let's check out the derivative of the composition of two functions: Here is the derivative of $\sin[x^2]$:

In[20]:=
```
Clear[f,x]; f[x_] = Sin[x^2]; f'[x]
```
Out[20]=
```
         2
2 x Cos[x ]
```

Or:

In[21]:=
```
D[Sin[x^2],x]
```
Out[21]=
```
         2
2 x Cos[x ]
```

This catches your eye because the derivative of $\sin[x]$ is $\cos[x]$ and the derivative of x^2 is $2\,x$. It seems that the derivative of $\sin[x^2]$ is manufactured from the derivative of $\sin[x]$ and the derivative of x^2. Here is the derivative of $\left(x^2 + \sin[x]\right)^8$:

In[22]:=
```
D[(x^2 + Sin[x])^8,x]
```
Out[22]=
```
                       2            7
8 (2 x + Cos[x]) (x  + Sin[x])
```

This catches your eye again because the derivative of x^8 is $8\,x^7$ and the derivative of $x^2 + \sin[x]$ is $2\,x + \cos[x]$. It seems that the derivative of $\left(x^2 + \sin[x]\right)^8$ is manufactured from the derivative of x^8, the derivative of $\sin[x]$, and the derivative of x^2. Here is the derivative of $f[g[x]]$:

In[23]:=
```
Clear[f,g]; D[f[g[x]],x]
```
Out[23]=
```
f'[g[x]] g'[x]
```

Very interesting and of undeniable importance. This formula, which says that the derivative of

$$h[x] = f[g[x]]$$

is

$$h'[x] = f'[g[x]]\, g'[x],$$

is called the *chain rule*. The chain rule tells you how to build the derivative of $f[g[x]]$ from the derivatives of $f[x]$ and $g[x]$. Here is the chain rule in action: If

$$h[x] = \sin[x^2],$$

then

$$h'[x] = \cos[x^2]\, 2\,x$$

in accordance with:

In[24]:=
```
D[Sin[x^2],x]
```

Out[24]=

```
        2
2 x Cos[x ]
```

And if

$$h[x] = \left(x^2 + \sin[x]\right)^8,$$

then

$$h'[x] = 8\left(x^2 + \sin[x]\right)^7 (2\,x + \cos[x])$$

in accordance with:

In[25]:=

```
D[(x^2 + Sin[x])^8,x]
```

Out[25]=

```
                     2          7
8 (2 x + Cos[x]) (x  + Sin[x])
```

B.2.a) Give an explanation of why the derivative of $f[g[x]]$ is $f'[g[x]]\,g'[x]$.

Answer:

Put $h[x] = f[g[x]]$. Recall that $h[x]$ grows $h'[x]$ times as fast as x. But

→ $f[g[x]]$ grows $f'[g[x]]$ times as fast as $g[x]$ and

→ $g[x]$ grows $g'[x]$ times as fast as x.

As a result, $f[g[x]]$ grows $f'[g[x]]\,g'[x]$ times as fast as x. This explains why the instantaneous growth rate of $f[g[x]]$ is $f'[g[x]]\,g'[x]$. In other words, the derivative of $f[g[x]]$ with respect to x is $f'[g[x]]\,g'[x]$. This rule is called the chain rule, and it is important.

B.2.b) Give the derivative of $\sin[5\,x]$. Check with *Mathematica.*

Answer: The derivative of $\sin[5\,x]$ is $\cos[5\,x]\,5 = 5\cos[5\,x]$. Check:

In[26]:=

```
Clear[x]; D[Sin[5 x],x]
```

Out[26]=

```
5 Cos[5 x]
```

Got it.

B.2.c) Give the derivative of $\sin[x^4]$. Check with *Mathematica.*

Answer: The derivative of $\sin[x^4]$ is $\cos[x^4]\left(4\,x^3\right)$. Check:

In[27]:=
```
Clear[x]; D[Sin[x^4],x]
```
Out[27]=
```
     3     4
4 x  Cos[x ]
```
Nailed it.

B.2.d) Give the derivative of $(g[x])^3$. Check with *Mathematica.*

Answer: The derivative of x^3 is $3\,x^2$. This tells you that the derivative of $(g[x])^3$ is $3\,(g[x])^2\,g'[x]$. Check:

In[28]:=
```
Clear[g,x]; D[g[x]^3,x]
```
Out[28]=
```
        2
3 g[x]  g'[x]
```
Check this out for $g[x] = \sin[x]$:

In[29]:=
```
D[Sin[x]^3,x]
```
Out[29]=
```
                 2
3 Cos[x] Sin[x]
```
This is the same as $3\,g[x]^2\,g'[x]$.

B.2.e) Give the derivative of $f[x^3y^2]$ with respect to y. Check with *Mathematica.*

Answer: The derivative of $f[x^3y^5]$ with respect to y calculates the instantaneous growth rate of $f[x^3y^5]$ as y changes. So x is regarded as a constant. The derivative of $f[x^3y^5]$ with respect to y is $f'[x^3y^5]\,5\,x^3\,y^4$. Check:

In[30]:=
```
Clear[f,x,y]; D[f[x^3 y^5],y]
```
Out[30]=
```
   3  4      3  5
5 x  y  f'[x  y ]
```
Got it. This means, for example, that the derivative of $\sin[x^3y^5]$ with respect to y is $\cos[x^3y^5]\,5\,x^3\,y^4$.

B.2.f) Give the derivative of $(e^x - x^2)^7$. Check with *Mathematica.*

Answer: The derivative of $(e^x - x^2)^7$ is $7\,(e^x - x^2)^6\,(e^x - 2\,x)$.

In[31]:=
```
Clear[x]; D[(E^x - x^2)^7,x]
```
Out[31]=
```
       x          x    2 6
7 (E  - 2 x) (E  - x )
```
Got it.

B.3) General rules for taking derivatives

B.3.a) Let's see what *Mathematica* thinks about this:

In[32]:=
```
Clear[f,g,x]; D[f[x] + g[x],x] == D[f[x],x] + D[g[x],x]
```
Out[32]=
```
True
```
Good. Check it out with a couple of examples:

In[33]:=
```
Clear[f,g,x]; f[x_] = 4 x^3; g[x_] = -9 x^2; D[f[x] + g[x],x]
```
Out[33]=
```
                2
-18 x + 12 x
```
In[34]:=
```
f'[x] + g'[x]
```
Out[34]=
```
                2
-18 x + 12 x
```
Yep, it works.

In[35]:=
```
Clear[f,g,x]; f[x_] = 4/x; g[x_] = Cos[x^2]; D[f[x] + g[x],x]
```
Out[35]=
```
-4                 2
-- - 2 x Sin[x ]
 2
x
```
In[36]:=
```
f'[x] + g'[x]
```
Out[36]=
```
-4                 2
-- - 2 x Sin[x ]
 2
x
```
It works again.

Explain why this will work every time you try it.

Answer: The instantaneous growth rate of the sum is the sum of the instantaneous growth rates.

If you prefer a more technical explanation, then read on.

Remember:

→ $f'[x]$ is the limiting case of $(f[x+h]-f[x])/h$ as h closes in on 0.

→ $g'[x]$ is the limiting case of $(g[x+h]-g[x])/h$ as h closes in on 0.

→ $D[f[x]+g[x],x]$ is the limiting case of $((f[x+h]+g[x+h])-(f[x]+g[x]))/h$ as h closes in on 0.

Note that:

$$\frac{(f[x+h]+g[x+h])-(f[x]+g[x])}{h}=\frac{f[x+h]-f[x]}{h}+\frac{g[x+h]-g[x]}{h}.$$

So

$$\begin{aligned}D[f[x]+g[x],x] &= \lim_{h\to 0}\frac{f[x+h]-f[x]}{h}+\frac{g[x+h]-g[x]}{h}\\ &= \lim_{h\to 0}\frac{f[x+h]-f[x]}{h}+\lim_{h\to 0}\frac{g[x+h]-g[x]}{h}\\ &= f'[x]+g'[x].\end{aligned}$$

B.3.b) Let's see what *Mathematica* thinks about this:

```
In[37]:=
  Clear[f,c,x]; D[c f[x],x] == c D[f[x],x]

Out[37]=
  True
```

No problem. Check it out with a couple of examples:

```
In[38]:=
  Clear[f,c,x]; f[x_] = E^(3 x);
  D[c f[x],x]

Out[38]=
      3 x
  3 c E

In[39]:=
  c f'[x]

Out[39]=
      3 x
  3 c E
```

Yep, it works.

```
In[40]:=
  Clear[f,c,x]; c = 5; f[x_] = Cos[x]; D[c f[x],x]

Out[40]=
  -5 Sin[x]
```

In[41]:=
```
c f'[x]
```
Out[41]=
```
-5 Sin[x]
```

It works again.

> Explain why this will work every time you try it.

Answer: If c is a constant, then $g[x] = c\,f[x]$ grows at a rate c times faster than $f[x]$ grows. Consequently, the instantaneous growth rate of $g[x] = c\,f[x]$ is c times the instantaneous growth rate of $f[x]$. In short, $D[c\,f[x], x] = c\,f'[x]$.

B.3.c) Let's see what *Mathematica* thinks about this:

In[42]:=
```
Clear[f,g,x]
D[ f[x] g[x],x] == f'[x] g'[x]
```
Out[42]=
```
g[x] f'[x] + f[x] g'[x] == f'[x] g'[x]
```

No, apparently *Mathematica* thinks that the derivative of $f[x]\,g[x]$ is $f'[x]\,g[x] + f[x]\,g'[x]$. Check it out with a couple of examples:

In[43]:=
```
Clear[f,g,x]; f[x_] = E^x; g[x_] = Sin[x];
D[f[x] g[x],x]
```
Out[43]=
```
 x           x
E  Cos[x] + E  Sin[x]
```

In[44]:=
```
f'[x] g[x] + f[x] g'[x]
```
Out[44]=
```
 x           x
E  Cos[x] + E  Sin[x]
```

Yep, it works.

In[45]:=
```
Clear[f,g,x]; f[x_] = x^2; g[x_] = Cos[x];
D[f[x] g[x],x]
```
Out[45]=
```
                 2
2 x Cos[x] - x  Sin[x]
```

In[46]:=
```
f'[x] g[x] + f[x] g'[x]
```
Out[46]=
```
                 2
2 x Cos[x] - x  Sin[x]
```

It works again. This time, try it with $f[x] = x^2$ and $g[x] =$ any other function.

In[47]:=
```
Clear[f,g,x]
f[x_] = x^2; D[f[x] g[x],x]
```
Out[47]=
```
              2
2 x g[x] + x  g'[x]
```
In[48]:=
```
f'[x] g[x] + f[x] g'[x]
```
Out[48]=
```
              2
2 x g[x] + x  g'[x]
```
There it is: $f'[x]\,g[x] + f[x]\,g'[x]$. It works again.

> Explain why this will work every time you try it.

Answer: That the derivative of $f[x]\,g[x]$ is $f'[x]\,g[x]+f[x]\,g'[x]$ is not transparent in terms of instantaneous growth rates. So instead of explaining this in everyday language, you'll probably want to resort to a technical explanation. Here is one such. Look at:

In[49]:=
```
Clear[f,g,x]
first[x_] = Expand[(f[x] + g[x])^2]/4
```
Out[49]=
```
    2                      2
f[x]  + 2 f[x] g[x] + g[x]
--------------------------
            4
```
In[50]:=
```
second[x_] = Expand[-(f[x] - g[x])^2]/4
```
Out[50]=
```
     2                      2
-f[x]  + 2 f[x] g[x] - g[x]
---------------------------
             4
```
Look at the formulas for first$[x]$ and second$[x]$ to see that

$$\text{first}[x] + \text{second}[x] = f[x]\,g[x]:$$

In[51]:=
```
Expand[first[x] + second[x]]
```
Out[51]=
```
f[x] g[x]
```
This means

$$f[x]\,g[x] = \frac{(f[x] + g[x])^2}{4} - \frac{(f[x] - g[x])^2}{4}.$$

As a result,

$$D[f[x]\, g[x], x] = D\left[\frac{(f[x]+g[x])^2}{4}, x\right] + D\left[-\frac{(f[x]-g[x])^2}{4}, x\right].$$

The point of all this is that the chain rule tells you what $D[(f[x]+g[x])^2/4, x]$ and $D[-(f[x]-g[x])^2/4, x]$ are. In fact, you can type them in by hand:

In[52]:=
```
derivfirst = Expand[(1/4) 2 (f[x] + g[x]) (f'[x] + g'[x])]
```

Out[52]=
```
f[x] f'[x]   g[x] f'[x]   f[x] g'[x]   g[x] g'[x]
---------- + ---------- + ---------- + ----------
    2            2            2            2
```

In[53]:=
```
derivsecond = Expand[(-1/4) 2 (f[x] - g[x]) (f'[x] - g'[x])]
```

Out[53]=
```
   f[x] f'[x]     g[x] f'[x]   f[x] g'[x]   g[x] g'[x]
-(----------) + ---------- + ---------- - ----------
       2            2            2            2
```

$D[f[x]\, g[x], x]$ is the sum of derivfirst and derivsecond. As you can see, some terms cancel and some don't. The result of adding derivfirst and derivsecond is $f'[x]\, g[x] + f[x]\, g'[x]$:

In[54]:=
```
Expand[derivfirst + derivsecond]
```

Out[54]=
```
g[x] f'[x] + f[x] g'[x]
```

Now you know where the "product rule" $D[f[x]\, g[x], x] = f'[x]\, g[x] + f[x]\, g'[x]$ comes from.

■ B.4) Using the logarithm to calculational advantage

B.4.a) How do you know that $f'[x] = f[x]\, D[\log[f[x]], x]$? What calculational advantage does this give you?

Answer: Because $D[\log[x], x] = 1/x$, the chain rule tells you that $D[\log[f[x]], x] = f'[x]/f[x]$. Multiply both sides by $f[x]$ to see that

$$f[x]\, D[\log[f[x]], x] = \frac{f[x]\, f'[x]}{f[x]} = f'[x].$$

This is a welcome calculational advantage in the cases in which taking the derivative of $f[x]$ is not easy, but taking the derivative of $\log[f[x]]$ is easy. Some old timers call this technique "logarithmic differentiation." Try this out: Put

$$f[x] = \frac{e^x}{x^5}.$$

Note that $f[x]$ is not easy to differentiate, but

$$\log[f[x]] = \log[e^x] - \log[x^5]$$
$$= x\log[e] - 5\log[x] = x - 5\log[x]$$

is easy to differentiate. In fact, $D[\log[f[x]], x] = 1 - 5/x$. The formula, $f'[x] = f[x]\, D[\log[f[x]], x]$ tells you that $f'[x]$ is given by:

In[55]:=
```
Expand[((E^x)/x^5) (1 - 5/x)]
```

Out[55]=
```
      x    x
  -5 E    E
  ----- + --
    6      5
   x      x
```

Check:

In[56]:=
```
Expand[D[(E^x)/x^5,x]]
```

Out[56]=
```
      x    x
  -5 E    E
  ----- + --
    6      5
   x      x
```

Right on the money.

B.4.b.i) Look at *Mathematica*'s calculation of the derivative of $f[x] = x^t$:

In[57]:=
```
Clear[x,t,f]; f[x_] = x^t; f'[x]
```

Out[57]=
```
   -1 + t
t x
```

> Use the identity
>
> $$f[x]\, D[\log[f[x]], x] = f'[x]$$
>
> to explain the basic formula
>
> $$D[x^t, x] = t x^{t-1}$$
>
> which some folks call the "power rule."

Answer: Put $f[x] = x^t$. Take the natural logarithm of both sides:

$$\log[f[x]] = \log[x^t] = t\log[x].$$

Now use

$$f'[x] = f[x]\, D[\log[f[x]], x] = x^t \frac{t}{x} = t\, x^{t-1}.$$

And that's it!

Some may object that this explanation does not cover the case when $x = 0$ because log[0] makes no sense. This objection has some merit. If you want to see how to deal with it, then go to the electronic version.

B.4.b.ii) Look at:

In[58]:=
```
Clear[x,f]; f[x_] = 1 + x + x^2 + x^3 + x^4 + x^5 + x^6
```
Out[58]=
```
            2    3    4    5    6
1 + x + x  + x  + x  + x  + x
```
In[59]:=
```
f'[x]
```
Out[59]=
```
                2      3      4      5
1 + 2 x + 3 x  + 4 x  + 5 x  + 6 x
```

> Explain the *Mathematica* calculation of $f'[x]$.

Answer: This is the power rule applied to each term.

B.4.c) This problem appears only in the electronic version.

■ B.5) The instantaneous percentage growth rate

B.5.a.i)

> Take any positive function $f[x]$ and look at the function per$[x, h]$ specified by
>
> $$f[x + h] = f[x] + h f[x] \operatorname{per}[x, h].$$
>
> What does per$[x, h]$ measure?

Answer: 100 per$[x, h]$ measures the average percentage growth rate of $f[x]$ per unit on the x-axis as x advances from x to $x + h$.

B.5.a.ii) Here is a formula for per$[x, h]$ for any given positive function $f[x]$:

In[60]:=
```
Clear[f,x,h,r,per]
Solve[f[x + h] == f[x] + h f[x] per[x,h],per[x,h]]
```
Out[60]=
```
                   -f[x] + f[h + x]
{{per[x, h] -> ------------------}}
                       h f[x]
```

This tells you that

$$\text{per}[x,h] = \frac{f[x+h]-f[x]}{h f[x]}.$$

And because the limiting case of $(f[x+h]-f[x])/h$ as h closes in on 0 is $f'[x]$, this tells you that the limiting case of $\text{per}[x,h]$ as h closes in on 0 is $f'[x]/f[x]$.

What does $100\, f'[x]/f[x]$ measure?

Answer: $100\,\text{per}[x,h]$ measures the average percentage growth rate of $f[x]$ per unit on the x-axis as x advances from x to $x+h$. As h closes in on 0, the limiting case of $100\,\text{per}[x,h]$ is

$$100\,\frac{f'[x]}{f[x]}.$$

So $100\, f'[x]/f[x]$ measures the instantaneous percentage growth rate of $f[x]$ per unit on the x-axis. Some folks like to call

$$100\,\text{per}[x] = 100\,\frac{f'[x]}{f[x]}$$

by the name "percentage derivative," but this name is not in common use.

B.5.b) Why would anyone care about the instantaneous percentage growth rate of a positive function $f[x]$?

Answer: It puts two functions on a level playing field. For instance, if you take $f[x] = x^{15}$ and $g[x] = e^x$, then you get:

In[61]:=

```
Clear[f,g,x]; f[x_] = x^15;
g[x_] = E^x; N[{f'[x],g'[x]}]/.x->20
```

Out[61]=

```
         19             8
{2.4576 10  , 4.85165 10 }
```

This tells you that when $x = 20$, then $f[x] = x^{15}$ is growing more than 10^{10} times faster than $g[x] = e^x$. But when you look at the two instantaneous percentage growth rates you get:

In[62]:=

```
Clear[f,g,x]; f[x_] = x^15; g[x_] = E^x;
N[{100 f'[x]/f[x],100 g'[x]/g[x]}]/.x->20
```

Out[62]=

```
{75., 100.}
```

This tells you that when $x = 20$, then the instantaneous percentage growth rate of $g[x] = e^x$ is actually greater than the instantaneous percentage growth rate of $f[x] = x^{15}$. Check out what happens at $x = 50$:

In[63]:=
```
N[{f'[x],g'[x]}]/.x->50
```

Out[63]=
```
          24              21
{9.15527 10  , 5.18471 10  }
```

At $x = 50$, the power function $f[x]$ is growing about 1.8×10^3 times faster than the exponential function $g[x]$.

In[64]:=
```
N[{100 f'[x]/f[x],100 g'[x]/g[x]}]/.x->50
```

Out[64]=
```
{30., 100.}
```

At $x = 50$, the power function $f[x]$ is growing at a slower percentage than the exponential function $g[x]$. In fact, if you go with $x = 1000$, you get:

In[65]:=
```
N[{100 f'[x]/f[x],100 g'[x]/g[x]}]/.x->1000
```

Out[65]=
```
{1.5, 100.}
```

In percentage terms, the power function $f[x]$ is not growing fast at all, but the exponential maintains its steady percentage growth. This is one reason that the exponential is guaranteed to catch and pass the power function.

■ B.6) Exponential growth dominates power growth and power growth dominates logarithmic growth

The instantaneous percentage growth rate can do a lot to explain why exponential growth always dominates power growth in the global scale. Here are an exponential function $e^{0.05x}$ and a power function x^{25} together with their instantaneous percentage growth rates:

In[66]:=
```
Clear[expon,power,exppercentgrowth,powerpercentgrowth]
expon[x_] = E^(0.05 x);
power[x_] = x^25;
exponpercentgrowth[x_] = 100 Chop[expon'[x]/expon[x]]
```

Out[66]=
```
5.
```

In[67]:=
```
powerpercentgrowth[x_] = 100 power'[x]/power[x]
```

Out[67]=
```
2500
----
 x
```

Plot the two percentage growth rates:

In[68]:=
```
Plot[{exponpercentgrowth[x],powerpercentgrowth[x]},
{x,1,10000},PlotStyle->{{Red,Thickness[0.01]},{Blue}},
AxesLabel->{"x","% growth"}]
```

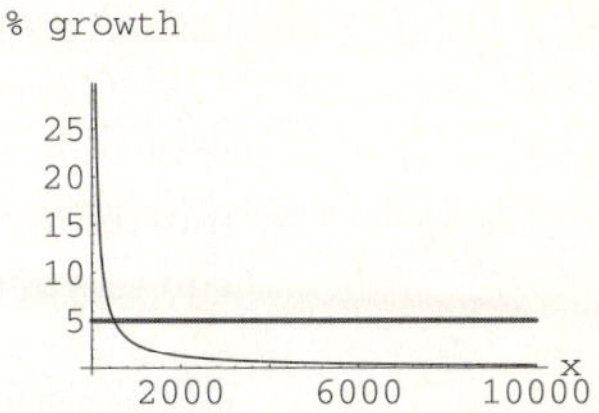

The percentage growth rate of the power function dies off as x gets big, but the percentage growth rate of the exponential function stays constant. This is why the exponential function eventually dominates the power function. In fact, if you look at the plot, you can see that the exponential function has to pass the power function eventually. Check it out with a plot:

In[69]:=
```
Plot[{expon[x],power[x]},{x,3500,4200},
PlotStyle->{{Red,Thickness[0.01]},{Blue}},
AxesLabel->{"x",""},PlotRange->All];
```

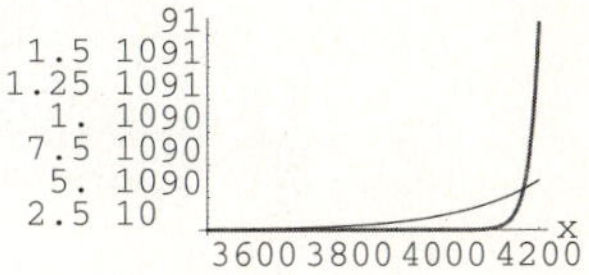

The plot of the exponential function is the thicker of the two plots.

That's the exponential making its move and leaving the poor power function in its dust. Once it gets ahead of the power function, it can never fall behind because its percentage growth rate is bigger than the percentage growth rate of the power function. A spectacular win in a battle of titans.

B.6.a) Take any exponential function e^{rx} with $r > 0$ and a power function x^k with $k > 0$ and look at their instantaneous percentage growth rates:

In[70]:=
```
Clear[expon,power,r,k,exppercentgrowth,powerpercentgrowth]
expon[x_] = E^(r x); power[x_] = x^k;
exponpercentgrowth[x_] = 100 expon'[x]/expon[x]
```

Out[70]=
```
100 r
```

In[71]:=
```
powerpercentgrowth[x_] = 100 power'[x]/power[x]
```

Out[71]=
```
100 k
-----
  x
```

> Use their instantaneous percentage growth rates to explain why e^{rx} dominates x^k in the global scale.

Answer: The instantaneous percentage growth rate of e^{rx} is $100\,r\%$; this is a fixed positive instantaneous percentage growth rate because r is given to be positive. The instantaneous percentage growth rate of x^k is $100\,k/x\%$; this is not a fixed positive instantaneous percentage growth rate. In fact, the global scale of $100\,k/x$ is 0. Consequently, the instantaneous percentage growth rate of x^k eventually slips way underneath the instantaneous percentage growth rate of e^{rx}. This is why, no matter what positive r and k you go with, the power function x^k is dominated by the exponential function e^{rx} in the global scale.

B.6.b) Saying that no matter what positive r and k you go with, then e^{rx} dominates x^k in the global scale is the same as saying

$$\lim_{x \to \infty} \frac{x^k}{e^{rx}} = 0.$$

Use this to explain why no matter what positive r you go with, then

$$\lim_{x \to \infty} \log \frac{[x]}{x^r} = 0;$$

so that power growth always dominates logarithmic growth in the global scale.

Answer: Here is a sneaky way of explaining this: Put $x = e^t$. Saying $x \to \infty$ is the same as saying $t \to \infty$. Consequently,

$$\begin{aligned}
\lim_{x \to \infty} \frac{\log[x]}{x^r} &= \lim_{t \to \infty} \frac{\log[e^t]}{(e^t)^r} && \text{(because } x = e^t\text{)} \\
&= \lim_{t \to \infty} \frac{\log[e^t]}{e^{rt}} && \text{(because } (e^t)^r = e^{rt}\text{)} \\
&= \lim_{t \to \infty} \frac{t}{e^{rt}} && \text{(because } \log[e^t] = t\text{)} \\
&= 0
\end{aligned}$$

because no matter what positive r and k you go with, then

$$\lim_{t \to \infty} \frac{t^k}{e^{rt}} = 0.$$

That's all folks!

Tutorials

■ T.1) Practicing with the chain rule

This problem appears only in the electronic version.

■ T.2) Practicing with the chain rule, the product rule, and the power rule

This problem appears only in the electronic version.

■ T.3) Linear dimension: Length, area, volume, and weight

The volume measurement, $V[r]$, and the surface area measurement, $S[r]$, of a sphere of radius r are given by: $V[r] = 4\pi r^3/3$ and $S[r] = 4\pi r^2$. This says that $V[r]$ is proportional to r^3 and $S[r]$ is proportional to r^2. For other three-dimensional objects, the formulas for volume and surface area are not so easy to come by, but the idea of proportionality survives. Here is the idea: A linear dimension of a given solid or shape is any length between specified locations on the solid. The radius of a sphere or the radius of a circle is a linear dimension. The total length of a solid, the total width, or the total height of a solid are all examples of linear dimensions. The diameter of the finger loop on a coffee cup is a linear dimension of the cup. Next take a given shape for a solid. If the shape stays the same but the linear dimensions change, then it is still true that the volume is proportional to the cube of any linear dimension and it is still true that the surface area is proportional to the square of any linear dimension. Here is a shape in the xy-plane:

You will learn more about the ParametricPlot instruction after just a few more lessons. If you can't wait, then feel free to play with this instruction.

In[1]:=

```
Clear[x,y,t]
x[t_] = 4 t ( 2 - t) E^(t/4);
y[t_] = 2 - Sin[Pi t]; a = 0; b = 2;
little = ParametricPlot[{x[t],y[t]},{t,a,b},Axes->None,
PlotRange->{{0,17},{0,10}},AspectRatio->Automatic,
PlotStyle->{{Red,Thickness[0.007]}}];
```

Here is the same shape with all its linear dimensions increased by a factor of 3:

In[2]:=

```
big = ParametricPlot[{3 x[t],3 y[t]},{t,a,b},Axes->None,
PlotRange->{{0,17},{0,10}},AspectRatio->Automatic,
PlotStyle->{{Red,Thickness[0.007]}}];
```

Here they are together:

In[3]:=

```
Show[little,big,AspectRatio->Automatic];
```

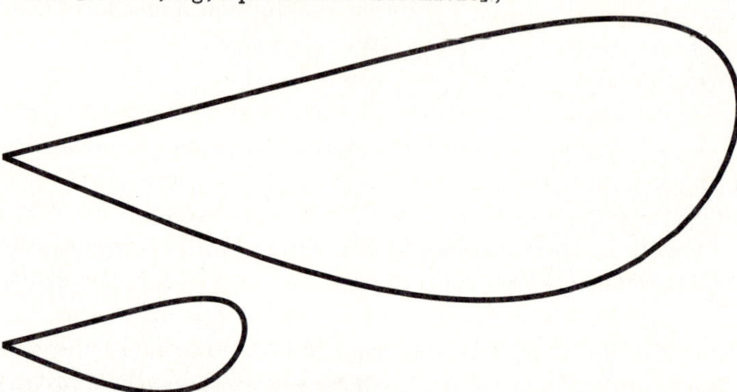

T.3.a) How does the area measurement of the larger blob above compare to the area measurement of the smaller blob?

Answer: Both are the same shape, but the linear dimensions of the larger blob are three times the linear dimensions of the smaller blob. The upshot: The area measurement of the larger blob is $3^2 = 9$ times the area of the smaller blob.

T.3.b) The idea of linear dimension leads to some intriguing biological implications. A giant mouse with linear dimension 10 times larger than the usual mouse would not be viable because the volume of its body would be larger than the volume of the usual mouse by a factor of 10^3, but the surface area of some of its critical supporting organs like lungs, intestines, and skin would be larger only by a factor of 10^2. That big mouse would be hungry and out of breath at all times! Similarly, there will never be a 12-foot-tall basketball player at Indiana or even at Duke. The approximate size of an adult mammal is dictated by its shape! The same common sense applies to buildings and other structures. An architect or engineer does not design a 200-foot-tall building by taking a proven design for a 20-foot-tall building and multiplying all the linear dimensions by 10. Now it's time for a calculation.

A crystal grows in such a way that all the linear dimensions increase by 25%. How do the new surface area and the new volume compare to the old surface area and volume?

Answer:

In[4]:=

```
Clear[newsurfacearea,oldsurfacearea]
newsurfacearea = (1.25)^2 oldsurfacearea
```

Out[4]=

```
1.5625 oldsurfacearea
```

An increase of the linear dimensions by 25% increases the surface area by about 56%. The percentage increase in volume is:

In[5]:=
```
Clear[newvolume,oldvolume]
newvolume = (1.25)^3 oldvolume
```

Out[5]=
```
1.95312 oldvolume
```

An increase of the linear dimensions by 25% increases the volume by about 95%.

T.3.c.i) Calculus&*Mathematica* thanks Ruth Reynolds, owner of Pioneer Bernese Mountain Dog Kennel in Greenwood, Florida for the data used in this problem.

Dogs and other animals grow so that a linear dimension of their bodies is given by what a lot of folks call a logistic function $b\,c\,e^{at}/(b-c+c\,e^{at})$ where t measures time in years elapsed since the birth of the animal. A good linear dimension for a dog is the height of the dog's body at the dog's shoulders:

In[6]:=
```
Clear[height,a,b,c,t]
height[t_] =  b c E^(a t)/(b - c + c E^(a t))
```

Out[6]=
```
        a t
   b c E
-------------
           a t
b - c + c E
```

To see what the parameters a, b, and c mean, look at:

In[7]:=
```
height[0]
```

Out[7]=
```
c
```

This tells you that c measures the dog's height at birth. The global scale of height$[t]$ $= b\,c\,e^{at}/(b-c+c\,e^{at})$ is $b\,c\,e^{at}/c\,e^{at} = b$. This tells you that b measures the dog's mature height. For a typically magnificent Bernese Mountain Dog bitch, as owned by the actor Robert Redford, $b = 24$ inches and $c = 4$ inches; so for the Bernese Mountain Dog, height$[t]$ is:

In[8]:=
```
b = 24.0; c = 4.5;
height[t]
```

Out[8]=
```
           a t
   108. E
--------------
             a t
19.5 + 4.5 E
```

The parameter a is related to how fast the dog grows. At one year, a typical Bernese Mountain Dog has achieved about 95% of its mature height. This gives you an equation to solve to get a:

In[9]:=
```
equation = height[1] == 0.95 24.5
```
Out[9]=
```
         a
   108. E
   ----------- == 23.275
             a
19.5 + 4.5 E
```
In[10]:=
```
Solve[equation,a]
```
Out[10]=
```
{{a -> 4.9353}}
```
Now you've got the height function for the typical Bernese Mountain Dog bitch:

In[11]:=
```
a = 4.9353;
height[t]
```
Out[11]=
```
          4.9353 t
   108. E
   -------------------
              4.9353 t
19.5 + 4.5 E
```
Here's a plot:

In[12]:=
```
heightplot =
Plot[{height[t],b,c},{t,0,2},PlotStyle->
{{Thickness[0.01]},{Brown},{Brown}},PlotRange->All,
AxesLabel->{"t","height[t]"},
PlotLabel->"Bernese Mountain Dog bitch"];
```
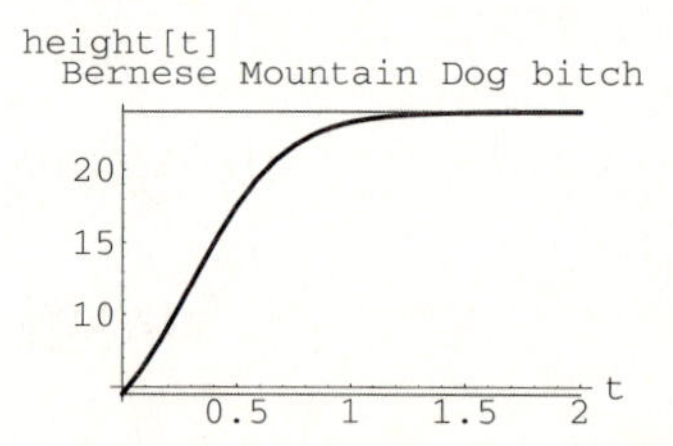

Looks OK.

Given that the typical mature Bernese Mountain Dog bitch as described above weighs 85 pounds, give an approximate plot of the bitch's weight as a function of time for the first three years and give a critique of the plot.

Answer: If you assume the bitch maintains the same shape throughout the growing process, then you can say:

→ weight$[t]$ is proportional to the volume of the bitch's body at time t and

→ the volume of the bitch's body at time t is proportional to height$[t]^3$.

So

$$\text{weight}[t] = k\,(\text{height}[t])^3 = k\left(\frac{b\,c\,e^{at}}{b - c + c\,e^{at}}\right)^3.$$

The global scale of weight[t] is $k\,(b\,c\,e^{at}/c\,e^{at})^3 = k\,b^3$. For the typical Bernese Mountain Dog bitch under study here, k is given by:

In[13]:=
```
Solve[85 == b^3 k,k]
```

Out[13]=
```
{{k -> 0.00614873}}
```

The weight of the typical Bernese Mountain Dog bitch under study here t years after her birth is:

In[14]:=
```
Clear[weight]; weight[t_] = 0.00614873 height[t]^3
```

Out[14]=
```
            14.8059 t
 7745.63 E
------------------------
              4.9353 t 3
(19.5 + 4.5 E          )
```

Here comes a plot:

In[15]:=
```
weightplot = Plot[{weight[t],85},{t,0,2},
PlotStyle->{{Thickness[0.01]},{Brown},{Brown}},
PlotRange->All, AxesLabel->{"t","weight[t]"},
PlotLabel->"Bernese Mountain Dog bitch"];
```

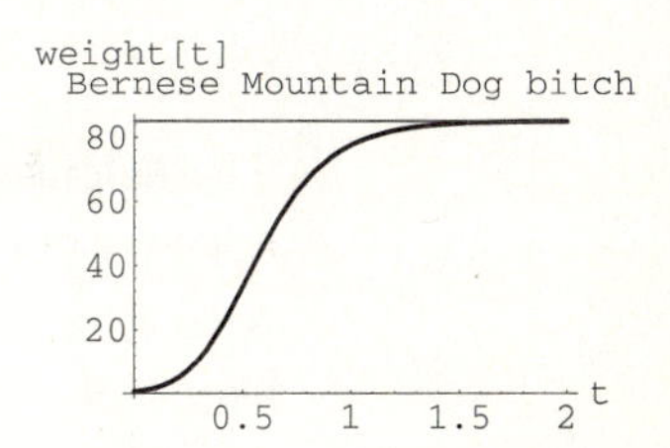

See the height plot and the weight plot side by side:

In[16]:=
```
Show[GraphicsArray[{heightplot,weightplot}]];
```

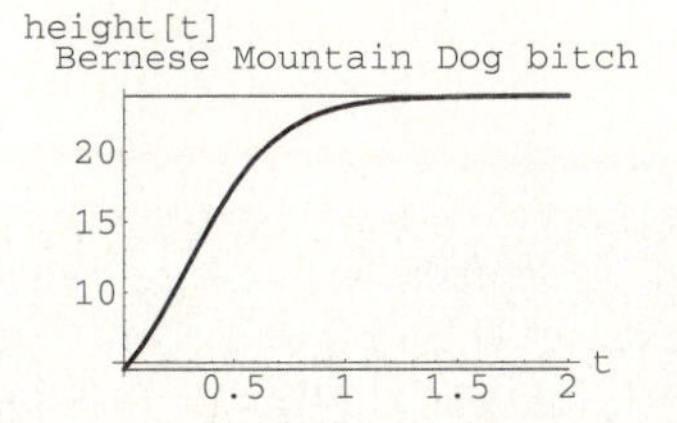

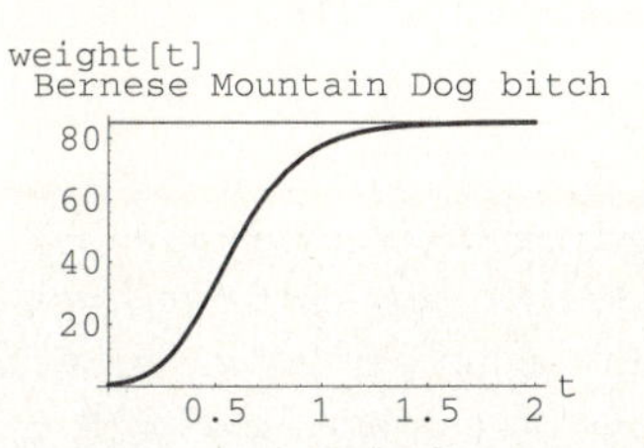

Somewhat interesting. Now comes the bad news: The dogs do not maintain the same shape throughout their growing years; they maintain only approximately the same shape as they grow. The upshot: The weight plot should be regarded only as a rather good approximation of the true story. One way to check it is to see what it predicts the birth weight of a Bernese Mountain Dog pup is:

In[17]:=
```
weight[0]
```

Out[17]=
```
0.560303
```

In ounces:

In[18]:=

```
weight[0] 16
```

Out[18]=

```
8.96485
```

Not bad. The typical birth weight of a Bernese Mountain Dog pup is 14 to 20 ounces. The approximation above is off, but not by very much.

T.3.c.ii) Here are plots of the instantaneous growth rates of the weight and height of the Bernese Mountain Dog bitch studied above. The thick plot is weight.

In[19]:=

```
Plot[{weight'[t],height'[t]},{t,0,2},
PlotStyle->{Thickness[0.01],Red},
AxesLabel->{"t",""}];
```

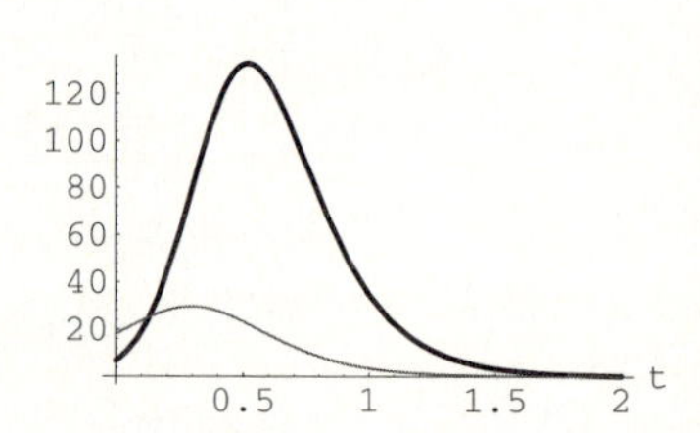

The height spurt happens before the weight spurt. This tells you that leggy adolescent animals are mathematical facts rather than anecdotal observations.

Now look at plots of the instantaneous percentage growth rates of the weight and height of the Bernese Mountain Dog bitch studied above.

In[20]:=

```
Plot[{100 weight'[t]/weight[t],
100 height'[t]/height[t]},{t,0,2},
PlotRange->All,PlotStyle->{Thickness[0.01],Red},
AxesLabel->{"t",""}];
```

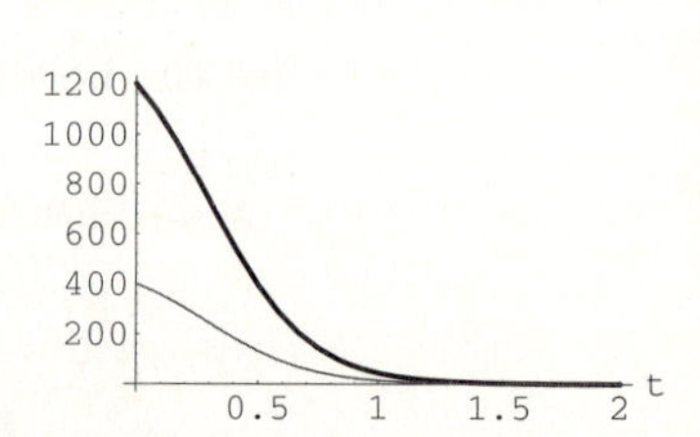

> This plot looks a little suspicious. It makes the strong suggestion that the instantaneous percentage growth rate of the weight is three times the instantaneous percentage growth rate of the height. Is this an accident?

Answer: Get off it. In mathematics, there are no accidents.

Remember

$$\text{weight}[t] = k\,\text{height}[t]^3.$$

So by the chain rule

$$\text{weight}'[t] = 3\,k\,\text{height}[t]^2\,\text{height}'[t].$$

Consequently, the instantaneous percentage growth rate of the weight is given by

$$\begin{aligned}\frac{100\,\text{weight}'[t]}{\text{weight}[t]} &= \frac{100\,3\,k\,\text{height}[t]\,\text{height}'[t]^2}{k\,\text{height}[t]^3}\\ &= \frac{3\,100\,\text{height}'[t]}{\text{height}[t]}\\ &= \text{instantaneous percentage growth rate of the height.}\end{aligned}$$

The upshot:

No matter what the height function is, the instantaneous percentage growth rate of the weight is three times the instantaneous percentage growth rate of the height. A new piece of biological insight brought to you by the chain rule.

Give It a Try

Experience with the starred ($\star$) problems will be especially beneficial for understanding later lessons.

■ G.1) Practicing with the chain rule★

Here are four derivative formulas which all successful practitioners of calculus know cold:

a) $D[e^{f(x)}, x] = e^{f(x)} f'[x]$:

In[1]:=

```
Clear[f,x]
D[E^f[x],x]
```

Out[1]=

```
 f[x]
E     f'[x]
```

This formula comes from the chain rule and the fact that $D[e^x, x] = e^x$.

b) $D[\log[f[x]], x] = f'[x]/f[x]$:

In[2]:=

```
D[Log[f[x]],x]
```

Out[2]=

```
f'[x]
-----
f[x]
```

This formula comes from the chain rule and the fact that $D[\log[x], x] = 1/x$.

c) $D[\sin[f[x]], x] = \cos[f[x]]\, f'[x]$:

In[3]:=
```
D[Sin[f[x]],x]
```

Out[3]=
```
Cos[f[x]] f'[x]
```

This formula comes from the chain rule and the fact that $D[\sin[x], x] = \cos[x]$.

d) $D[\cos[f[x]], x] = -\sin[f[x]]\, f'[x]$:

In[4]:=
```
D[Cos[f[x]],x]
```

Out[4]=
```
-(Sin[f[x]] f'[x])
```

This formula comes from the chain rule and the fact that $D[\cos[x], x] = -\sin[x]$.

Use one of these formulas to explain the *Mathematica* output from each of the following *Mathematica* instructions. In each case, say what $f[x]$ and $f'[x]$ are. The first one is worked out as a freebie for your learning pleasure.

Rapid hand calculation for its own sake is not the goal of this course, but if you expect to be successful in understanding the myriad of applications of the derivative and the chain rule, you should be in a position to do these derivatives without much work or thought. Besides, if you can't take derivatives like these by hand, people will make fun of you.

G.1.a) The derivative with respect to x of e^{3x}:

Sample answer: e^{3x} is $e^{f(x)}$ with $f[x] = 3\,x$. Here $f'[x] = 3$. The derivative of $e^{f(x)}$ is $e^{f(x)} f'[x]$. So the derivative of e^{3x} is $e^{3x}\, 3 = 3\, e^{3x}$.

G.1.b) The derivative with respect to x of e^{-x^2}:

G.1.c) The derivative with respect to x of $e^{\sin(x)}$:

G.1.d) The derivative with respect to x of $e^{\sin(4x)}$:

G.1.e) The derivative with respect to x of $\sin[7\,x]$:

G.1.f) The derivative with respect to t of $\sin[\pi\,(t - a)]$:

G.1.g) The derivative with respect to x of $\log[1 + 4\,x]$:

G.1.h) The derivative with respect to x of $\log[1 + 4\,x^2]$:

G.1.i) The derivative with respect to x of $3\cos[x^2]$:

G.1.j) The derivative with respect to t of $\sqrt{\pi}\, e^{-t^2}$:

G.1.k) The derivative with respect to x of $\log[\cos[x]]$:

G.1.l) The derivative with respect to t of $\sin[e^{2t}]$:

G.1.m) The derivative with respect to x of $\sin[\log[x]]$:

G.1.n) The derivative with respect to x of $\sin[e^{ax}]$:

G.1.o) The derivative with respect to x of $\sin[b\, e^{ax}]$:

G.1.p) The derivative with respect to y of $a\, e^{-b(y-x)^2}$:

G.1.q) The derivative with respect to t of $\sin[b\, e^{ax}]$:

■ G.2) Practicing with the chain rule, the product rule, and the power rule★

Type the derivative with respect to x for each of the following functions and then let *Mathematica* check your answer. You should be able to do these by hand faster than you can type them into your computer.

G.2.a) $x^5 + 6$

G.2.b) $3\,x^5 - 4\,x$

G.2.c) $\frac{8}{x}$

G.2.d) $e^{2x}\,\sin[3\,x]$

G.2.e) $\left(x^3+1\right)^8\left(x^4+1\right)^7$

G.2.f) $\frac{\cos[\pi\,x]}{x^2}$

G.2.g) $a\,x^9 e^{-xy} + b\,y$

G.2.h) $x\log[x] - x$

G.2.i) $\frac{1}{\sqrt{1-x^2}}$

G.2.j) $\cos[5\,x]^3$

G.2.k) $\sqrt{x}$

G.2.l) $1 + x + \frac{x^2}{2} + \frac{x^3}{6} + \frac{x^4}{24} + \frac{x^5}{120}$

G.2.m)

$$1 - x + \frac{x^2}{2} - \frac{x^3}{6} + \frac{x^4}{24} - \frac{x^5}{120}$$

G.2.n)

$e^{xy} \sin[\pi\, x^2\, y]$

■ G.3) Global scale★

G.3.a)

Put $f[x] = \left(2\,x^4 + 50 \log[x]\right) / \left(x^4 + 3\,x^2 + 1\right)$. What do you say is the limiting value

$$\lim_{x \to \infty} f[x]?$$

G.3.b)

What do you say is the limiting value

$$\lim_{x \to \infty} \frac{x^{0.8} + 4\, \log[x]}{3\,x^{0.8} + 2\, \log[x]}?$$

Illustrate with a plot.

G.3.c)

What do you say is the limiting value

$$\lim_{x \to \infty} \frac{45\,x^8 - 123 \cos[x] + 6\,x^6}{e^{0.04x}}?$$

Illustrate with a plot.

G.3.d)

What do you say is the limiting value

$$\lim_{x \to \infty} \frac{\log[x]}{x^{0.3}}?$$

Illustrate with a plot.

G.3.e)

Rank the following functions in order of dominance as $x \to \infty$:

x^{52}, $0.0004\, e^{0.01x}$, $e^{0.02x}/x$, $x\, \log[x]$,
$89\,x^2$, $\sqrt{x}$, $100\, \log[x]$, $17\,x$, $0.08\,x^3$,
$0.0000013\, e^{2x}$, $100\,x^{0.004}$.

■ G.4) Exponential functions and their constant percentage growth rate★

G.4.a.i) Put $f[x] = e^{0.5x}$ and calculate the instantaneous percentage growth rate of $f[x]$ at any point x.

G.4.a.ii) Keep $f[x] = e^{0.5x}$.

Multiple choice:

a) Every time x advances by one unit, $f[x]$ goes up by 50%.

b) Every time x advances by one unit, $f[x]$ goes up by more than 50%.

c) Every time x advances by one unit, $f[x]$ goes up by less than 50%.

G.4.b.i) Take positive numbers a and r and put $f[x] = a\, e^{rx}$. Calculate in terms of r the instantaneous percentage growth rate of $f[x]$ at any point x.

G.4.b.ii) Keep $f[x] = a\, e^{rx}$ with a and $r > 0$.

Multiple choice:

a) Every time x advances by one unit, $f[x]$ goes up by $100\, r$ percent.

b) Every time x advances by one unit, $f[x]$ goes up by more than $100\, r$ percent.

c) Every time x advances by one unit, $f[x]$ goes up by less than $100\, r$ percent.

G.4.c) Come up with a function $f[x]$ with the properties that:

→ $f[0] = 10$ and

→ the instantaneous percentage growth rate of $f[x]$ is a steady 25%.

G.4.d.i) Put $f[x] = e^{-0.5x}$ and calculate the instantaneous percentage growth rate of $f[x]$ at any point x.

G.4.d.ii) Keep $f[x] = e^{-0.2x}$.

Multiple choice:

a) Every time x advances by one unit, $f[x]$ goes down by 20%.

b) Every time x advances by one unit, $f[x]$ goes down by more than 20%.

c) Every time x advances by one unit, $f[x]$ goes down by less than 20%.

G.4.e.i) Take positive numbers a and r and put $f[x] = a\,e^{-rx}$ and calculate in terms of r the instantaneous percentage growth rate of $f[x]$ at any point x.

G.4.e.ii) Keep $f[x] = a\,e^{-rx}$ with a and $r > 0$.

Multiple choice:

a) Every time x advances by one unit, $f[x]$ goes down by $100\,r$ percent.

b) Every time x advances by one unit, $f[x]$ goes down by more than $100\,r$ percent.

c) Every time x advances by one unit, $f[x]$ goes down by less than $100\,r$ percent.

G.4.f) Come up with a function $f[x]$ with the properties that:

→ $f[0] = 4000$ and

→ the instantaneous percentage growth rate of $f[x]$ is a steady -25%.

■ G.5) Relating the plots of $f[x]$ and $f'[x]$

You can't get too much of a good thing.

G.5.a.i) Given $f[x] = \sin[x^2]$, *Mathematica* calculates the derivative $f'[x]$ as follows:

In[5]:=

```
Clear[x,f]
f[x_] = Sin[x^2];
f'[x]
```

Out[5]=

```
          2
2 x Cos[x ]
```

Use the chain rule to confirm the *Mathematica* output.

G.5.a.ii) Here is a plot of $f[x] = \sin[x^2]$ and its derivative $f'[x] = \cos[x^2]2\,x$ on the same axes for $-4 \le x \le 4$.

In[6]:=
```
Clear[f,x]; f[x_] = Sin[x^2];
Plot[{f[x],f'[x]},{x,-4,4},
PlotStyle->Thickness[0.01],
AxesLabel->{"x",""}];
```

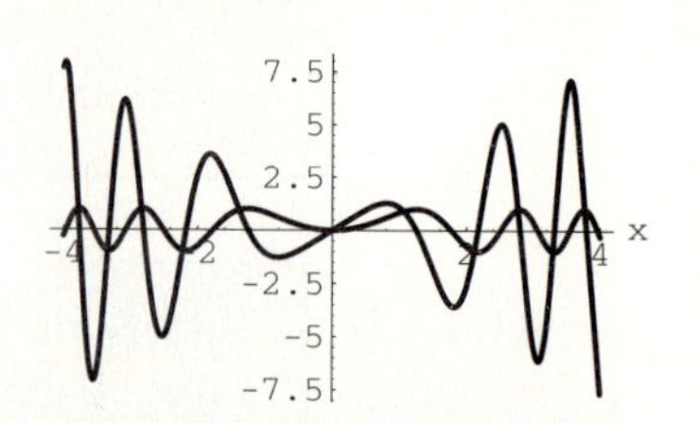

> Identify the plot of $f'[x]$ and discuss the relations between the two plots. Pay special attention to:
>
> → what $f[x]$ is doing when $f'[x]$ is positive.
>
> → what $f[x]$ is doing when $f'[x]$ is negative.
>
> → what $f[x]$ is doing when $f'[x]$ is changing sign.
>
> → what $f'[x]$ is doing when $f[x]$ is going up.
>
> → what $f'[x]$ is doing when $f[x]$ is going down.
>
> → what $f'[x]$ is doing when $f[x]$ is at the top of a crest or at the bottom of a dip.

G.5.b.i) Given $f[x] = \cos[x^2]^2$, *Mathematica* calculates the derivative $f'[x]$ as follows:

In[7]:=
```
Clear[x,f]
f[x_] = Cos[x^2]^2;
f'[x]
```

Out[7]=
```
            2        2
-4 x Cos[x ] Sin[x ]
```

> Use the chain rule to confirm the *Mathematica* output.

G.5.b.ii) Here is a plot of $f[x] = \cos[x^2]^2$ and its derivative $f'[x] = -2\sin[x^2]\cos[x^2]\,2\,x$ on the same axes for $-3 \le x \le 3$.

In[8]:=
```
Clear[f,x]
f[x_] = Cos[x^2]^2;
Plot[{f[x],f'[x]},{x,-3,3},
PlotStyle->Thickness[0.01],AxesLabel->{"x",""}];
```

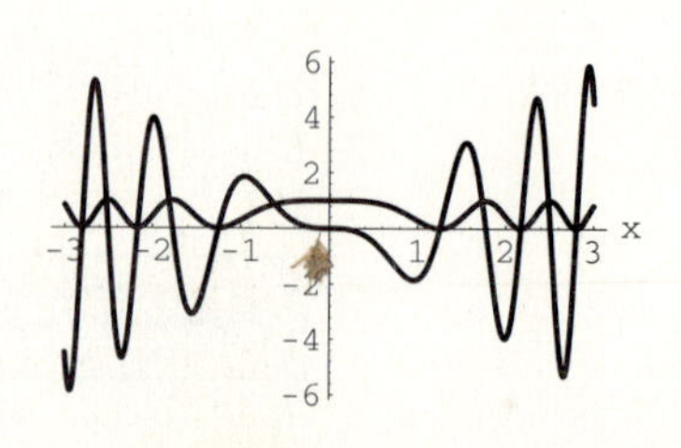

Identify the plot of $f'[x]$ and discuss the relations between the two plots. Pay special attention to:

→ what $f[x]$ is doing when $f'[x]$ is positive.

→ what $f[x]$ is doing when $f'[x]$ is negative.

→ what $f[x]$ is doing when $f'[x]$ is changing sign.

→ what $f'[x]$ is doing when $f[x]$ is going up.

→ what $f'[x]$ is doing when $f[x]$ is going down.

→ what $f'[x]$ is doing when $f[x]$ is at the top of a crest or at the bottom of a dip.

G.5.c.i) Here is *Mathematica*'s calculation of the derivative of $f[x] = (3\,x^2 + x + 2)\,/\,(x^2 + 1)$:

In[9]:=

```
Clear[f,x]
f[x_] = (3 x^2 + x + 2)/(x^2 + 1);
f'[x]
```

Out[9]=

```
                          2
1 + 6 x     2 x (2 + x + 3 x )
-------  -  ------------------
     2               2 2
1 + x          (1 + x )
```

Write $f[x]$ in the form $f[x] = (3\,x^2 + x + 2)\,(x^2 + 1)^{-1}$ and use the product rule to confirm *Mathematica*'s calculation of $f'[x]$.

G.5.c.ii) Here is a plot of $f[x] = (3\,x^2 + x + 2)\,/\,(x^2 + 1)$ together with the plot of its derivative for $-3 \leq x \leq 3$:

In[10]:=

```
Clear[f,x]
f[x_] = (3 x^2 + x + 2)/(x^2 + 1);
Plot[{f[x],f'[x]},{x,-3,3},
PlotStyle->{{Blue,Thickness[0.01]},
{Red}},AxesLabel->{"x",""}];
```

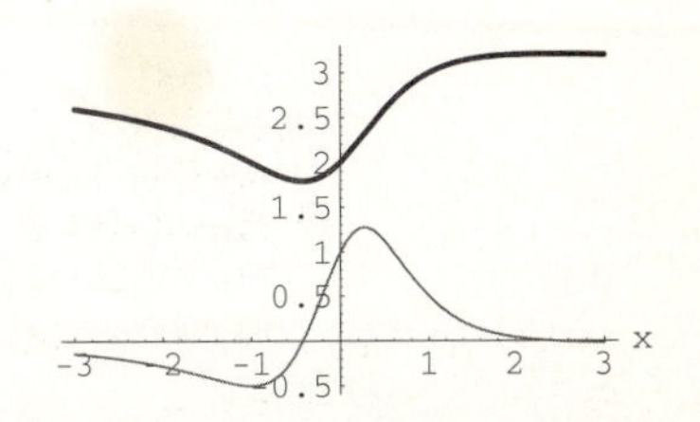

Describe the relations between the two plots.

What is the value of $f'[x]$ at the point in $[-3, 3]$ at which $f[x]$ is as low as it can be for $-3 \leq x \leq 3$? How does the plot of the derivative on the right explain the lazy growth of $f[x]$ as x goes from 2 to 3?

Estimate the point x at which $f[x]$ is going up most rapidly.

G.5.d.i) Here is a plot of $f[x] = e^{-x}\sin[6\,x]$ together with plots of e^{-x} and $-e^{-x}$ for $0 \leq x \leq 4$:

In[11]:=
```
Clear[f,x]
f[x_] = E^(-x) Sin[6 x];
Plot[{f[x],E^(-x),-E^(-x)},{x,0,4},
PlotStyle->{{Blue,Thickness[0.01]},Red,Red},
PlotRange->All,AxesLabel->{"x",""}];
```

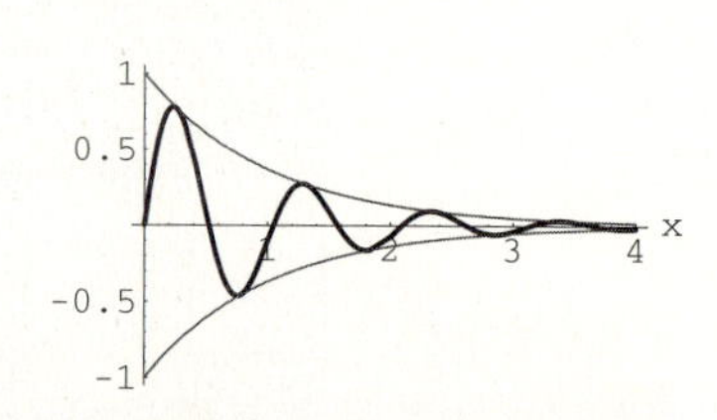

> Explain why you think the plot of $f[x] = e^{-x}\sin[6\,x]$ is mashed between the plots of e^{-x} and $-e^{-x}$. What is the global scale
>
> $$\lim_{x\to\infty} e^{-x}\sin[6\,x]?$$

G.5.d.ii) Here is *Mathematica*'s calculation of the derivative of $f[x] = e^{-x}\sin[6\,x]$:

In[12]:=
```
Clear[f,x]
f[x_] = E^(-x) Sin[6 x];
f'[x]
```

Out[12]=
```
6 Cos[6 x]   Sin[6 x]
---------- - --------
     x           x
    E           E
```

> Use the product rule to confirm *Mathematica*'s calculation.

G.5.d.iii) Here is a plot of $f[x] = e^{-x}\sin[6\,x]$ together with the plot of its derivative for $0 \leq x \leq 3$:

In[13]:=
```
Clear[f,x]
f[x_] = E^(-x) Sin[6 x];
Plot[{f[x],f'[x]},{x,0,3},
PlotStyle->{{Blue,Thickness[0.01]}, Red},
PlotRange->All,AxesLabel->{"x",""}];
```

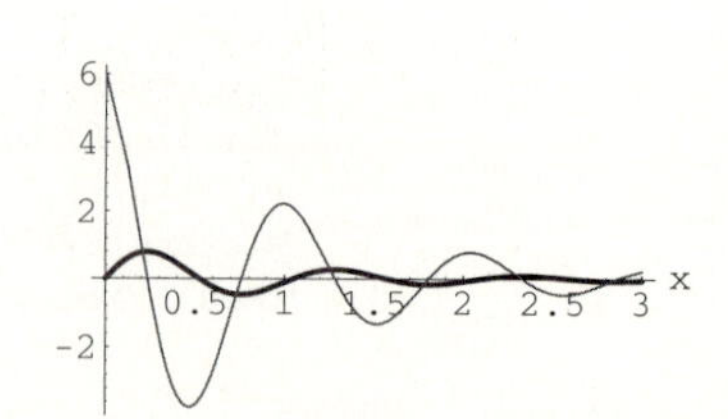

The plot of $f[x]$ is thicker than the plot of $f'[x]$.

> Describe the relations between the two plots.
>
> What is the value of $f'[x]$ at the point in $[0, 3]$ at which $f[x]$ is as low as it can be for $0 \leq x \leq 3$?

What is the value of $f'[x]$ at the point in $[0,3]$ at which $f[x]$ is as high as it can be for $0 \leq x \leq 3$?

How does the plot of the derivative tell you the point x in $[0,3]$ at which $f[x]$ the instantaneous growth rate is as big as possible?

G.5.e.i) Here is a plot of $(x^2+x)\,e^{-x^2}$ for $-4 \leq x \leq 4$:

In[14]:=

```
Clear[f,x]
f[x_] = (x^2 + x) E^(-x^2);
Plot[f[x],{x,-4,4},
PlotStyle->{{Blue,Thickness[0.01]}},
PlotRange->All,AxesLabel->{"x",""}];
```

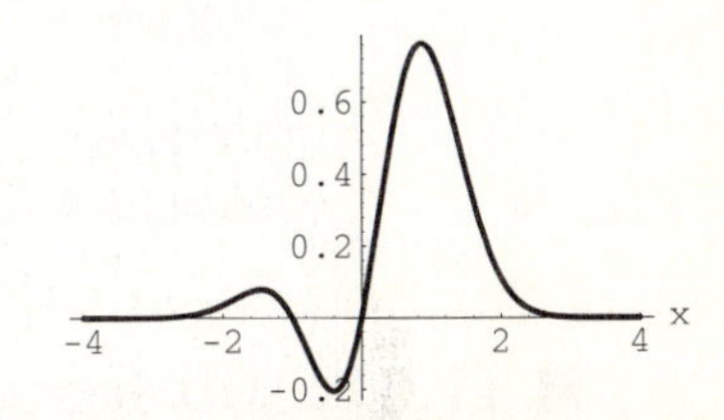

Use what you know about exponential growth to explain what's happening to the left and to the right.

G.5.e.ii) Here is *Mathematica*'s calculation of the derivative of $f[x] = (x^2+x)\,e^{-x^2}$:

In[15]:=

```
Clear[f,x]
f[x_] = (x^2 + x) E^(-x^2);
f'[x]
```

Out[15]=

```
                     2
1 + 2 x   2 x (x + x )
------- - ------------
   2           2
  x           x
 E           E
```

Use the product rule and the chain rule to confirm *Mathematica*'s calculation.

G.5.e.iii) Here is a plot of $(x^2+x)e^{-x^2}$ together with the plot of its derivative for $-4 \leq x \leq 4$:

In[16]:=

```
Clear[f,x]
f[x_] = (x^2 + x) E^(-x^2);
Plot[{f[x],f'[x]},{x,-4,4},
PlotStyle->{{Blue,Thickness[0.01]},Red},
PlotRange->All,AxesLabel->{"x",""}];
```

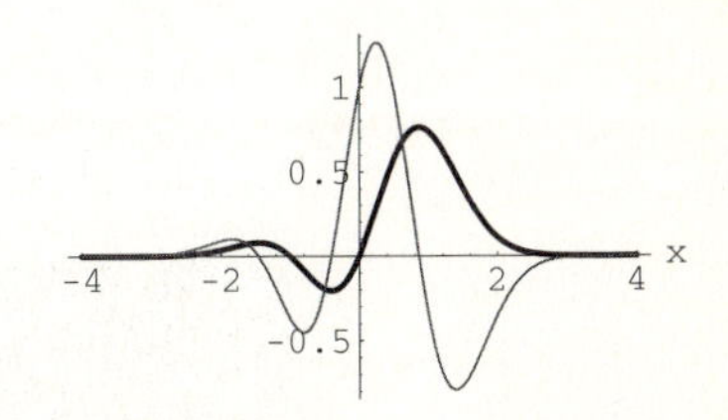

The plot of $f[x]$ is thicker than the plot of $f'[x]$.

Describe the relations between the two plots.

What is the value of $f'[x]$ at the point in $[-4,4]$ at which $f[x]$ is as low as it can be for $-4 \leq x \leq 4$?

What is the value of $f'[x]$ at the point in $[-4,4]$ at which $f[x]$ is as high as it can be for $-4 \leq x \leq 4$?

How does the plot of the derivative give you a pretty good estimate of the point x in $[-4,4]$ at which $f[x]$ the instantaneous growth rate is as big as possible? Approximately where is this point? Why is the plot of the derivative on the left and right in harmony with the lazy growth of $f[x]$ as x goes from -4 to -3 and as x grows from 3 to 4?

G.6) 100 log[$f[x]$] and the instantaneous percentage growth rate

G.6.a) Look at this:

In[17]:=

```
Clear[f,x]
100 D[Log[f[x]],x]
```

Out[17]=

```
100 f'[x]
---------
  f[x]
```

True or False: If $f[x]$ is a positive function, then the instantaneous percentage growth rate of $f[x]$ is exactly the same as the instantaneous growth rate of $100 \log[f[x]]$.

G.6.b) Look at this:

In[18]:=

```
Clear[f,k,x]
100 D[Log[k f[x]],x]
```

Out[18]=

```
100 f'[x]
---------
  f[x]
```

True or False: If $f[x]$ is a positive function, then the instantaneous percentage growth rate of $h[x] = k\, f[x]$ is exactly the same as the instantaneous percentage growth rate of $f[x]$.

G.6.c) Look at this:

In[19]:=
```
Clear[f,g,x]
100 D[Log[f[x] + g[x]],x]
```

Out[19]=
```
100 (f'[x] + g'[x])
-------------------
    f[x] + g[x]
```

True or False: If $f[x]$ and $g[x]$ are positive functions, then the instantaneous percentage growth rate of $f[x] + g[x]$ is exactly the same as the instantaneous percentage growth rate of $f[x]$ plus the instantaneous percentage growth rate of $g[x]$.

G.6.d) Look at this:

In[20]:=
```
Clear[f,g,x]
100 D[Log[f[x] g[x]],x]
```

Out[20]=
```
100 (g[x] f'[x] + f[x] g'[x])
-----------------------------
          f[x] g[x]
```

In[21]:=
```
Simplify[100 D[Log[f[x] g[x]],x]]
```

Out[21]=
```
100 f'[x]   100 g'[x]
--------- + ---------
  f[x]        g[x]
```

True or False: If $f[x]$ and $g[x]$ are positive functions, then the instantaneous percentage growth rate of $h[x] = f[x]\, g[x]$ is exactly the same as the instantaneous percentage growth rate of $f[x]$ plus the instantaneous percentage growth rate of $g[x]$.

G.6.e) Look at this:

In[22]:=
```
Clear[f,g,x]
100 D[Log[f[x]/g[x]],x]
```

Out[22]=
```
             f'[x]   f[x] g'[x]
100 g[x] (------- - ----------)
              g[x]          2
                       g[x]
-------------------------------
              f[x]
```

In[23]:=

```
Simplify[100 D[Log[f[x]/g[x]],x]]
```

Out[23]=

$$\frac{100\ f'[x]}{f[x]} - \frac{100\ g'[x]}{g[x]}$$

> True or False: If $f[x]$ and $g[x]$ are positive functions, then the instantaneous percentage growth rate of $h[x] = f[x]/g[x]$ is exactly the same as the instantaneous percentage growth rate of $f[x]$ minus the instantaneous percentage growth rate of $g[x]$.

G.6.f) Look at this:

In[24]:=

```
Clear[f,k,x]
100 D[Log[k f[x]^2],x]
```

Out[24]=

$$\frac{200\ f'[x]}{f[x]}$$

> True or False: If $f[x]$ is a positive function, then the instantaneous percentage growth rate of $h[x] = k\,f[x]^2$ is exactly the same as twice the instantaneous percentage growth rate of $f[x]$.

G.6.g) Look at this:

In[25]:=

```
Clear[f,r,x]
100 D[Log[k f[x]^r],x]
```

Out[25]=

$$\frac{100\ r\ f'[x]}{f[x]}$$

> True or False: If $f[x]$ is a positive function, then the instantaneous percentage growth rate of $h[x] = k\,f[x]^r$ is exactly the same as r times the instantaneous percentage growth rate of $f[x]$.

G.6.h)

> Try to use properties of $\log[x]$ like
>
> $\rightarrow$ $D[\log[x], x] = 1/x$,
>
> $\rightarrow$ $\log[u\,v] = \log[u] + \log[v]$,

→ $\log[u/v] = \log[u] - \log[v]$, and

→ $\log[u^v] = v \log[u]$

to explain the statements you marked true.

■ G.7) Linear dimension: Length, area, volume, and weight*

G.7.a) Here is a true scale plot of a solid:

In[26]:=

```
Clear[x,y,z,s,t]
x[s_,t_] = Sin[s] Cos[t]; y[s_,t_] = Sin[s] Sin[t];
z[s_,t_] = (0.6 ( 1 - Cos[s])) Cos[s];
little =  ParametricPlot3D[{x[s,t],y[s,t],z[s,t]},
{s,0,Pi},{t,0,2 Pi},PlotRange->{{-3,3},{-3,3},{-3,3}},
ViewPoint->{2.747, 1.594, 1.168},Boxed->False,Axes->None];
```

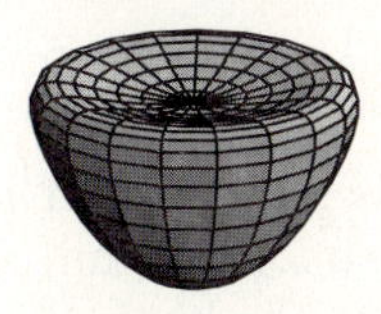

Here is a true scale plot of the same solid with all its linear dimensions increased by a factor of 2.5:

In[27]:=

```
big = ParametricPlot3D[
{2.5 x[s,t], 2.5 y[s,t],2.5 z[s,t]},{s,0,Pi},{t,0,2 Pi},
PlotRange->{{-3,3},{-3,3},{-3,3}},
ViewPoint->{2.747, 1.594, 1.168},Boxed->False,Axes->None];
```

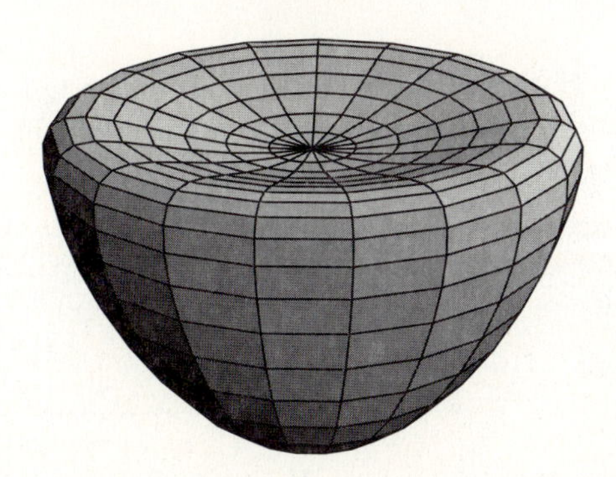

Here they are side by side:

In[28]:=
```
Show[GraphicsArray[{little,big}]];
```

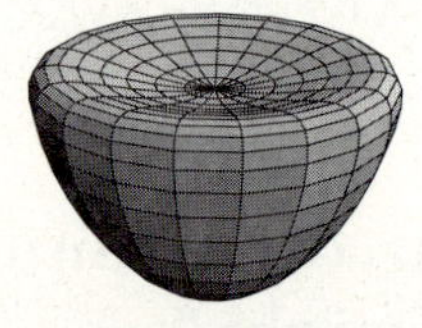

> The volume measurement of the larger solid is what percent of the volume measurement of the smaller solid? The surface area measurement of the larger solid is what percent of the surface area measurement of the smaller solid?

G.7.b)

> A crystal grows in such a way that all the linear dimensions increase by 150%. How do the new surface area and the new volume compare to the old surface area and volume?

G.7.c.i) Calculus&*Mathematica* thanks Terri Birk, D.V.M., owner of St. Joseph Animal Hospital, St. Joseph, Illinois and breeder of Haflinger horses, for supplying data for this problem.

Horses and other animals grow so that a linear dimension of their bodies is given by what a lot of folks call a logistic function $b\,c\,e^{at}/\,(b - c + c\,e^{at})$ where t measures time in years elapsed since the birth of the animal. A good linear dimension for a horse is the height of the horse's body at the horse's shoulders:

In[29]:=
```
Clear[height,a,b,c,t]
height[t_] = b c E^(a t)/(b - c + c E^(a t))
```

Out[29]=
```
       a t
  b c E
-----------
           a t
b - c + c E
```

A magnificent Haflinger horse, as owned by the Queen of England, is born with a height of 36 inches and eventually grows to a height of 59 inches. It achieves 95% of its adult height at age 2.

> Use these data to determine the values of the parameters a, b, and c and then plot height$[t]$ for the Haflinger horse for the first four years of its life.
>
> Give a plot.

G.7.c.ii) Given that the typical mature Haflinger horse as described above weighs 1000 pounds, give an approximate plot of the horse's weight as a function of time for the first four years and give a critique of the plot. The Haflinger weighs about 150 pounds at birth.

Show the height plot beside the weight plot and discuss what you see. Does the horse grow up before it grows out?

G.7.c.iii) Give plots of the instantaneous growth rates of the weight and height of the Haflinger horse studied above for the first four years. Discuss the timing between the height spurt and the weight spurt. Then give plots of the instantaneous percentage growth rates of the weight and height of the Haflinger horse studied above. Experienced horse people often say that a new colt or filly drops out of the womb growing fast, with the growth rate tapering off as the horse grows up. Is this opinion reflected in your plot?

G.7.d) Repeat all parts of part G.7.c) above, but this time use data from your own body or the body of your sister, brother, mother, father, or friend. You will need to know:

→ height at birth,

→ height at maturity,

→ % of mature height achieved at a given age,

→ weight at maturity.

Explain why the weight plot may be way off for beer-swilling couch potatoes.

G.7.e) Explain the following statement: As a bear grows up, the instantaneous percentage growth rate of its weight is about 1.5 times the instantaneous percentage growth rate of the number of hairs in its fur.

■ G.8) Interest compounded every instant versus interest compounded every month

G.8.a.i) When banks announce an interest rate of $100\,r$ percent per year compounded every instant, this means that when you deposit $\$a$ into an account and never touch the money, then the amount in dollars in your account t years after the initial deposit is:

In[30]:=

```
Clear[balance,t,a,r]
balance[t] = a E^(r t)
```

Out[30]=

```
  r t
a E
```

> Measure the instantaneous percentage growth rate of balance[t] for instantaneous compounding and discuss the meaning of the result.

G.8.a.ii) When banks announce an interest rate of $100\,r$ percent per year compounded every month, this means that when you deposit \$$a$ into an account and never touch the money, then you calculate the amount in dollars in your account t years after the initial deposit as follows: Start with the advertised yearly rate of $100\,r$ percent; convert this to $100\,r/12$ percent, which the bank pays you each month. The balance t years after the deposit is given by balance[t] $= a\,e^{s(r)t}$ where $s[r]$ is set so that

$$100\left(\frac{\text{balance}[t+1/12]}{\text{balance}[t]} - 1\right) = 100\frac{r}{12}:$$

In[31]:=

```
Clear[balance,t,a,s]
balance[t_] = a E^(s[r] t)
```

Out[31]=

```
  t s[r]
a E
```

In[32]:=

```
Simplify[100(balance[t + 1/12]/balance[t] - 1)]
```

Out[32]=

```
          s[r]/12
100 (-1 + E       )
```

This tells you that:

$$e^{s(r)/12} - 1 = r/12;$$

$$e^{s(r)/12} = 1 + r/12;$$

$$s[r]/12 = \log[1 + r/12];$$

$$s[r] = 12\log[1 + r/12];$$

$$s[r] = \log[(1 + r/12)^{12}];$$

In[33]:=

```
s[r_] = Log[(1 + r/12)^12];
balance[t]
```

Out[33]=

```
         r   12 t
a ((1 + --)     )
        12
```

> Measure the instantaneous percentage growth rate of balance[t] for monthly compounding.

G.8.a.iii) In part G.8.a.i) above, you measured the instantaneous percentage growth rate of balance[t] for instantaneous compounding at an advertised interest rate of $100\,r$ percent. In part G.8.a.ii) above, you measured instantaneous percentage growth rate of balance[t] for monthly compounding. If you did everything right, then both measurements came out in terms of r.

> Plot these two measurements on the same axes as functions of r for $0 \leq r \leq 0.15$ and discuss what the plot reveals about compounding every instant versus compounding every month for advertised interest rates between 0% and 15%.

LESSON 1.05

Using the Tools

Basics

■ B.1) Finding maximum and minimum values

You can tell what happens to

$$f[x] = e^{-x^2}\left(x^4 - 2\sin[x]\right)$$

as x leaves $x = 1.0$ by looking at $f'[1.0]$:

In[1]:=
```
Clear[f,x]
f[x_] = E^(-x^2) (x^4 - 2 Sin[x]); f'[1.0]
```

Out[1]=
```
1.57647
```

Positive. This tells you that as x leaves 1.0 and advances a little bit, then $f[x]$ increases. This also tells you that as x leaves 1.0 and decreases a little bit, then $f[x]$ decreases. Check with a plot:

In[2]:=
```
a = 1.0;
localfplot = Plot[f[x],{x,a - 0.2, a + 0.2},
PlotStyle->{{Red,Thickness[0.01]}},
DisplayFunction->Identity];
points = {Graphics[{PointSize[0.03],Point[{a,f[a]}]}],
Graphics[{PointSize[0.03],Point[{a,0}]}]};
line = Graphics[Line[{{a,0},{a,f[a]}}]];
Show[localfplot,points,line,AxesLabel->{"x","f[x]"},
DisplayFunction->$DisplayFunction];
```

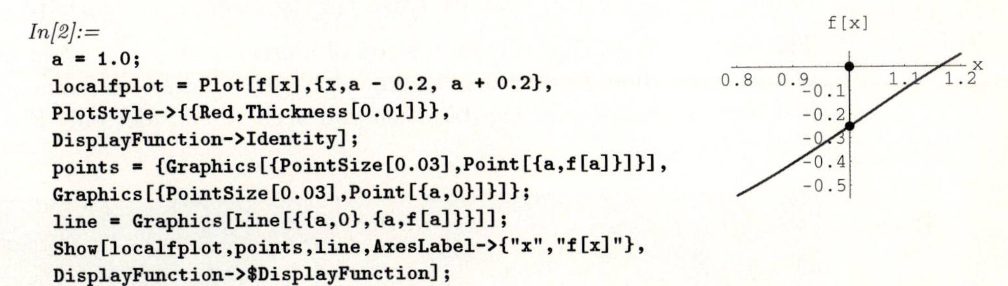

Yep.

→ As x leaves 1.0 and advances a little bit, $f[x]$ goes up.

→ But as x leaves 1.0 and decreases a bit, $f[x]$ goes down.

This was predicted in advance because $f'[1.0]$ is positive. Now see what $f[x]$ is doing at $x = 1.8$:

In[3]:=

```
f'[1.8]
```

Out[3]=

```
-0.27404
```

Negative. Check it out with a plot:

In[4]:=

```
a = 1.8;
Plot[f[x],{x,a - 0.2,a + 0.2},
PlotStyle->{{Red,Thickness[0.01]}},
AxesLabel->{"x","f[x]"},
AxesOrigin->{a,f[a]}];
```

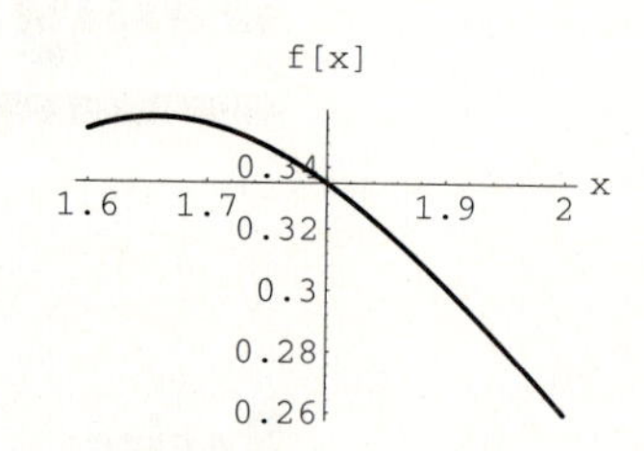

Lookin' good.

→ As x leaves 1.8 and advances a little bit, $f[x]$ goes down.

→ But as x leaves 1.8 and decreases a bit, $f[x]$ goes up.

This was predicted in advance because $f'[1.8]$ is negative. Try some other selections of x.

B.1.a.i) What's the moral?

Answer:

→ If $f'[a] > 0$, then you can increase $f[x]$ by pushing x slightly to the right of a.

→ If $f'[a] > 0$, then you can decrease $f[x]$ by pushing x slightly to the left of a.

→ If $f'[a] < 0$, then you can increase $f[x]$ by pushing x slightly to the left of a.

→ If $f'[a] < 0$, then you can decrease $f[x]$ by pushing x slightly to the right of a.

The upshot: If $\{a, f[a]\}$ sits at the top of a crest or at the bottom of a dip on the plot of $f[x]$, then $f'[a] = 0$. Consequently, if you want to make $f[x]$ as high as it can be or as low as it can be, then you need to consider only the x's for which $f'[x] = 0$.

B.1.a.ii) Give the x's that correspond to the highest crest and the lowest dip on the graph of $f[x] = e^{-x^2}\left(x^4 - 2\sin[x]\right)$.

Answer: Look at a plot:

In[5]:=
```
Clear[f,x]
f[x_] = E^(-x^2) (x^4 - 2 Sin[x]);
fplot = Plot[f[x],{x,-4,4},
PlotStyle->{{Red,Thickness[0.01]}},
AxesLabel->{"x","f[x]"}];
```

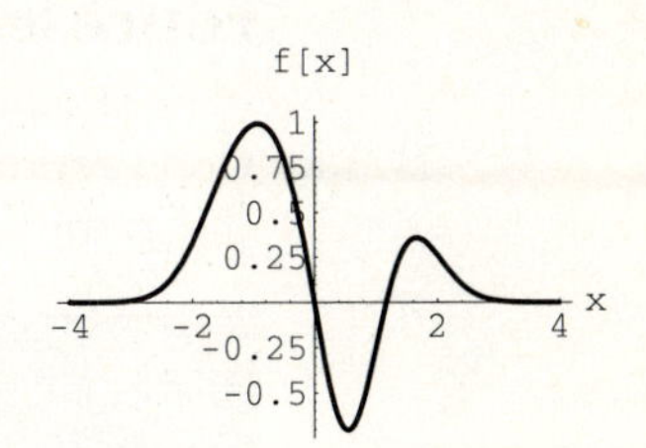

The global scale dominance of the e^{-x^2} term is already evident at both ends of the plot. You can be sure that the highest crest and the lowest dip are staring at you from this plot.

The highest crest is almost over $x = -1$.

In[6]:=
```
FindRoot[f'[x] == 0,{x,-1}]
```

Out[6]=
```
{x -> -0.939457}
```

The highest crest is at $\{-0.939457,\ f[-0.939457]\}$. Take a look:

In[7]:=
```
a = -0.939457;
points =
{Graphics[{PointSize[0.03],Point[{a,f[a]}]}],
Graphics[{PointSize[0.03],Point[{a,0}]}]};
line = Graphics[Line[{{a,0},{a,f[a]}}]];
Show[fplot,points,line];
```

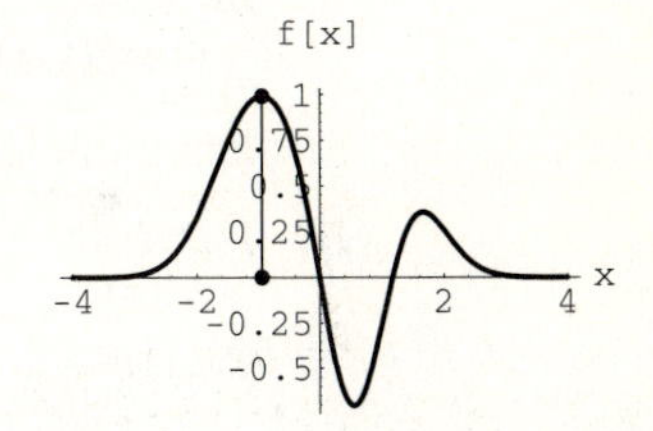

Nailed it. The lowest dip is almost under $x = 0.5$.

In[8]:=
```
FindRoot[f'[x] == 0,{x,0.5}]
```

Out[8]=
```
{x -> 0.548742}
```

The lowest dip is at $\{0.548742,\ f[0.548742]\}$. Take a look:

In[9]:=

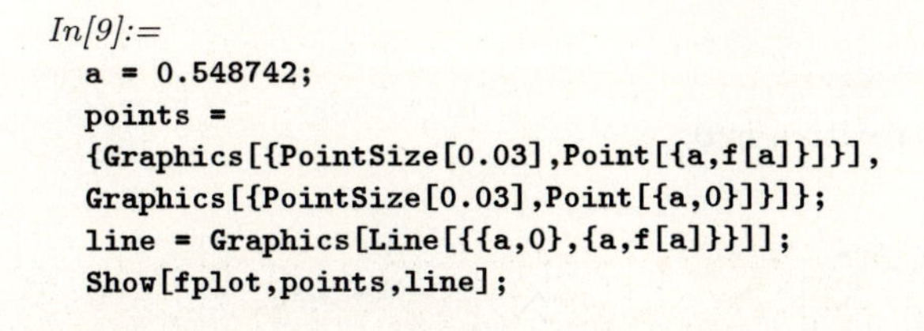
```
a = 0.548742;
points =
{Graphics[{PointSize[0.03],Point[{a,f[a]}]}],
Graphics[{PointSize[0.03],Point[{a,0}]}]};
line = Graphics[Line[{{a,0},{a,f[a]}}]];
Show[fplot,points,line];
```

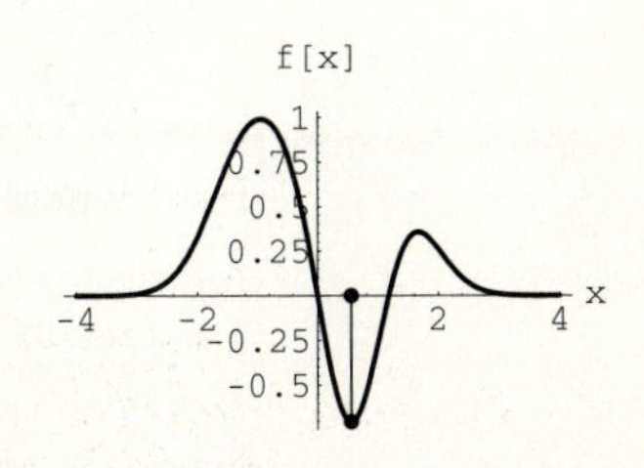

On the money.

■ B.2) Using the derivative to get a good representative plot

B.2.a) Good representative plots of functions try to exhibit all the dips and the crests of the graph and give a strong flavor of the global scale behavior.

> Why does a good representative plot of a function normally include all points at which its derivative is 0?

Answer: This way you can be sure that the curve does not change direction after it leaves the screen on the left and on the right.

B.2.b)

> Use the derivative to help you set up a good representative plot of
>
> $$f[x] = e^{-x/10}\left(48 + 24\,x + x^2\right).$$

Answer:

In[10]:=

```
Clear[f,x]; f[x_] = (E^(-x/10)) (48 + 24 x + x^2);
```

Examine the derivative:

In[11]:=

```
f'[x]
```

Out[11]=

```
                         2
24 + 2 x   48 + 24 x + x
-------- - --------------
  x/10           x/10
 E           10 E
```

Go for a common denominator:

In[12]:=

```
Together[f'[x]]
```

Out[12]=

```
              2
192 - 4 x - x
--------------
     x/10
 10 E
```

Find where the derivative is 0.

In[13]:=

```
Solve[192 - 4 x - x^2  == 0,x]
```

Out[13]=

```
{{x -> -16}, {x -> 12}}
```

The first shot for a good representative plot should begin to the left of $x = -16$ and end to the right of $x = 12$.

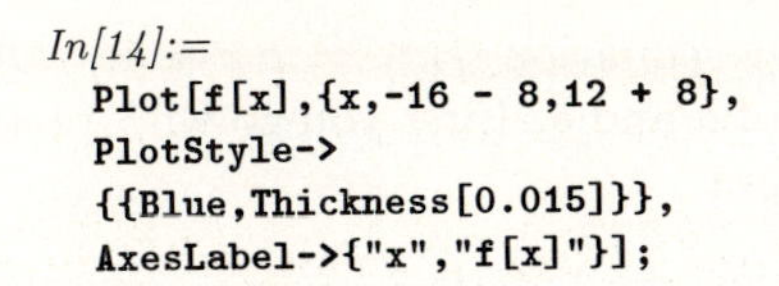

In[14]:=
```
Plot[f[x],{x,-16 - 8,12 + 8},
PlotStyle->
{{Blue,Thickness[0.015]}},
AxesLabel->{"x","f[x]"}];
```

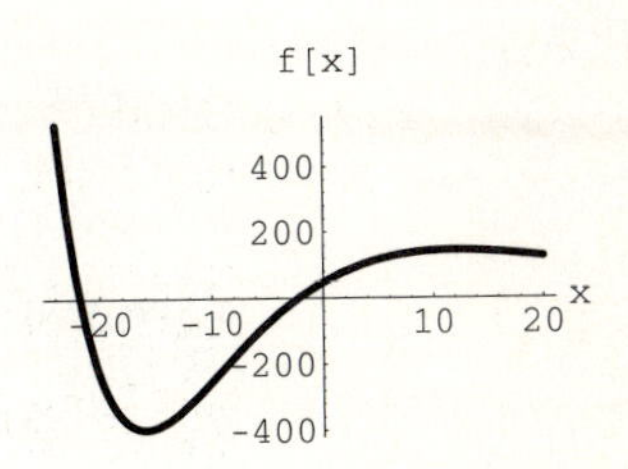

The plot changes direction at $x = -16$ and $x = 12$; it cannot change direction at any other point.

Look at the formula for $f[x] = e^{-x/10}\left(48 + 24\,x + x^2\right)$. The powerful $e^{-x/10}$ factor is dominating the quadratic factor on the right.

■ B.3) Using the derivative to fit data by curves

B.3.a) Here are some data and a plot of the data:

In[15]:=
```
data = {{0.23, 1.11}, {0.37, 1.22}, {0.90, 1.35},
{1.72, 1.58}, {1.84, 1.57}, {2.44, 1.74},
{3.21, 2.02}, {3.31, 1.95}, {3.98, 2.31},
{4.70, 2.64}, {4.79, 2.76}};
dataplot =
ListPlot[data,PlotStyle->PointSize[0.02]];
```

2.75
2.5
2.25
2 3 4
1.75
1.5
1.25

They don't line up in a straight line, but they are pretty close. Here is *Mathematica*'s compromise line fit of these data:

In[16]:=
```
Clear[fitter,x]; fitter[x_] = Fit[data,{1,x},x]
```

Out[16]=
```
1.00626 + 0.333981 x
```

Check out the plot:

In[17]:=
```
fitplot = Plot[fitter[x],{x,0,5},
PlotStyle->{{Red,Thickness[0.01]}},
DisplayFunction->Identity];
Show[fitplot,dataplot,
DisplayFunction->$DisplayFunction];
```

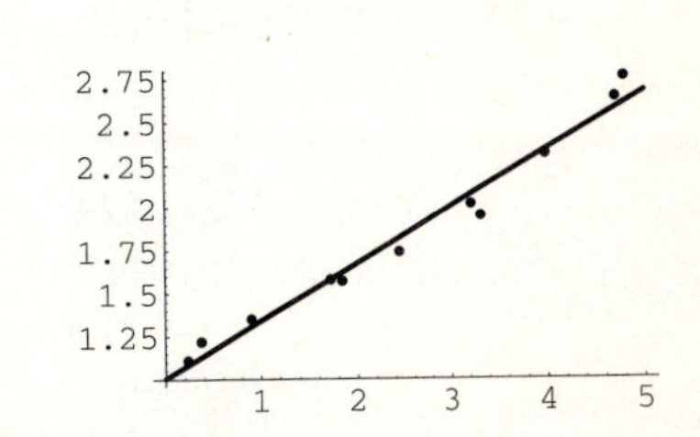

Nice job, *Mathematica*.

How did *Mathematica* do it?

Answer: Well, any line has the equation $f[x] = a\,x + b$. *Mathematica* was programmed to find good choices of a and b. First you go with prelimf$[x] = a\,x + b$ without specifiying what a and b are.

Then you take the data:

$$\{x_1, y_1\},\ \{x_2, y_2\},\ \{x_3, y_3\},\ \ldots,\ \{x_m, y_m\}$$

and form the square error function

$$\text{sqerror}[a, b] = (\text{prelimf}[x_1] - y_1)^2 + (\text{prelimf}[x_2] - y_2)^2 + (\text{prelimf}[x_3] - y_3)^2 + \cdots + (\text{prelimf}[x_m] - y_m)^2 .$$

The reason you use the squares is to prevent negative errors from canceling positive errors.

Here is how you get the square error function for the data given above:

In[18]:=

```
Clear[a,b,x,prelimf,sqerror,k]; prelimf[x_] = a x + b;
```

In[19]:=

```
sqerror[a_,b_] = Sum[(prelimf[data[[k,1]]] - data[[k,2]])^2,{k,1,Length[data]}]
```

Out[19]=

```
                   2                    2                   2
(-1.11 + 0.23 a + b)  + (-1.22 + 0.37 a + b)  + (-1.35 + 0.9 a + b)  +

                    2                    2                    2
 (-1.58 + 1.72 a + b)  + (-1.57 + 1.84 a + b)  + (-1.74 + 2.44 a + b)  +

                    2                    2                    2
 (-2.02 + 3.21 a + b)  + (-1.95 + 3.31 a + b)  + (-2.31 + 3.98 a + b)  +

                   2                    2
 (-2.64 + 4.7 a + b)  + (-2.76 + 4.79 a + b)
```

Now you search for the a and b that make sqerror$[a, b]$ as small as it can be. From what you know from the earlier Basics in this lesson, the a and b that make sqerror$[a, b]$ as small as it can be cannot be points for which $D[\text{sqerror}[a, b], a] \neq 0$ or $D[\text{sqerror}[a, b], b] \neq 0$.

This means that the a and b you want must make $D[\text{sqerror}[a, b], a] = 0$ and $D[\text{sqerror}[a, b], b] = 0$. Try it:

In[20]:=

```
equation1 = D[sqerror[a,b],a] == 0;
equation2 = D[sqerror[a,b],b] == 0;
bestaandb = Solve[{equation1,equation2},{a,b}]
```

Out[20]=

```
{{a -> 0.333981, b -> 1.00626}}
```

This tells you what a and b should be for a great fit:

In[21]:=

```
greatfit[x_] = prelimf[x]/.bestaandb[[1]]
```

Out[21]=

```
1.00626 + 0.333981 x
```

Compare:

In[22]:=

```
Clear[fitter,x]; fitter[x_] = Fit[data,{1,x},x]
```

Out[22]=

```
1.00626 + 0.333981 x
```

The very same thing. Fancy folks like to call this the "least squares fit." Now you know the calculus and the algebra behind the Fit instruction.

It's hardly ever a good idea to do this calculus and algebra by hand because the calculations are just too darn tedious. Doing them by hand is analogous to sawing many 4′ by 8′ sheets of plywood in half with a hand saw. You can do it in theory, but you're going to make lots of small mistakes and you'll be worn out at the end. Carpenters know when power tools are appropriate, and so should scientists.

B.3.b.i) Here are some data and a plot of the data:

In[23]:=

```
data = {{0.1, 4.5}, {0.30, 5.20}, {0.45, 6.16},
{0.65, 7.66}, {0.93, 8.62}, {1.27, 8.04},
{1.59, 7.02}, {1.82, 5.76}, {1.98, 4.44},
{2.15, 3.27}, {2.41, 2.38}, {2.75, 1.83},
{3.08, 1.65}, {3.34, 1.78}, {3.51, 2.14},
{3.67, 2.66}, {3.90, 3.23}, {4.22, 3.80},
{4.56, 4.32}, {4.84, 4.76}, {5.04, 5.12},
{5.19, 5.39}, {5.40, 5.59}, {5.69, 5.74},
{6.04, 5.83}};
dataplot = ListPlot[data,
PlotStyle->PointSize[0.02]];
```

It doesn't make a lot of sense to try to put a compromise line through the data; so you can have *Mathematica* put an interpolating function through the data:

In[24]:=

```
Clear[interpolator,x]
interpolator[x_] = Interpolation[data][x];
interpolplot = Plot[ interpolator[x],{x,0.1,6.04},
PlotStyle->{{Red,Thickness[0.01]}},
DisplayFunction->Identity];
Show[interpolplot,dataplot,
DisplayFunction->$DisplayFunction];
```

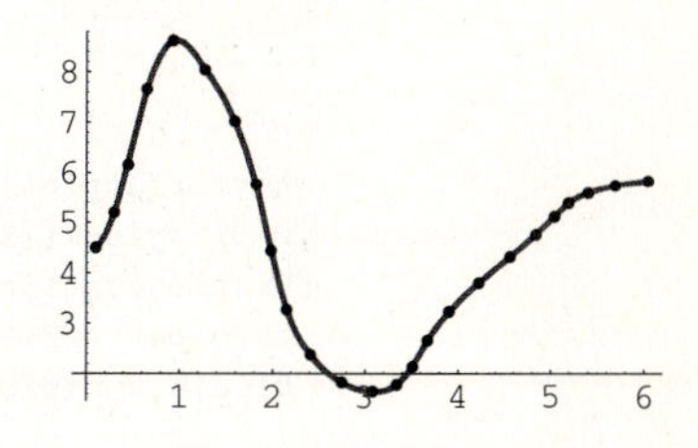

But the formula for this function is not available:

In[25]:=

```
interpolator[x]
```

Out[25]=

```
InterpolatingFunction[{0.1, 6.04}, <>][x]
```

> How do you use sine and cosine waves to come up with a compromise function, with a fairly simple formula, whose plot goes with the flow of the data?

Answer: Very easily. First look at the interval $[0, 6.04]$ over which the data are plotted. Put $L = 6.04$ and go for a fit in terms of sine and cosine waves:

In[26]:=
```
L = 6.04; Clear[fitter,x]
fitter[x_] = Fit[data,N[{1, Sin[Pi x/(L)],Cos[Pi x/(L)],
Sin[2 Pi x/(L)],Cos[2 Pi x/(L)]}],x]
```
Out[26]=
```
-2.52172 - 0.416701 Cos[0.520131 x] + 6.95178 Cos[1.04026 x] +
 11.2932 Sin[0.520131 x] + 1.49908 Sin[1.04026 x]
```
Try it out:

In[27]:=
```
fitplot = Plot[fitter[x],{x,0,L},
PlotStyle->{{Red,Thickness[0.01]}},
DisplayFunction->Identity];
Show[fitplot,dataplot,
DisplayFunction->$DisplayFunction];
```

Not half bad. You can get a better fit by using more sines and cosines:

In[28]:=
```
Clear[betterfitter,x]
betterfitter[x_] =
Fit[data,N[{1, Sin[Pi x/(L)],Cos[Pi x/(L)],Sin[2 Pi x/(L)],Cos[2 Pi x/(L)],
Sin[3 Pi x/(L)],Cos[3 Pi x/(L)],Sin[4 Pi x/(L)],Cos[4 Pi x/( L)]}],x]
```
Out[28]=
```
26.4797 + 3.312 Cos[0.520131 x] - 24.0394 Cos[1.04026 x] -
  4.15711 Cos[1.56039 x] + 2.563 Cos[2.08052 x] - 38.5552 Sin[0.520131 x] -
  3.73879 Sin[1.04026 x] + 12.9683 Sin[1.56039 x] + 2.41121 Sin[2.08052 x]
```
Not a terribly complicated formula. Check out the plot:

In[29]:=

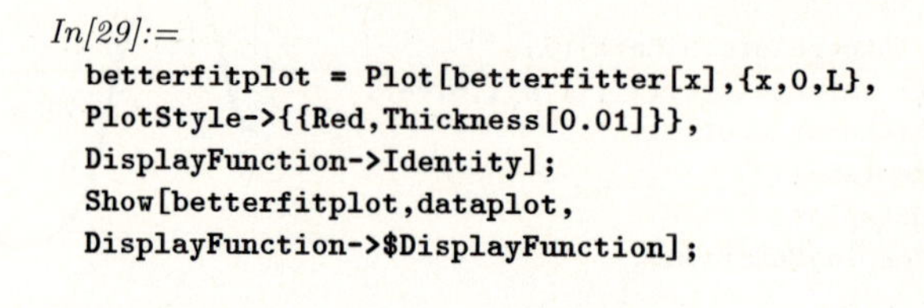

```
betterfitplot = Plot[betterfitter[x],{x,0,L},
PlotStyle->{{Red,Thickness[0.01]}},
DisplayFunction->Identity];
Show[betterfitplot,dataplot,
DisplayFunction->$DisplayFunction];
```

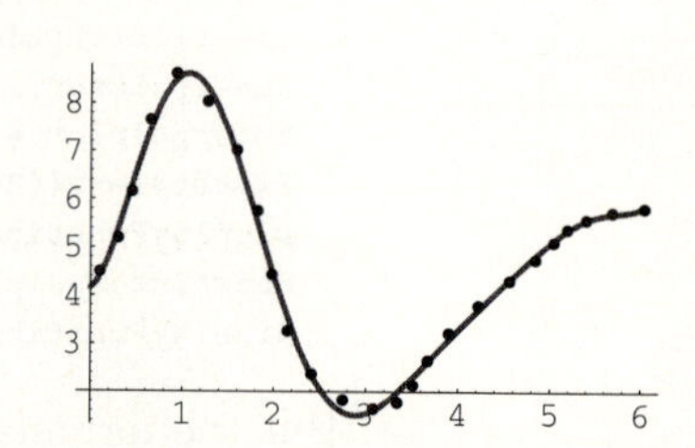

Absolutely stunning. The mathematician who had this fantastic idea was Jean Baptiste Fourier (1772–1830), who revolutionized mathematics with his theory that functions are built out of well-chosen sine and cosine waves the way that music is made from basic harmonics. The great mathematician Antoni Zygmund put it this way: "[Fourier's theory] has been a source of new ideas . . . for the last two centuries and is likely to be so in the years to come."

In any case, sines and cosines are a helluva lot more useful than most trigonometry books make you believe.

B.3.b.ii) What is the main drawback in trying to fit with sine and cosine waves?

Answer: Fitting with sine and cosine is not always good for looking beyond the data.

In[30]:=
```
extendedbetterfitplot =
Plot[betterfitter[x],{x,0,4 L},
PlotStyle->{{Red,Thickness[0.01]}},
DisplayFunction->Identity];
Show[extendedbetterfitplot,dataplot,
DisplayFunction->$DisplayFunction];
```

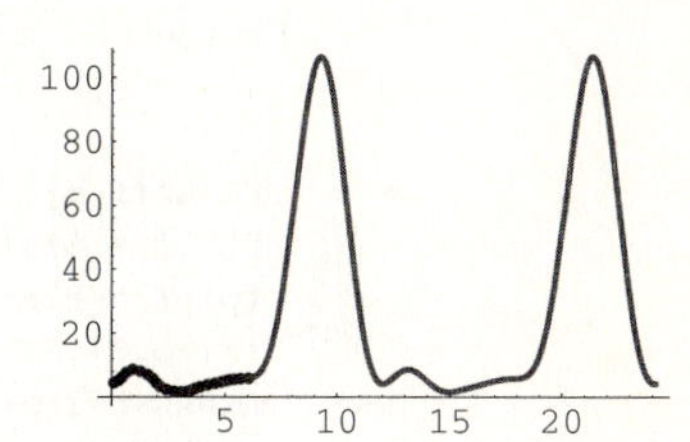

As you can see, part of the plot is unrelated to the data.

B.3.b.iii) How did *Mathematica* find the formulas for the good sine and cosine fit?

Answer: It used the same procedure that it used to go after a good line fit but it started with

$$\begin{aligned}\text{prelimf}[x] = a + b\sin\left[\frac{\pi\,x}{L}\right] + c\cos\left[\frac{\pi\,x}{L}\right] + d\sin\left[\frac{2\,\pi\,x}{L}\right] + e\cos\left[\frac{2\,\pi\,x}{L}\right]\\ + f\sin\left[\frac{3\,\pi\,x}{L}\right] + g\cos\left[\frac{3\,\pi\,x}{L}\right] + h\sin\left[\frac{4\,\pi\,x}{L}\right] + i\cos\left[\frac{4\,\pi\,x}{L}\right]\end{aligned}$$

without specifiying what a, b, c, d, e, f, g, h, and i are.

Then you gave it the data: $\{x_1, y_1\}$, $\{x_2, y_2\}$, $\{x_3, y_3\}, \ldots, \{x_m, y_m\}$ and *Mathematica* formed the square error function

$$\begin{aligned}\text{sqerror}[a,b,c,d,e,f,g,h,i] = (\text{prelimf}[x_1] - y_1)^2 + (\text{prelimf}[x_2] - y_2)^2\\ + (\text{prelimf}[x_3] - y_3)^2 + \cdots + (\text{prelimf}[x_m] - y_m)^2 .\end{aligned}$$

And finally *Mathematica* solved the equations

$$D[\text{sqerror}[a,b,c,d,e,f,g,h,i],a] = 0;$$
$$D[\text{sqerror}[a,b,c,d,e,f,g,h,i],b] = 0;$$

through

$$D[\text{sqerror}[a,b,c,d,e,f,g,h,i],i] = 0.$$

for a, b, c, d, e, f, g, h, and i. An impossible job for a human, but an easy job for a machine.

Tutorials

■ T.1) Highest and lowest points on the graph

T.1.a) Determine the highest and lowest points on the graph of the function

$$y = f[x] = \frac{\sin[x]}{3 - 2x + x^2}.$$

Answer: Size up the situation:

In[1]:=
```
Clear[f,x]
f[x_] = Sin[x]/(3 - 2x + x^2);
fplot = Plot[f[x],
{x,-6,8},PlotStyle->Thickness[0.01],
AxesLabel->{"x","f[x]"},PlotRange->All];
```

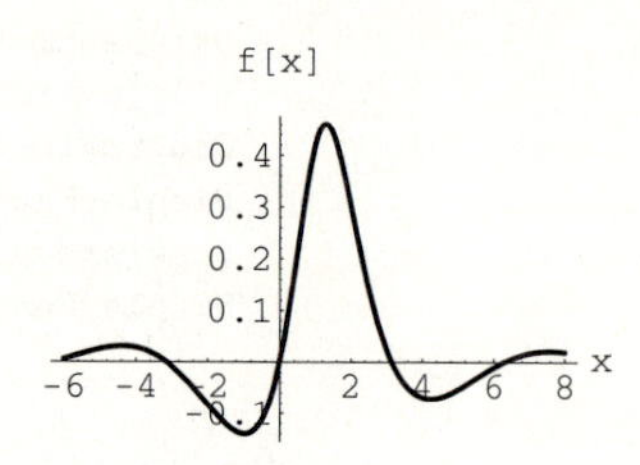

This plot suggests that the highest and lowest points on the graph of

$$f[x] = \frac{\sin[x]}{3 - 2x + x^2}$$

are somewhere between $x = -2$ and $x = 2$. The plot seems to dampen off quite rapidly. This is not surprising because $\sin[x]$ is condemned to oscillate between -1 and 1 while the remaining factor $1/\left(3 - 2x + x^2\right)$ shrinks to 0. Look at

$$\frac{1}{3 - 2x + x^2}, \quad \frac{-1}{3 - 2x + x^2} \quad \text{and} \quad f[x] = \frac{\sin[x]}{3 - 2x + x^2}$$

on the same plot:

In[2]:=
```
Plot[{1/(3 - 2 x + x^2),
f[x],-1/(3 - 2 x + x^2)},
{x,-12,12},PlotStyle->{{Thickness[0.01]},
{Red},{Thickness[0.01]}},
AxesLabel->{"x",""}];
```

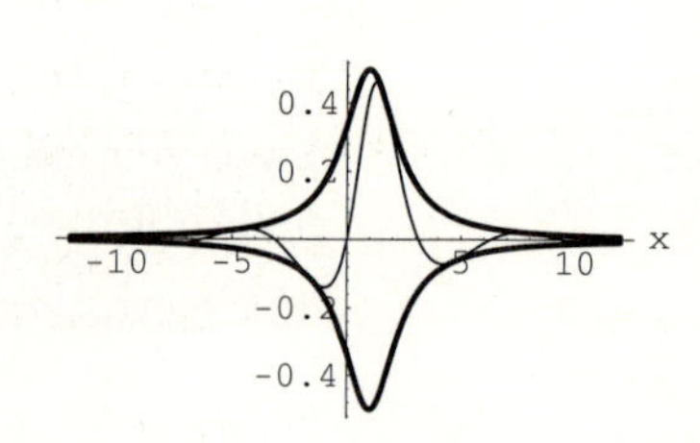

Sure enough. The factor

$$\frac{1}{3 - 2x + x^2}$$

is in control except near the origin. It squashes all the action in the

$$f[x] = \frac{\sin[x]}{3 - 2x + x^2}$$

curve except near the origin. This confirms the initial reaction.

The lowest point is nearly under $x = -1$:

In[3]:=
```
FindRoot[f'[x] == 0,{x,-1}]
```
Out[3]=
```
{x -> -0.981371}
```
In[4]:=
```
lowest = {x,f[x]}/.x->-0.981371
```
Out[4]=
```
{-0.981371, -0.140277}
```
The highest point is nearly over $x = 1.5$:

In[5]:=
```
FindRoot[f'[x] == 0,{x,1.5}]
```
Out[5]=
```
{x -> 1.29516}
```
In[6]:=
```
highest = {x,f[x]}/.x->1.29516
```
Out[6]=
```
{1.29516, 0.461043}
```
Confirm with a plot:

In[7]:=
```
highlow =
{Graphics[{Red,PointSize[0.04],Point[highest]}],
Graphics[{Red,PointSize[0.04],Point[lowest]}]};
Show[fplot,highlow];
```

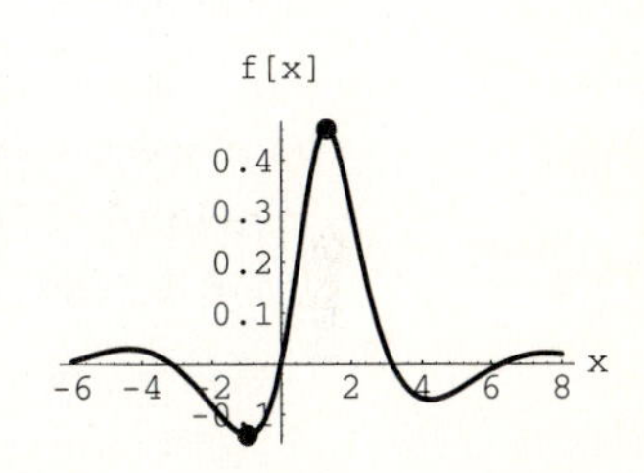

How sweet it is.

T.1.b) Determine the highest and lowest points on the graph of

$$f[x] = e^{-x^2}\left(x^8 + 8\,x^5 + 16\,x + 2\right).$$

Answer: The decay of e^{-x^2} is so devastatingly rapid that it will quickly squash any action that $x^8 + 8\,x^5 + 16\,x + 2$ might have in mind.

In[8]:=
```
Clear[x,f]
f[x_] = E^(-x^2) ( x^8 + 8 x^5 + 16 x + 2);
fplot = Plot[f[x],{x,-6,6},
PlotStyle->{{Thickness[0.01],Red}},
AxesLabel->{"x","f[x]"}, PlotRange->All];
```

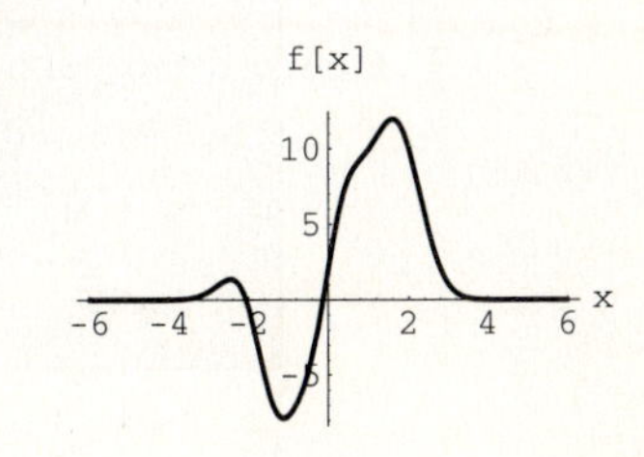

Interesting plot. The e^{-x^2} factor has completely wiped out all trace of the polynomial factor $x^8 + 8x^5 + 16x + 2$ to the right of $x = 4$ and to the left of $x = -4$.

The highest point sits nearly above $x = 1.8$.

In[9]:=
```
FindRoot[f'[x],{x,1.8}]
```
Out[9]=
```
{x -> 1.59545}
```
The highest point is:

In[10]:=
```
highest = {x, f[x]}/.x->1.59545
```
Out[10]=
```
{1.59545, 11.9388}
```
The lowest point sits nearly below $x = -1.0$.

In[11]:=
```
FindRoot[f'[x],{x,-1.0}]
```
Out[11]=
```
{x -> -1.11425}
```
The lowest point is:

In[12]:=
```
lowest = {x, f[x]}/.x->-1.11425
```
Out[12]=
```
{-1.11425, -7.85685}
```
Confirm with a plot:

In[13]:=
```
highlow = {Graphics[{Blue,PointSize[0.04],Point[highest]}],
Graphics[{Blue,PointSize[0.04],Point[lowest]}]};
Show[fplot,highlow];
```

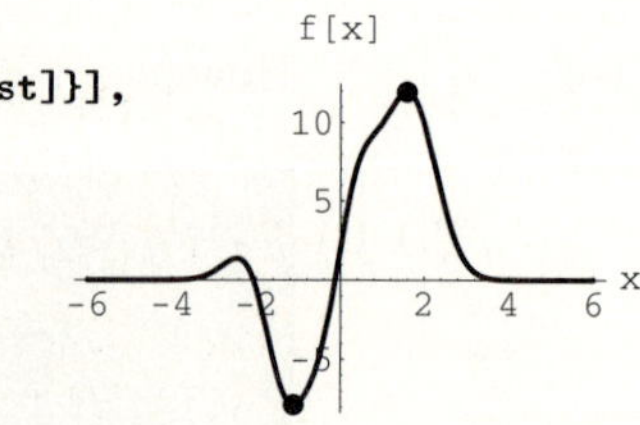

That's all folks.

T.1.c) Determine the highest and lowest values of

$$f[x] = \frac{x^7 - 58x^2 + 8}{2x^6 + 11}$$

for $-1 \le x \le 4$.

Answer: Plot the function for $-1 \leq x \leq 4$:

In[14]:=
```
Clear[f,x]
f[x_] = (x^7 - 58 x^2 + 8)/(2 x^6 + 11);
fplot = Plot[f[x],{x,-1,4},
PlotStyle->{{DarkGreen,Thickness[0.01]}},
PlotRange->All,AxesLabel->{"x","f[x]"}];
```

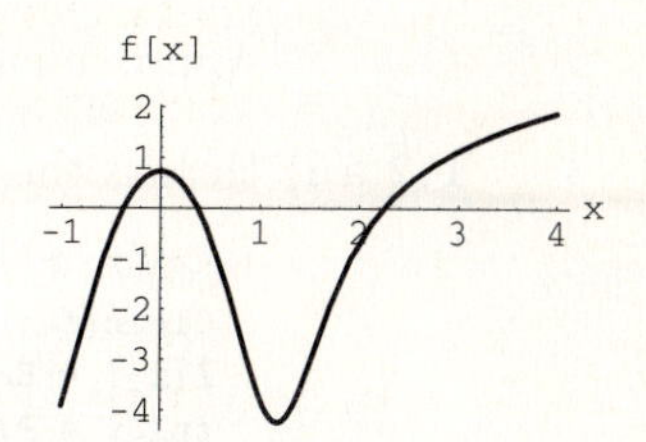

It's clear that the function takes its highest value for

$-1 \leq x \leq 4$ at $x = 4$.

In[15]:=
```
highest = {x,f[x]}/.x->4.0
```

Out[15]=
```
{4., 1.88516}
```

That dip nearly under $x = 1$ seems like the lowest value for $-1 \leq x \leq 4$.

In[16]:=
```
FindRoot[f'[x] == 0,{x,1.0}]
```

Out[16]=
```
{x -> 1.18316}
```

In[17]:=
```
lowest = {x,f[x]}/.x->1.18316
```

Out[17]=
```
{1.18316, -4.24268}
```

Just to be sure that the function isn't lowest at the left endpoint, $x = -1$, look at:

In[18]:=
```
{x,f[x]}/.x->-1.0
```

Out[18]=
```
{-1., -3.92308}
```

Good. The highest and lowest points for $-1 \leq x \leq 4$ are nailed down. Confirm with a plot:

In[19]:=
```
highlow =
{Graphics[{Blue,PointSize[0.04],Point[highest]}],
Graphics[{Blue,PointSize[0.04],Point[lowest]}]};
Show[fplot,highlow];
```

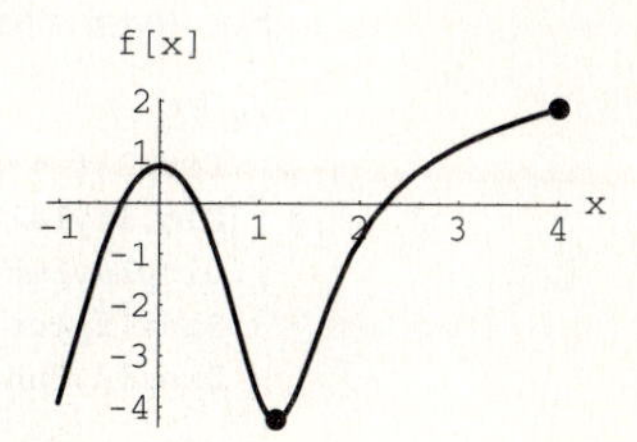

And you're out of here.

■ T.2) Approximations

Approximations are the stuff of a lot of advanced mathematics.

T.2.a.i) Here is the function $f[x] = e^{-x/2} \cos[x]$ and its plot on $-2 \leq x \leq 6$:

In[20]:=
```
Clear[f,x]
f[x_] = E^(-x/2) Cos[x];
fplot = Plot[f[x],{x,-2,6},
PlotStyle->Blue,
AxesLabel->{"x","f[x]"}];
```

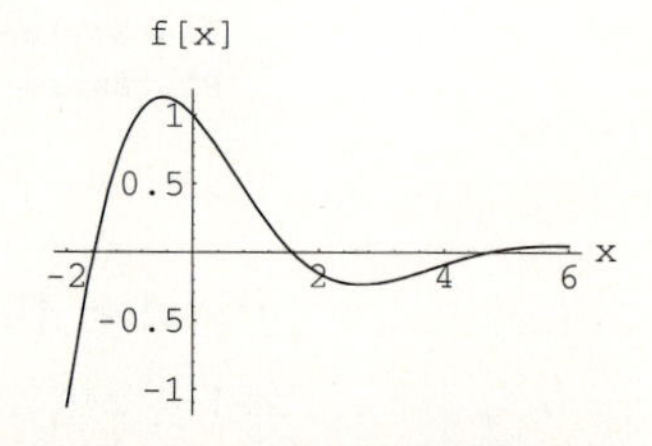

Try to get a decent approximation of this function on $[-2, 6]$ by a polynomial.

Answer: Generate some points on the curve:

In[21]:=
```
points = Table[N[{x,f[x]}],{x,-2,6}]
```

Out[21]=
```
{{-2., -1.1312}, {-1., 0.890808}, {0, 1.}, {1., 0.32771}, {2., -0.153092},
  {3., -0.220897}, {4., -0.088461}, {5., 0.0232844}, {6., 0.0478041}}
```

See the points:

In[22]:=
```
pointplot = ListPlot[points,
PlotStyle->PointSize[0.03],
DisplayFunction->Identity];
Show[fplot,pointplot,
DisplayFunction->$DisplayFunction];
```

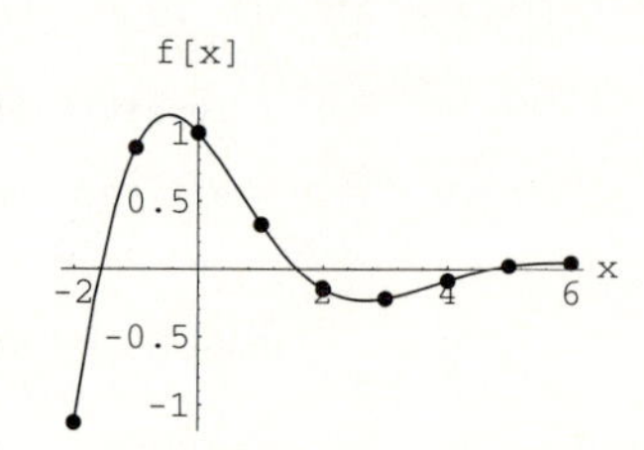

So far, so good. Fit the points with a polynomial:

In[23]:=
```
polyfit[x_] = Fit[points,{1,x,x^2,x^3,x^4},x]
```

Out[23]=
```
                                      2              3               4
0.985156 - 0.327088 x - 0.344774 x  + 0.142729 x  - 0.0134571 x
```

See everything:

In[24]:=
```
fitplot = Plot[polyfit[x],{x,-2,6},
PlotStyle->Red,
DisplayFunction->Identity];
Show[fplot,pointplot,fitplot,AxesLabel->{"x",""},
DisplayFunction->$DisplayFunction];
```

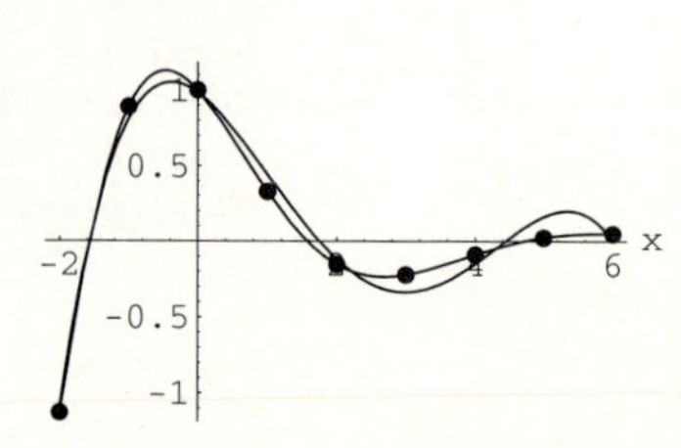

Not bad for beginners.

T.2.a.ii) What can you do to try to improve the quality of the approximation of

$$f[x] = e^{-x/2}\cos[x]$$

on $-2 \leq x \leq 6$ by a polynomial?

Answer: Use more points and use more powers of x. Here you go:

In[25]:=
```
morepoints = Table[N[{x,f[x]}],{x,-2,6,0.5}];
```

See the points:

In[26]:=
```
pointplot =
ListPlot[morepoints,PlotStyle->PointSize[0.02],
DisplayFunction->Identity];
Show[fplot,pointplot,
DisplayFunction->$DisplayFunction];
```

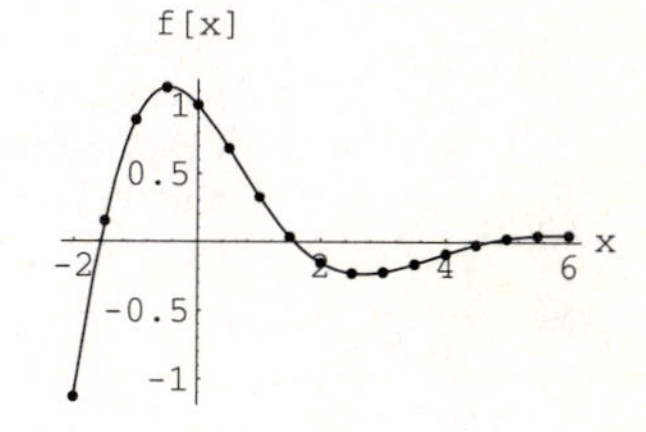

Fit the points with a polynomial using more powers of x:

In[27]:=
```
Clear[polyfit]
polyfit[x_] = Fit[morepoints,{1,x,x^2,x^3,x^4,x^5,x^6},x]
```

Out[27]=
```
                                    2               3
0.983426 - 0.44556 x - 0.347483 x  + 0.183083 x  -

               4                5                 6
  0.0166669 x  - 0.00245621 x  + 0.000330374 x
```

See everything:

In[28]:=
```
fitplot = Plot[polyfit[x],{x,-2,6},
PlotStyle->Red,
DisplayFunction->Identity];
Show[fplot,pointplot,fitplot,AxesLabel->{"x",""},
DisplayFunction->$DisplayFunction];
```

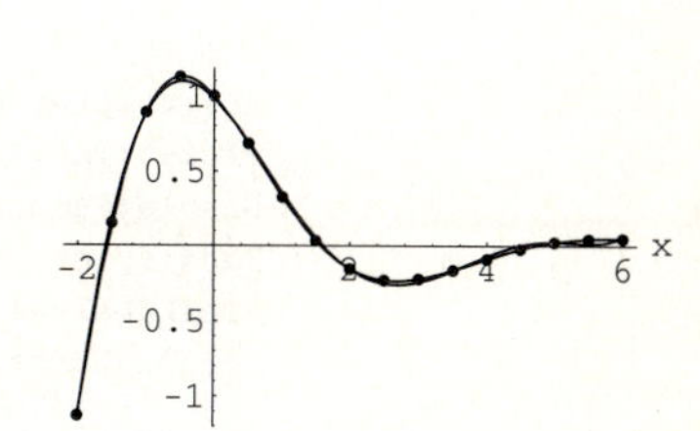

Nice approximation.

T.2.b.i) Here is the same function $f[x] = e^{-x/2}\cos[x]$ and its plot on $-2 \leq x \leq 6$:

In[29]:=
```
Clear[f,x]
f[x_] = E^(-x/2) Cos[x];
fplot = Plot[f[x],{x,-2,6},
PlotStyle->Blue,
AxesLabel->{"x","f[x]"}];
```

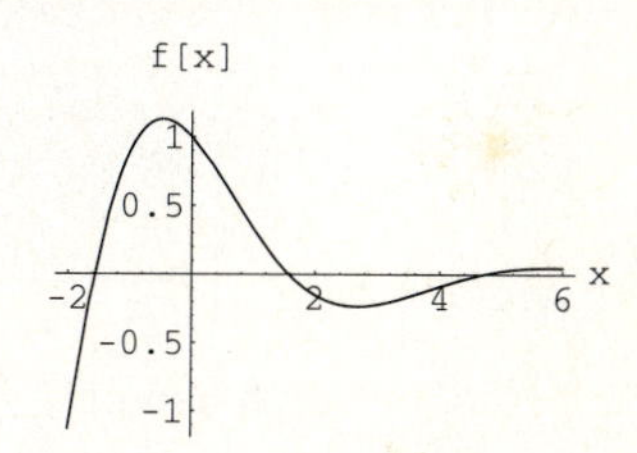

Try to get a decent approximation of this function on $-2 \leq x \leq 6$ by sines and cosines.

Answer: Generate some points on the curve:

In[30]:=
```
points = Table[N[{x,f[x]}],{x,-2,6}];
```

See the points:

In[31]:=
```
pointplot =
ListPlot[points,PlotStyle->PointSize[0.02],
DisplayFunction->Identity];
Show[fplot,pointplot,
DisplayFunction->$DisplayFunction];
```

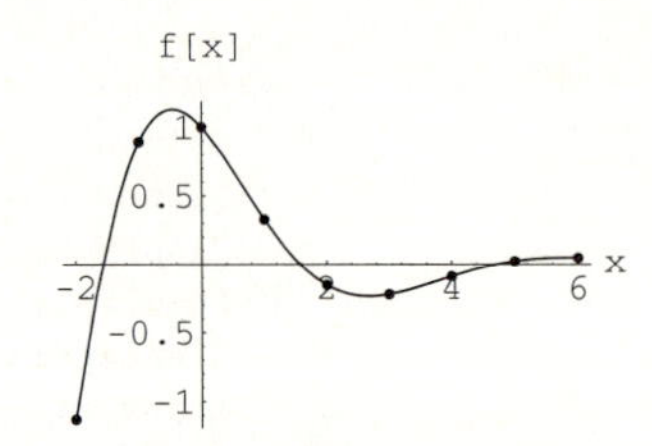

Fit the points with sine and cosine waves:

In[32]:=
```
L = 6; Clear[trigfit,x]
trigfit[x_] = Fit[points, N[{1,Sin[Pi x/L],Cos[Pi x/L],
Sin[2 Pi x/L],Cos[2 Pi x/L]}],x]
```

Out[32]=
```
-0.417196 + 0.611361 Cos[0.523599 x] + 0.860099 Cos[1.0472 x] +
  0.836134 Sin[0.523599 x] - 0.4487 Sin[1.0472 x]
```

See everything:

In[33]:=
```
trigfitplot = Plot[trigfit[x],{x,-2,6},
PlotStyle->Red,
DisplayFunction->Identity];
Show[fplot,pointplot,trigfitplot,
AxesLabel->{"x",""},
DisplayFunction->$DisplayFunction];
```

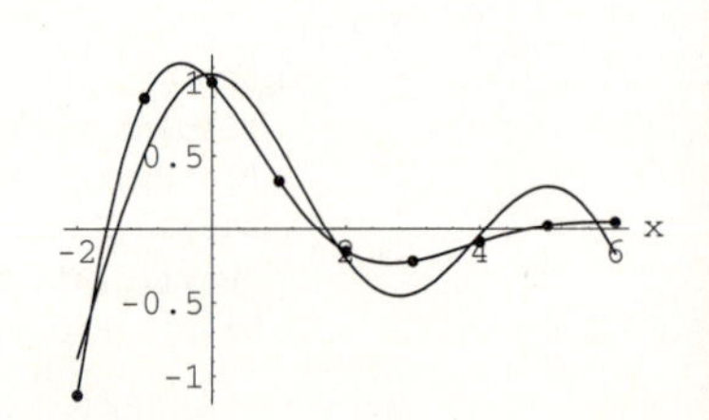

Not bad, but not good.

T.2.b.ii) What can you do to try to improve the quality of the approximation of

$$f[x] = e^{-x/2}\cos[x]$$

on $-2 \leq x \leq 6$ by sines and cosines?

Answer: Use more points and use more sines and cosines. Here you go:

In[34]:=
```
morepoints = Table[N[{x,f[x]}],{x,-2,6,0.5}];
```

See the points:

In[35]:=
```
pointplot =
ListPlot[morepoints,PlotStyle->PointSize[0.02],
DisplayFunction->Identity];
Show[fplot,pointplot,
DisplayFunction->$DisplayFunction];
```

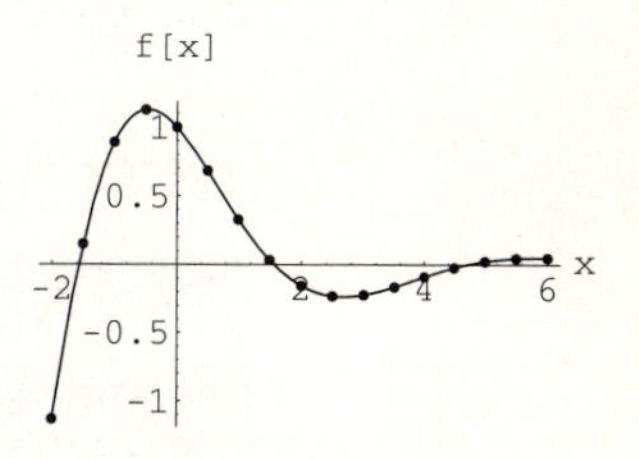

Fit the points using more sines and cosines:

In[36]:=
```
L = 6; Clear[trigfit]
trigfit[x_] = Fit[morepoints, N[{1,Sin[Pi x/(L)],Cos[Pi x/(L)],
Sin[2 Pi x/(L)],Cos[2 Pi x/(L)], Sin[3 Pi x/(L)],Cos[3 Pi x/(L)],
Sin[4 Pi x/(L)],Cos[4 Pi x/(L)]}],x];
```

See everything:

In[37]:=
```
fitplot = Plot[trigfit[x],{x,-2,6},
PlotStyle->Red,
DisplayFunction->Identity];
Show[fplot,pointplot,fitplot,
AxesLabel->{"x",""},
DisplayFunction->$DisplayFunction];
```

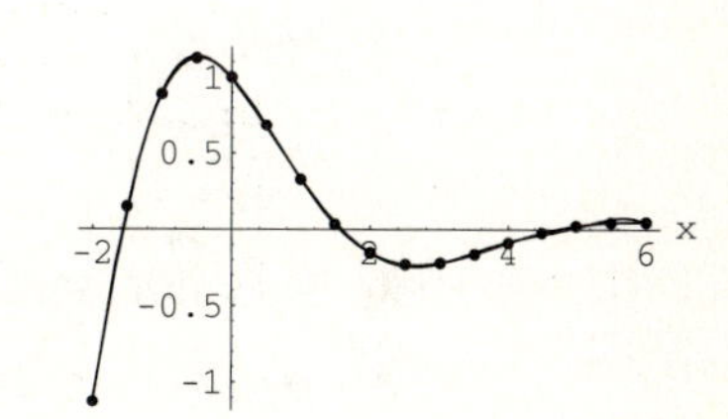

Darn good.

■ T.3) Fish gotta swim: The least energy

This problem was adapted from E. Batschlet, *Introduction to Mathematics for Life Scientists*, Springer-Verlag, New York, 1979. This book is highly recommended for supplementary browsing.

T.3.a) A river near the Puget Sound is flowing at a constant speed v_r mph relative to the river bank. A salmon swims upstream at a steady speed v_f mph relative to the water, intent on reaching a spawning point s miles up the river.

Given that $v_f > v_r$, the time t needed for the salmon to complete the journey is

$$t = \frac{s}{v_f - v_r}$$

because $s = (v_f - v_r)\,t$. The energy the salmon must expend to maintain its journey is determined by friction in the water and by the time t necessary to reach the spawning point. Fish biologists measuring this energy have come to the conclusion that this energy is jointly proportional to the time t needed to complete the journey and to $(v_f)^k$ for some empirical constant $k > 2$. This tells you that

$$\text{energy} = c\,(v_f)^k\,t = \frac{c\,(v_f)^k\,s}{v_f - v_r}$$

where c is a positive constant of proportionality.

> Calculate the steady speed v_f that makes the salmon's trip possible with the least energy E.

Answer:

In[38]:=
```
Clear[energy,vf,vr,t,c]; energy = c vf^k t;
```

You know that $t = s/\,(v_f - v_r)$. So in terms of v_f:

In[39]:=
```
energy = energy/.t-> (s/(vf - vr))
```

Out[39]=
```
       k
c s vf
--------
vf - vr
```

Now differentiate with respect to v_f and factor:

In[40]:=
```
Factor[D[energy,vf]]
```

Out[40]=
```
       -1 + k
c s vf         (-vf + k vf - k vr)
-----------------------------------
                   2
          (-vf + vr)
```

Remembering that c and s are positive and noticing that the denominator cannot be negative, you see that the sign of the derivative is the same as the sign of:

$$(-v_f + kv_f - kv_r) = (v_f\,(k-1) - kv_r) = (k-1)\left(v_f - \frac{k}{k-1}v_r\right).$$

Accordingly, because $k > 2$, the derivative $D[\text{energy}, v_f]$ is negative for

$$v_f < \left(\frac{k}{k-1}\right) v_r$$

and $D[\text{energy}, v_f]$ is positive for

$$v_f > \left(\frac{k}{k-1}\right) v_r.$$

This means that as a function of v_f, the energy decreases as v_f advances from v_r to $(k/(k-1))\,v_r$ and the energy increases as v_f advances from $(k/(k-1))\,v_r$ to ∞. Consequently, the least energy velocity is

$$v_f = \left(\frac{k}{k-1}\right) v_r.$$

Note that the optimal speed v_f of the fish does not depend on the distance s.

■ T.4) Designing a box

You are working for the MBA Box Co. An order comes in for cardboard boxes with a volume of 6 cubic feet each.

The lazy MBA boss says: "That's simple; we'll just make the boxes 3 feet long, 2 feet wide, and 1 foot high." You respond: "We can make them cheaper by using different dimensions." You are right because you know some calculus. You arrived at your answer by first laying out the information at hand:

You put

$x =$ length of the box in feet

$y =$ width of the box in feet

$z =$ height of the box in feet.

The material for the box consists of six rectangular pieces. Two of them (the top and bottom) have area $x\,y$ each; these cost \$0.40 per square foot.

In[41]:=
```
Clear[topbottom,x,y,z]; topbottom[x_,y_,z_] = 2 0.4 x y
```

Out[41]=
```
0.8 x y
```

Another two (the sides) have area $x\,z$ each; these cost \$0.30 per square foot.

In[42]:=
```
Clear[sides,x,y,z]; sides[x_,y_,z_] = 2 0.3 x z
```

Out[42]=
```
0.6 x z
```

And the other two (the ends) have area $y\,z$; these cost \$0.35 per square foot.

In[43]:=
```
Clear[ends,x,y,z]; ends[x_,y_,z_] = 2 0.35 y z
```

Out[43]=
```
0.7 y z
```

So the cost in dollars of the materials for making each box is:

In[44]:=
```
Clear[cost,x,y,z]; cost[x_,y_,z_] =
topbottom[x,y,z] + sides[x,y,z] + ends[x,y,z]
```

Out[44]=
```
0.8 x y + 0.6 x z + 0.7 y z
```

T.4.a.i) How did you know the boxes could have been made more cheaply by using dimensions other than the dimensions $x = 3$, $y = 2$, and $z = 1$ suggested by the silly boss?

Answer: The box is to contain 6 cubic feet so $x\,y\,z = 6$. This tells you that $z = 6/(x\,y)$.

In[45]:=
```
Clear[newcost]; newcost[x_,y_] = cost[x,y,6/(x y)]
```

Out[45]=
```
4.2   3.6
--- + --- + 0.8 x y
 x     y
```

Look at the derivative of newcost$[x, y]$ with respect to x:

In[46]:=
```
D[newcost[x,y],x]
```

Out[46]=
```
-4.2
---- + 0.8 y
  2
 x
```

Put in the boss's suggestion of $x = 3$ and $y = 2$:

In[47]:=
```
D[newcost[x,y],x]/.{x->3,y->2}
```

Out[47]=
```
1.13333
```

Positive. This tells you that holding $y = 2$ and making x a little smaller than 3 will result in a cheaper box.

This discredits the boss. To really rub it in, look at the derivative of newcost$[x, y]$ with respect to y:

In[48]:=
```
D[newcost[x,y],y]
```

Out[48]=
```
        3.6
0.8 x - ---
         2
        y
```

Put in the boss's suggestion of $x = 3$ and $y = 2$:

In[49]:=
```
D[newcost[x,y],y]/.{x->3,y->2}
```

Out[49]=
```
1.5
```

Positive. This tells you that holding $x = 3$ and making y a little smaller than 2 will result in a cheaper box. This is more than enough to discredit the boss.

So the MBA manager's quick answer of $x = 3$, $y = 2$ would have thrown company money away. Everyone can see this from the following plot:

In[50]:=
```
Plot3D[newcost[x,y],
{x,3 - 0.5,3 + 0.5},{y,2 - 0.5,2 + 0.5}];
```

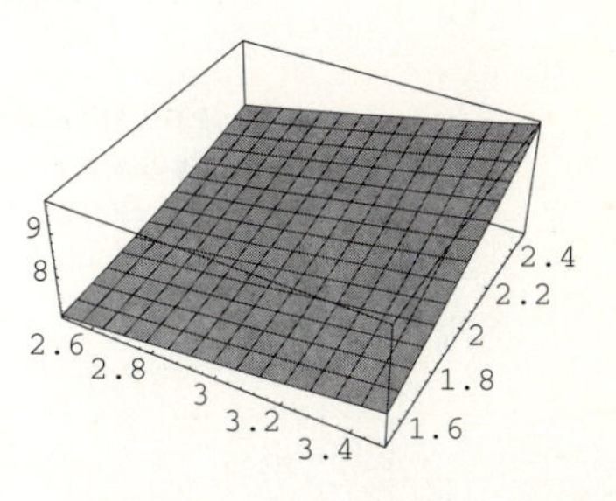

The plot indicates that $x = 2.5$ and $y = 1.5$ is a better choice than $x = 3$ and $y = 2$.

T.4.a.ii) What dimensions result in least cost? What is the least cost? How much more than the least cost does the MBA boss's design cost?

Answer: If either

```
D[newcost[x, y], x] /. {x->x0, y->y0}
```

or

```
D[newcost[x, y], y] /. {x->x0, y->y0}
```

is not 0 for a selection $\{x_0, y_0\}$ of measurements, then the same analysis you did above shows that you can change x from x_0 or you can change y from y_0 to reduce the cost of the box.

Consequently, the dimensions x and y that result in least cost must be found in the output from:

In[51]:=
```
bestxandy = N[Solve[{D[newcost[x,y],x] == 0, D[newcost[x,y],y]==0},{x,y}]]
```

Out[51]=
```
{{y -> 1.56827, x -> 1.82965},
 {y -> -0.784137 + 1.35817 I, x -> -0.914826 + 1.58453 I},
 {y -> -0.784137 - 1.35817 I, x -> -0.914826 - 1.58453 I}}
```

Throw out the solutions involving imaginary numbers. The length of the least cost box is:

In[52]:=
```
bestx = 1.83
```

Out[52]=

```
1.83
```

The width of the least cost box is:

In[53]:=

```
besty = 1.57
```

Out[53]=

```
1.57
```

Take a look at a plot:

In[54]:=

```
Plot3D[newcost[x,y],{x,bestx -0.2, bestx + 0.2},
{y,besty - 0.2,besty + 0.2},
ViewPoint->{2.747, 1.594, 1.168}];
```

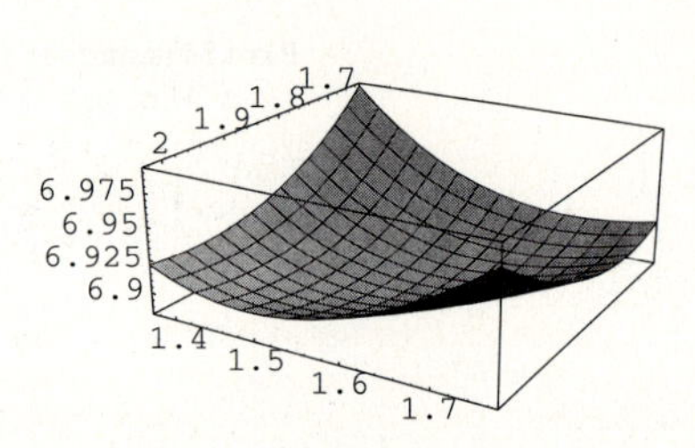

That depression in the center of the plot of newcost$[x, y]$ at $x = 1.83$ and $y = 1.57$ is visual substantiating evidence.

The height z of the least cost box in terms of x and y is $z = 6/\,(x\,y)$. So the height of the least cost box is given by:

In[55]:=

```
bestz = 6/(x y)/.{x->bestx,y->besty}
```

Out[55]=

```
2.08834
```

The cost in dollars of the least expensive box is:

In[56]:=

```
cost[bestx,besty,bestz]
```

Out[56]=

```
6.88656
```

The cost in dollars of the MBA boss's box is:

In[57]:=

```
cost[3.0,2.0,1.0]
```

Out[57]=

```
8.
```

You've saved the company about

In[58]:=

```
8.00 - 6.89
```

Out[58]=

```
1.11
```

dollars per box. Keep this in mind when you ask for a pay raise.

T.4.a.iii) For the best choices of x, y, and z, the corresponding costs of the components in dollars are:

In[59]:=
```
topbottom[bestx,besty,bestz]
ends[bestx,besty,bestz]
sides[bestx,besty,bestz]
```
Out[59]=
```
2.29848
```
Out[59]=
```
2.29508
```
Out[59]=
```
2.29299
```

Is this output an accident?

Answer: Naw. In fact, the small discrepancies are due to roundoff errors.

T.4.a.iv) How do you know that these results are not accidents, and how can you be sure that the small discrepancies are all roundoff errors?

Answer: Do the problem with symbols instead of numbers: The material for the box consists of six rectangular pieces. Two of them (the top and bottom) have area $x\,y$ each; these cost $\$t$ per square foot.

In[60]:=
```
Clear[topbottom,t,x,y,z]; topbottom[x_,y_,z_] = 2 t x y
```
Out[60]=
```
2 t x y
```

Another two (the sides) have area $x\,z$ each; these cost $\$s$ per square foot.

In[61]:=
```
Clear[sides,s,x,y,z]; sides[x_,y_,z_] = 2 s x z
```
Out[61]=
```
2 s x z
```

And the other two (the ends) have area $y\,z$; these cost $\$e$ per square foot (this e is not the usual number e, of course; it is just a symbol).

In[62]:=
```
Clear[ends,e,x,y,z]; ends[x_,y_,z_] = 2 e y z
```
Out[62]=
```
2 e y z
```

So the cost in dollars of the materials for making each box is:

In[63]:=
```
Clear[cost,x,y,z]; cost[x_,y_,z_] =
topbottom[x,y,z] + sides[x,y,z] + ends[x,y,z]
```
Out[63]=
```
2 t x y + 2 s x z + 2 e y z
```

But $x\,y\,z = 6$; so $z = 6/\,(x\,y)$:

In[64]:=

```
Clear[newcost]; newcost[x_,y_] = cost[x,y,6/(x y)]
```

Out[64]=

```
12 e   12 s
---- + ---- + 2 t x y
 x      y
```

In[65]:=

```
bestxandy = Solve[{D[newcost[x,y],x] == 0, D[newcost[x,y],y] == 0},{x,y}]
```

Out[65]=

```
           1/3  2/3          1/3  2/3
          6    s            6    e
{{y -> ------------, x -> ------------},
         1/3  1/3           1/3  1/3
        e    t             s    t

            2/3  1/3  2/3              2/3  1/3  2/3
       (-1)     6    s            (-1)     6    e
 {y -> ------------------, x -> ------------------},
                1/3  1/3                 1/3  1/3
               e    t                   s    t

            4/3  1/3  2/3              4/3  1/3  2/3
       (-1)     6    s            (-1)     6    e
 {y -> ------------------, x -> ------------------}}
                1/3  1/3                 1/3  1/3
               e    t                   s    t
```

Throw out the solutions involving imaginary numbers. Here come the dimensions of the least cost box:

In[66]:=

```
{bestx,besty,bestz} = {x,y,6/(x y)}/.bestxandy[[1]]
```

Out[66]=

```
  1/3  2/3    1/3  2/3    1/3  2/3
 6    e      6    s      6    t
{----------, ----------, ----------}
  1/3  1/3    1/3  1/3    1/3  1/3
 s    t      e    t      e    s
```

The cost in dollars of the least expensive box is:

In[67]:=

```
cost[bestx,besty,bestz]
```

Out[67]=

```
 5/3  1/3  1/3  1/3
6    e    s    t
```

For the best choices of x, y, and z, the corresponding costs of the components in dollars are:

In[68]:=

```
topbottom[bestx,besty,bestz]
ends[bestx,besty,bestz]
sides[bestx,besty,bestz]
```

Out[68]=

```
     1/3  1/3  1/3  1/3
288     e    s    t
```

Out[68]=

```
     1/3  1/3  1/3  1/3
288     e    s    t
```

Out[68]=

```
    1/3  1/3  1/3  1/3
288    e    s    t
```

Exactly the same. Think of it! The component costs for the cheapest box are all exactly the same no matter what the costs of the materials are.

T.4.a.v) Wouldn't you have saved a lot of trouble if originally you had calculated the following

```
topbottom[x_,y_,z_] = 2  0.4 x y
sides[x_,y_,z_] = 2  0.3 x z
ends[x_,y_,z_] = 2  0.35 y z
cost[x_,y_,z_] = topbottom[x,y,z] + sides[x,y,z] + ends[x,y,z]
```

and then solved the system

$D[\mathrm{cost}[x,y,z],x] = 0,$

$D[\mathrm{cost}[x,y,z],y] = 0,$

$D[\mathrm{cost}[x,y,z],z] = 0?$

Answer: Try it and see:

In[69]:=

```
Clear[topbottom,sides,ends,cost,x,y,z]
topbottom[x_,y_,z_] = 2 0.4 x y;
sides[x_,y_,z_] = 2 0.3 x z;
ends[x_,y_,z_] = 2 0.35 y z;
cost[x_,y_,z_] = topbottom[x,y,z] + sides[x,y,z] + ends[x,y,z];
Solve[{D[cost[x,y,z],x] == 0,D[cost[x,y,z],y] == 0,
D[cost[x,y,z],z] == 0},{x,y,z}]
```

Out[69]=

```
{{x -> 0, y -> 0, z -> 0}}
```

Pure bull manure. Reason: When you try to do the problem this way, you never incorporate the essential information that $x\,y\,z = 6$.

■ T.5) Largest and smallest

T.5.a) If a one-quart can of fruit punch has least possible surface area, then what is the ratio of its height to the diameter of its base?

Answer: Put

V = volume of a one quart can

h = its height

r = radius of its base.

Then $V = \pi r^2 h$. Also:

area of the top = area of the bottom $= \pi r^2$.

The area of the cylindrical side is the height times the circumference of the top. So:

In[70]:=
```
Clear[V,S,r,h]; SurfArea = 2 Pi r^2 + 2 Pi r h
```

Out[70]=
```
                   2
2 h Pi r + 2 Pi r
```

Next solve $V = \pi r^2 h$ for h.

In[71]:=
```
Solve[V == Pi r^2 h,h]
```

Out[71]=
```
            V
{{h -> -------}}
            2
        Pi r
```

Substitute into the expression $V = \pi r^2 h$ immediately above. The result is surface area as a function of r.

In[72]:=
```
Clear[SA]; SA[r_] = SurfArea/.h->V/(Pi r^2)
```

Out[72]=
```
        2   2 V
2 Pi r  + ---
             r
```

The physical setup tells you that there is a least surface area and that this must happen when the derivative $SA'[r]$ is 0.

In[73]:=
```
optimalr = Solve[SA'[r] == 0,r]
```

Out[73]=
```
          1/3                       1/3  1/3                   2/3  1/3
         V                      (-1)    V                  (-1)    V
{{r -> ----------}, {r -> -(----------------)}, {r -> ----------------}}
               1/3                     1/3                       1/3
         (2 Pi)                  (2 Pi)                    (2 Pi)
```

The best radius is:

In[74]:=
```
bestr = optimalr[[1,1,2]]
```

Out[74]=
```
  1/3
 V
----------
      1/3
(2 Pi)
```

The corresponding height is:

In[75]:=
```
besth = V/(Pi r^2)/.r->bestr
```

Out[75]=
```
 4  1/3  1/3
(--)    V
 Pi
```

and the ratio of its height to the diameter of its base is:

In[76]:=
```
Simplify[besth/(2 bestr)]
```

Out[76]=
```
1
```

For the least surface area, the height equals the diameter of the base.

T.5.b) Find the rectangle with largest area among those rectangles whose diagonals are d units long.

Answer:

In[77]:=
```
Clear[d,base,height]; area = base height
```

Out[77]=
```
base height
```

The Pythagorean theorem says that the diagonal has length $d = \sqrt{\text{base}^2 + \text{height}^2}$. So height $= \sqrt{d^2 - \text{base}^2}$. Substitute this into the area expression to find area as a function of the base:

In[78]:=
```
areafunctofbase = area/.height->Sqrt[d^2 - base^2]
```

Out[78]=
```
                     2    2
base Sqrt[-base  + d ]
```

It's not surprising that area $= 0$ when base $= 0$ or when base $= d$. At the true optimum value, the curve changes from increasing to decreasing, so the derivative

```
D[areafunctofbase, base]
```

must be 0 when the base is at its optimum value.

In[79]:=
```
bestbase = Solve[D[areafunctofbase,base]==0,base]
```

Out[79]=
```
                  d                            d
{{base -> -------}, {base -> -(-------)}}
               Sqrt[2]                    Sqrt[2]
```

Take the positive root. The base of the largest rectangle is $d/\sqrt{2}$. Then the height of the largest rectangle is:

In[80]:=
```
bestheight = Sqrt[d^2 - base^2]/.base->d/Sqrt[2]
```

Out[80]=
```
        2
Sqrt[d ]
--------
Sqrt[2]
```

Because d is positive, this is the same as:

In[81]:=

```
Clear[x]; bestheight/.Sqrt[(x_)^2]->x
```

Out[81]=

```
   d
-------
Sqrt[2]
```

A square! Some folks will argue that a square is not a rectangle. But those in the know say that a square is just a funny-shaped rectangle.

Give It a Try

Experience with the starred ($\star$) problems will be especially beneficial for understanding later lessons.

■ G.1) Good representative plots*

G.1.a) If your plot of a function includes all points at which the derivative is 0, explain why you can be sure that your plot cannot miss any of the dips and the crests of the graph of the function.

G.1.b) Give a good representative plot of

$$f[x] = e^{-x/5}\,(x+3)^2.$$

Say what $f[x]$ is doing as $x \to \infty$.

G.1.c) Give a good representative plot of

$$f[x] = \frac{\log[x]}{x}.$$

Say what $f[x]$ is doing as $x \to \infty$.

G.1.d) Give a good representative plot of

$$f[x] = \frac{x^2+24}{x^2+1}$$

Say what the global scale of $f[x]$ looks like.

■ G.2) Highest and lowest points on the graph★

G.2.a) Find the highest point on the graph of

$$f[x] = e^{-x^2}\left(2 + \cos[x] + \frac{1}{2}\sin[x]\right).$$

Is there a lowest point on the graph?

G.2.b) Find the highest point on the graph of

$$f[x] = -577 + 736\,x - 324\,x^2 + 60\,x^3 - 4\,x^4.$$

Is there a lowest point on this graph?

G.2.c) Find as accurately as you can the highest and lowest points on the graph of

$$f[x] = x\left(\frac{240 - 7\,x^2}{240 + 3\,x^2}\right)$$

for $-6 \le x \le 6$.

■ G.3) Approximations by polynomials and sine and cosine waves★

G.3.a) Here is the plot for $-2 \le x \le 2$ of the function

$$f[x] = \left(x^2 + x\right) e^{-x^2}:$$

In[1]:=

```
Clear[f,x]
f[x_] = (x^2 + x) E^(-x^2);
Plot[f[x],{x,-2,2},
PlotStyle->{{Blue,Thickness[0.01]}},
PlotRange->All,AxesLabel->{"x",""}];
```

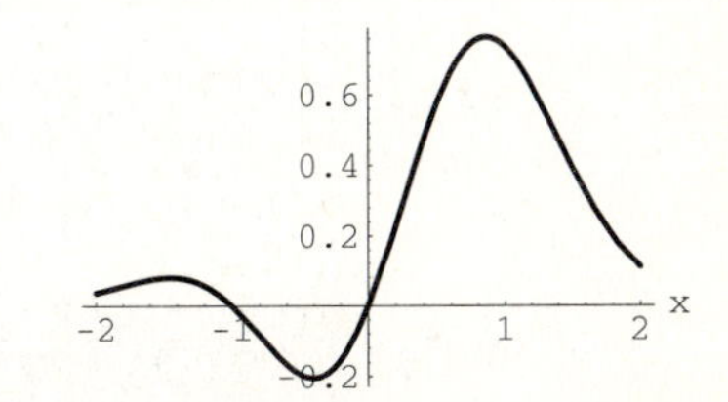

G.3.a.i) Try to get a decent approximation to this function on $[-2, 2]$ by a polynomial.

G.3.a.ii) Try to get a decent approximation to this function on $[-2, 2]$ by sines and cosines.

G.3.b) Here is an oddball function that:

→ runs with $2x$ for $x < 0.5$

→ runs with x^2 for $0.5 \leq x \leq 1$

→ runs with e^x for $1 < x$.

And here is its plot on $[0, 1.5]$:

In[2]:=

```
Clear[f,x]
f[x_] := 2 x/; x < 0.5;
f[x_] := x^2/;0.5 <= x <= 1;
f[x_] := E^(x)/;1 < x ;
fplot = Plot[f[x],{x,0,1.5},
PlotStyle->{{Thickness[0.01],Blue}},
AxesLabel->{"x","f[x]"}];
```

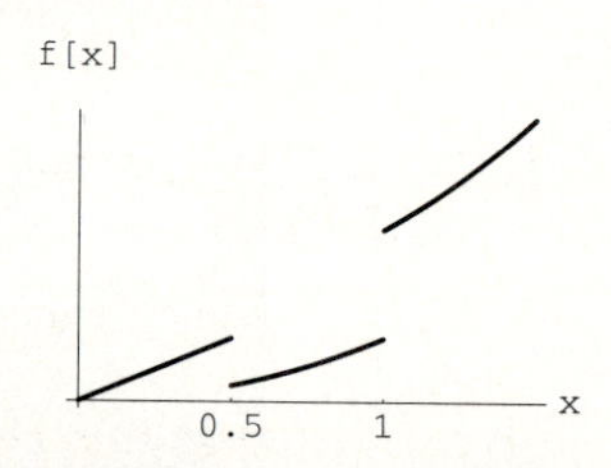

> Attempt to get a decent approximation of this function on $[0, 1.5]$ by a polynomial. Attempt to get a decent approximation of this function on $[0, 1.5]$ by sines and cosines. Why do you think this function is rather resistant to approximation by polynomials or by sines and cosines?
>
> Which approximation tool do you prefer?

■ G.4) Oil slicks

An oil tanker hits a reef and dumps V cubic feet of oil onto the surface of a calm sea. The oil spreads in a circular pattern of uniform thickness $h[t]$ and radius $r[t]$ at t hours after the spill.

Given that V cubic feet of oil were spilled, you can see that $V = \pi\, r[t]^2\, h[t]$ because the volume of a cylinder is the area of the base times the height.

Laboratory experiments indicate that, after half an hour, $h[t]$ is proportional to $1/\sqrt{t}$ for $t > 0$. This means that, for $t \geq 0.5$, the height is $h[t] = K/\sqrt{t}$ where K is a positive constant.

G.4.a)

> Find an expression for the instantaneous growth rate $r'[t]$ of the radius $r[t]$ in terms of V, K, and t (remember that V and K do not change with time).

G.4.b) In a huge oil spill, 10^5 cubic feet of oil is dumped almost instantly, and one hour later it is noticed that the oil on the sea is 10^{-3} feet thick (i.e., $h[1] = 10^{-3}$).

> Find the value of the constant K for this spill.

Then plot the radius $r[t]$ and $r'[t]$ on the same axes for $0.5 \leq t \leq 36$ hours.

Discuss how the plot indicates why action must be taken in the first few hours.

When is the slick spreading fast? When is it spreading more slowly?

Does it make a whole lot of difference whether the cleaning up begins at 18 hours after the spill or 24 hours after the spill?

■ G.5) The second derivative $f''[x]^\star$

Some folks like to look at the derivative of the derivative, which they call the "second derivative." The second derivative is the instantaneous growth rate of the instantaneous growth rate. As such, the second derivative is a measurement tool in its own right.

Here are three ways to get second derivatives from *Mathematica*:

In[3]:=
```
Clear[f,x]; f[x_] = x^2 Log[x];
```

In[4]:=
```
{f''[x], D[D[f[x],x],x], D[f[x],{x,2}]}
```

Out[4]=
```
{3 + 2 Log[x], 3 + 2 Log[x], 3 + 2 Log[x]}
```

Play with this for other functions.

Look at the following plots of $f[x]$ and $f''[x]$ with an eye toward saying what the sign of $f''[x]$ tells you about the shape of the plot of $f[x]$ (in all cases the plot of $f[x]$ is thicker than the plot of $f''[x]$).

In[5]:=
```
Clear[f,x]
f[x_] = x^3 + x^2 - 2 x - 2;
Plot[{f[x],f''[x]},{x,-2.5,2.5},
PlotStyle->{{Blue,Thickness[0.01]},Red},
AxesLabel->{"x",""}];
```

In[6]:=
```
Clear[f,x]
f[x_] = Erf[x];
Plot[{f[x],f''[x]},{x,-2,2},
PlotStyle->{{Blue,Thickness[0.01]},Red},
AxesLabel->{"x",""}];
```

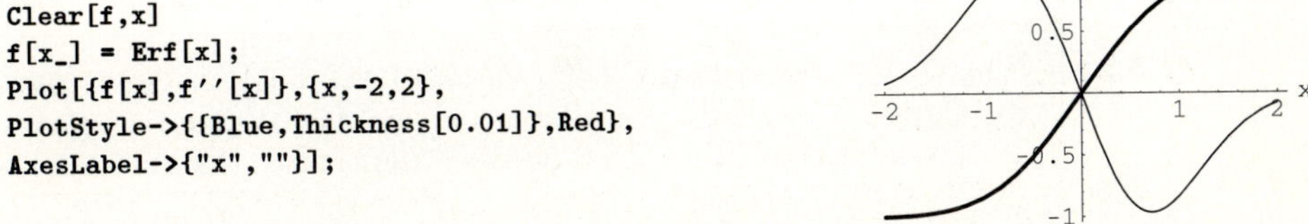

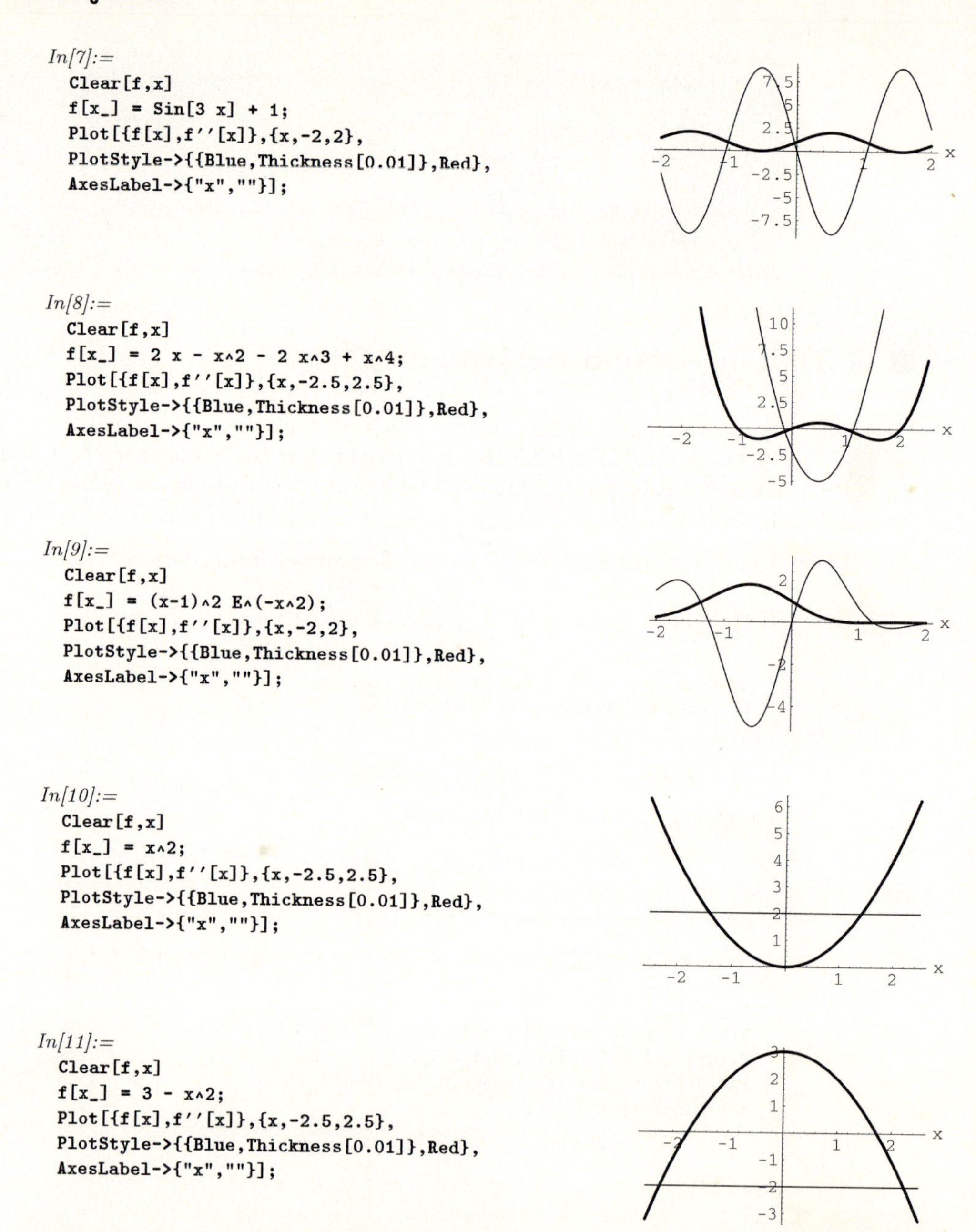

In[7]:=
```
Clear[f,x]
f[x_] = Sin[3 x] + 1;
Plot[{f[x],f''[x]},{x,-2,2},
PlotStyle->{{Blue,Thickness[0.01]},Red},
AxesLabel->{"x",""}];
```

In[8]:=
```
Clear[f,x]
f[x_] = 2 x - x^2 - 2 x^3 + x^4;
Plot[{f[x],f''[x]},{x,-2.5,2.5},
PlotStyle->{{Blue,Thickness[0.01]},Red},
AxesLabel->{"x",""}];
```

In[9]:=
```
Clear[f,x]
f[x_] = (x-1)^2 E^(-x^2);
Plot[{f[x],f''[x]},{x,-2,2},
PlotStyle->{{Blue,Thickness[0.01]},Red},
AxesLabel->{"x",""}];
```

In[10]:=
```
Clear[f,x]
f[x_] = x^2;
Plot[{f[x],f''[x]},{x,-2.5,2.5},
PlotStyle->{{Blue,Thickness[0.01]},Red},
AxesLabel->{"x",""}];
```

In[11]:=
```
Clear[f,x]
f[x_] = 3 - x^2;
Plot[{f[x],f''[x]},{x,-2.5,2.5},
PlotStyle->{{Blue,Thickness[0.01]},Red},
AxesLabel->{"x",""}];
```

G.5.a) Play with some more functions of your own choice and then say what you think the sign of $f''[x]$ tells you about the shape of the plot of $f[x]$.

G.5.b) Look at a plot of $\log[x]$:

In[12]:=
```
Clear[f,x]
f[x_] = Log[x];
Plot[f[x],{x,0.5,9},
PlotStyle->{{Blue,Thickness[0.01]}},
AxesLabel->{"x","Log[x]"}];
```

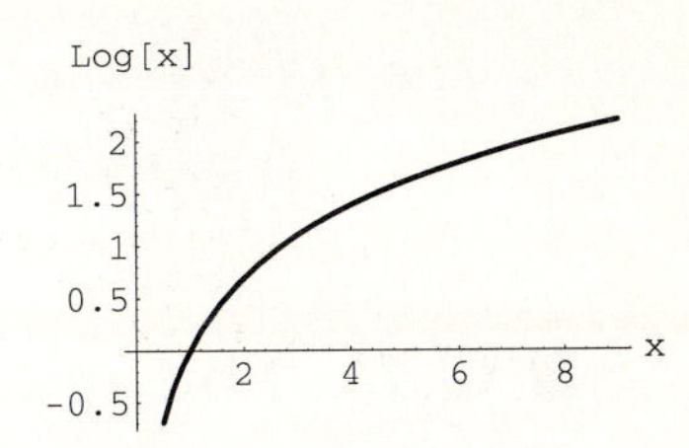

This $f[x]$ goes up as x advances from left to right; but $f[x]$ goes up more and more slowly as x advances. What does this tell you about the sign of $f''[x]$?

Confirm your answer with a plot of $f''[x]$.

G.5.c) At a point x_0 at which $f'[x]$ is as high or as low as $f'[x]$ can be, what do you expect $f''[x_0]$ to be?

■ G.6) Driving the big Mack trucks

Calculus&*Mathematica* is pleased to acknowledge that the idea for this problem comes from *Calculus for a New Century*, ACM-GCLA Calculus Reform Project, Robert Fraga (editor), 1990.

G.6.a) You are the chief dispatcher for the C&M Trucking Company, which sends Mack trucks on the straight shot between Chicago and New Orleans on Interstate 57. You know that:

→ The run between the two cities is 750 miles.

→ Running at a steady 50 miles per hour, the Mack gets 4 miles per gallon.

→ For each mile per hour increase in speed, the big Mack loses 1/10 of a mile per gallon in its mileage.

→ The driver team gets 27 dollars per hour.

→ Keeping the truck on the road costs an extra 12 dollars per hour over and above the cost of the fuel.

→ Diesel fuel for the Mack costs $1.19 per gallon.

Come up with a function $f[x]$ that measures the total cost of running the Mack from Chicago to New Orleans at a steady speed of x miles per hour.

G.6.b) Use *Mathematica* to calculate $f'[x]$ and plot it over a reasonable interval like $40 \leq x \leq 80$.

Use your plot to estimate the number s such that $f'[x] < 0$ for $40 < x < s$ and $f'[x] > 0$ for $s < x < 80$.

G.6.c) Approximately what steady speed should you tell your drivers to hold in order to make the run at least cost? Discuss how you arrived at your answer.

■ G.7) The space shuttle Challenger and its O-rings

Calculus&*Mathematica* thanks Edward Tufte of Yale University for suggesting this problem in a lecture at the University of Illinois.

Intriguing discussions of this and related issues appear in Tufte's book, *Visual Explanations*, Graphics Press, 1992. Tufte is ahead of everyone else at using graphics to visualize information. You will probably enjoy his books *The Visual Display of Quantitative Information*, Graphics Press, 1983, and *Envisioning Information*, Graphics Press, 1990.

Everyone was horrified to see the space shuttle Challenger explode shortly after it was launched on January 28, 1986. In early 1987, a fascinated nation watched the great scientist Richard Feynman show a congressional hearing how failure of O-rings led to the disaster.

Here are data on O-ring failures that were available to NASA prior to the launch of the Challenger. The data are given in the form $\{x, y\}$ where

$x =$ outdoor temperature in Fahrenheit at the time of a launch;

$y =$ number of damaged O-rings for the same launch.

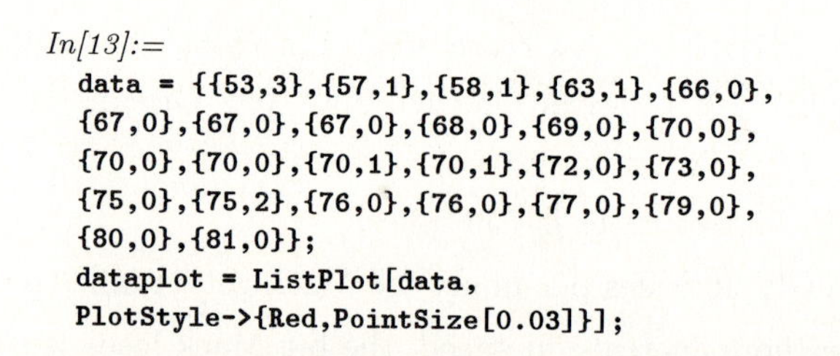

In[13]:=
```
data = {{53,3},{57,1},{58,1},{63,1},{66,0},
{67,0},{67,0},{67,0},{68,0},{69,0},{70,0},
{70,0},{70,0},{70,1},{70,1},{72,0},{73,0},
{75,0},{75,2},{76,0},{76,0},{77,0},{79,0},
{80,0},{81,0}};
dataplot = ListPlot[data,
PlotStyle->{Red,PointSize[0.03]}];
```

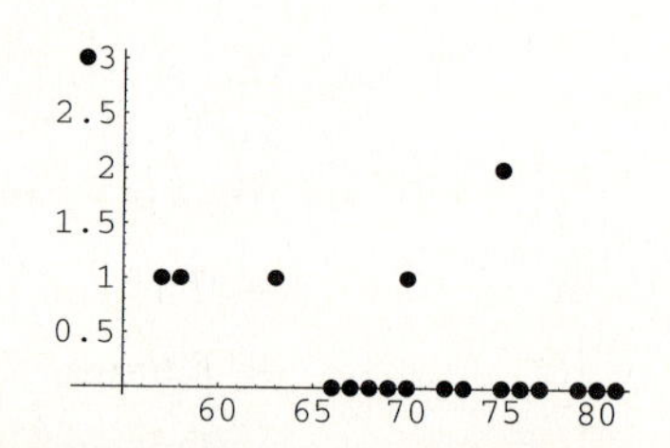

G.7.a) Fit the data with a compromise polynomial that you believe goes with the flow of the data for temperatures ranging from 0 through 80.

Plot your polynomial together with the data and discuss what the plot suggests about launching space shuttles when the temperature is under 40 degrees Fahrenheit.

G.7.b) When the Challenger was launched on January 28, 1986, the outdoor temperature was a freezing 31 degrees Fahrenheit. If the big shots at NASA had seen the plot you did above, then do you think that they would have gone ahead with the launch of the Challenger?

■ G.8) Management analysis

Some of these problems were suggested by Sherman Stein's book *Calculus in the First Three Dimensions*, McGraw-Hill, New York, 1967.

G.8.a) You are on the payroll of the C&M Wholesaling Co. of Peoria, Illinois. One of C&M's products is the Delco Freedom 74-60 battery, which it buys from General Motors and sells to independent retailers like Don's Automotive in Homer, Illinois. Sales records indicate that C&M can expect to sell B individual Delco Freedom 74-60 batteries each year.

Until recently, C&M just ordered all B batteries once each year and held them in inventory until they were sold. But while reading *Business Week* magazine, the C&M big shots learned of the Japanese method for running with low inventories but frequent reordering as demand indicates. Your boss, knowing that you know calculus, assigns you to look at the problem and to write a report for the C&M big shots to look at.

This is a real opportunity for you to make some points with the big brass at C&M. You don't want to screw it up; so you lay out the facts at hand:

→ The current policy of ordering all B batteries at once means, on the average, that the C&M Co. maintains an inventory of $B/2$ batteries at any given time of the year. This means a lot of money is lost to costs of maintaining an inventory.

It costs S dollars per year to store each battery; so the current policy spends $SB/2$ dollars per year just to maintain the inventory.

→ If C&M decides to order additional batteries each day or each hour, then the reorder costs will get out of hand.

You decide to see what happens if C&M orders B batteries per year in lots of x batteries per order.

The cost of placing an order for x batteries is

$$F + Px$$

where F is the fixed cost (secretarial time, telephone, fax costs, and other overhead) for each order and P is the fixed cost (packing, shipping, and handling) for each battery ordered by C&M. Consequently, if C&M orders B batteries per year in lots of x batteries per order, there will be B/x orders for a total ordering cost of

$$(F + Px)\,\frac{B}{x}$$

dollars per year.

When C&M orders in lots of x batteries, C&M Co. expects to maintain an inventory of $x/2$ batteries at any given time of the year. So the inventory costs now become

$$\frac{Sx}{2} \text{ dollars per year.}$$

The overall cost in dollars per year for this strategy is:

In[14]:=

```
Clear[cost,F,P,B,S,x]
cost[x_] = S x/2 + (F + P x)(B/x)
```

Out[14]=

```
S x   B (F + P x)
--- + -----------
 2         x
```

You analyze the derivative:

In[15]:=

```
Factor[cost'[x]]
```

Out[15]=

```
              2
-2 B F + S x
-------------
       2
    2 x
```

And you say if

$$0 < x < \sqrt{\frac{2\,FB}{S}},$$

then increasing x a bit will lower cost$[x]$, but if

$$\sqrt{\frac{2\,FB}{S}} < x,$$

then decreasing x a bit will lower cost$[x]$. You are right.

G.8.a.i) Explain why you are right.

G.8.a.ii) Then you say that if C&M makes the size of x as close to $\sqrt{2\,FB/S}$ as possible, then C&M will be minimizing its costs from inventory and reordering. Again you are right.

Explain why you are right.

G.8.a.iii) If C&M goes with $x = \sqrt{2\,FB/S}$, then what is the ratio

$$\frac{S\,x/2}{FB/x}$$

of the inventory costs to fixed costs?

G.8.a.iv) You go back to your bosses and say that a simple strategy for setting the size of x is to make sure that the inventory costs $Sx/2$ should be approximately equal to the fixed costs FB/x.

Explain why you are right and why your bosses should be impressed with you.

For more reading, see the book *Analysis of Inventory Systems* by G. Hadley and T. M. Whitin, Prentice-Hall, 1963.

G.8.b) The C&M Express Company is setting up some warehouses to make deliveries around the Chicago area. To fit the layout of the roads, they chop Chicagoland into identical squares each of area x. A warehouse will be built at the center of each square, and deliveries destined for a location within a square will be made from the warehouse located within the same square. The engineers tell us that to cover each square from a warehouse located at the center, the transportation costs per item are proportional to $\sqrt{x}$ and the warehouse cost per item is proportional to $1/x$.

So the overall cost of servicing a square is $a\sqrt{x} + b/x$ where a and b are positive constants.

G.8.b.i) How should you set x in terms of a and b to make the overall cost of servicing a square $a\sqrt{x} + b/x$ as small as it can be?

G.8.b.ii) What should the ratio

$$\frac{a\sqrt{x}}{b/x} = \frac{\text{transportation cost}}{\text{warehouse cost}}$$

be if x is selected so that the cost of servicing each square is as small as it can be?

■ G.9) Up then down for x^t/e^x★

G.9.a) Plot x^3/e^x for $0 \leq x \leq a$ where a is chosen to be large enough to see the curve go up and then down on $[0, a]$.

Factor the derivative to find the exact turning point x at which the curve changes direction.

Explain why the curve cannot change direction at any other point.

G.9.b) Do the same for x^6/e^x.

G.9.c) Do the same for x^{12}/e^x.

G.9.d.i) Given a positive number t, factor the derivative of x^t/e^x to explain why the curve $y = x^t/e^x$ first goes up as x advances from 0 and grows until x reaches a point x_t after which the curve goes down.

Find the exact value of the turning point x_t in terms of t.

G.9.d.ii) How does the result of part G.9.d.i) reflect the fact that in the global scale as $x \to \infty$, the exponential growth of e^x dominates the power growth of x^t?

G.10) Other max-min problems

G.10.a) Explain the statement: Of all rectangles with a fixed perimeter, the square measures out to the largest area.

G.10.b) Calculus&*Mathematica* thanks Mrs. Jodie Melton, Postmaster, U.S. Post Office, Homer, Illinois 61849 for her help in setting up this problem.

The U.S. Postal Service carries small packages, but when the boxes get big, the post office folks take out a tape measure and measure the length x of the box, the width y of the box, and the height z of the box with all measurements in inches.

If $x + 2\,y + 2\,z > 100$, then the box is rejected; otherwise it is accepted.

What measurements x, y, and z give rise to the acceptable box with the biggest volume?

G.10.c) C. J. Pennycuick, in his article, "The mechanics of bird migration" (*Ibis*, 111, pp. 525–556, 1969), determined the following formula for the power which a bird has to maintain to fly at speed v:

$$\text{power}[v] = \frac{w^2}{2\,d\,S\,v} + \frac{d\,A\,v^3}{2}:$$

Here d is air density, w is the weight of the bird, and S and A are certain parameters connected with the bird's shape and size.

Use the derivative power$'[v]$ to determine a formula for velocity v that makes power$[v]$ as small as possible.

G.10.d) A bin with an open top, rectangular sides, rectangular ends, and a rectangular base is to be built using A square feet of material.

Give a formula, in terms of A, for the dimensions of the bin with the largest possible volume.

G.10.e) You need to scoop 160 tons of coal into a railroad car. To do that, workers can be hired for $40 a day. Each worker you hire will shovel 16 tons of coal per day, but you have to provide each worker with a $15 shovel. You also have to pay $53 per day for parking the railroad car during the time the car is being loaded.

How do you get this job done with the least amount of money?

G.10.f) This problem involves no calculus.

Surfers and other beach persons, as well as residents of Illinois and Iowa, often wonder whether haze, smog, or curvature of the earth cause lack of visibility of distant objects. It's time to remove this discussion from bars and beaches into the far more secure setting of calculation:

For this problem, agree that the earth is a sphere of radius 3940 miles and agree that there are 5280 feet in a mile. Look at the graphics:

In[16]:=

```
Show[Graphics[{Blue,Circle[{0, -1}, 1,{Pi/3, Pi/2}],
GrayLevel[0],Line[{{0,-1},{.35,0}}],
Text["C",{-.04,-1.03}], Text["R",{-.04,-.50}],
Text["R",{ .23,-0.50}], Text["Q",{ .35, .05}],
Text["P",{ .35,-0.14}]}],
AxesOrigin->{0,0},AxesLabel->{"x","y"},
PlotRange->{{-.2,.5},{-1.1,.1}},Axes->True,
AspectRatio->Automatic,Ticks->None];
```

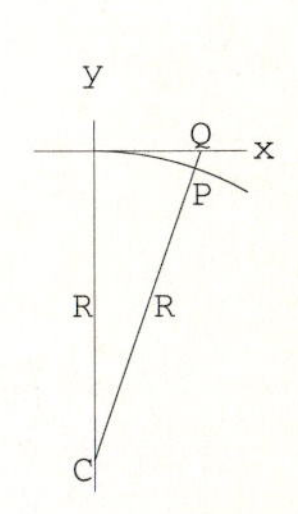

Interpret this diagram as follows: You are positioned at the origin on the surface of the earth, and the x-axis represents your line of sight. C represents the center of the earth, and R is the radius of the earth. At a horizontal distance x miles from your position is the point Q. At that distance you can see nothing of height less than the length of the line segment PQ.

G.10.f.i) Give the formula for length PQ of the shortest visible object (in miles) as a function of x measured in miles.

G.10.f.ii) Give a plot of the height of the shortest visible object as a function of its horizontal distance from your position. Discuss the plot.

G.10.f.iii) The Empire State Building, built in 1930–1931, is 1250 feet high and was for a long time the tallest building in the world.

From about how far out in the ocean is the Empire State Building in the line of sight of a submarine's periscope positioned on the surface of the ocean on a clear day?

■ G.11) At what age is the Bernese Mountain Dog growing the fastest?

G.11.a) Back in the last lesson, you saw that the height of a typical Bernese Mountain Dog bitch t years after her birth is given approximately by:

In[17]:=
```
Clear[height,t]
height[t_] = (108 E^(4.9353 t))/(19.5 + 4.5 E^(4.9353 t))
```

Out[17]=
```
        4.9353 t
   108 E
----------------------
               4.9353 t
19.5 + 4.5 E
```

A plot:

In[18]:=
```
Plot[height[t],{t,0,2},
PlotStyle->{{Thickness[0.01],Brown}},
AxesLabel->{"t","height[t]"}];
```

Here is a plot of the instantaneous growth rate of height[t]:

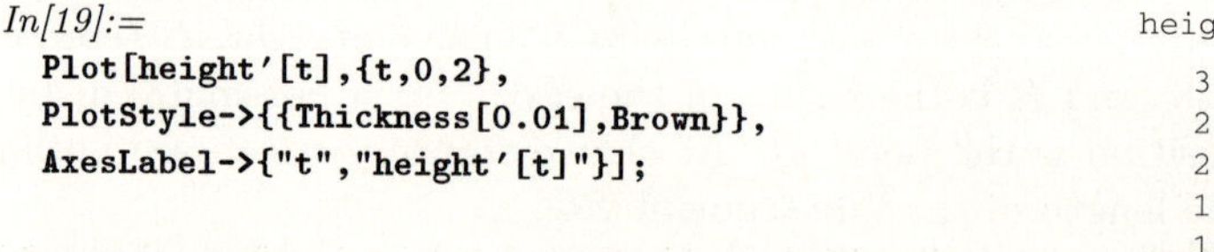

In[19]:=
```
Plot[height'[t],{t,0,2},
PlotStyle->{{Thickness[0.01],Brown}},
AxesLabel->{"t","height'[t]"}];
```

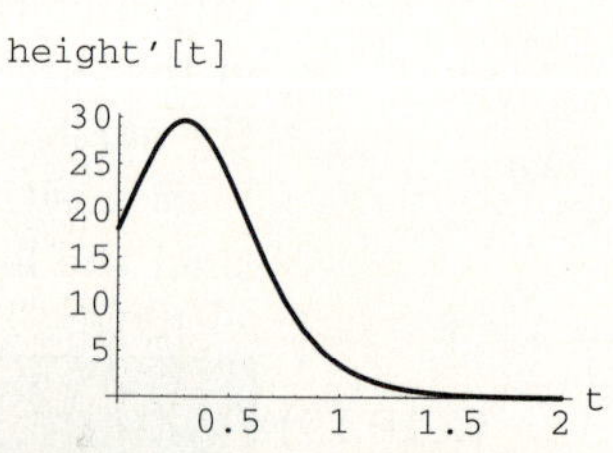

At what point is the bitch's instantaneous growth rate as big as it can be?

Is this the same point at which the bitch is growing fastest?

How tall is she at the instant she is growing fastest?

How does this height compare to her eventual mature height?

Do you think your answer is an accident?

LESSON 1.06

The Differential Equations of Calculus

Basics

■ B.1) The differential equation $y'[x] = r\,y[x]$

B.1.a) Given a number r, to solve the differential equation

$$y'[x] = r\,y[x],$$

you want a function $y[x]$ with the property that

$$y'[x] = r\,y[x]$$

for all x's. This means

$$\frac{100\,y'[x]}{y[x]} = 100\,r$$

no matter what x is.

> How does this give away a formula for $y[x]$?

Answer: The equation is $y'[x] = r\,y[x]$ for all x's. This means $100\,y'[x]/y[x] = 100\,r$ no matter what x is.

This tells you that $y[x]$ grows with a steady instantaneous percentage growth rate of $100\,r$ percent. And this tells you that $y[x]$ is the exponential function $y[x] = k\,e^{rx}$ for some number k. Try it out:

In[1]:=

```
Clear[x,y,k,r];  y[x_] = k E^(r x);  y'[x] == r y[x]
```

Out[1]=
```
True
```

No problem-o. You can get *Mathematica* to spit this out directly:

In[2]:=
```
Clear[y,x]; DSolve[y'[x] == r y[x],y[x],x]
```

Out[2]=
```
            r x
{{y[x] -> E     C[1]}}
```

This means $C[1]$ can be any constant. You can reconcile the solution $y[x] = k\,e^{rx}$ above with the *Mathematica* solution $y[x] = C[1]e^{rx}$ by taking $k = C[1]$.

B.1.b) Now that you know that $y[x] = k\,e^{3x}$ solves the differential equation

$$y'[x] = 3\,y[x],$$

what data do you need to nail down what the value of k is?

Answer: You know $y[x] = k\,e^{3x}$; so you have one unknown (namely k) to determine. You need one data point on $y[x]$ to do this. For instance, if you know $y[1.71]$ is to be equal to 18.06, then you determine k by solving for k as follows:

In[3]:=
```
Clear[x,k]; Solve[k (E^(3 x)/.x->1.71) == 18.06,k]
```

Out[3]=
```
{{k -> 0.106853}}
```

This gives you the specific solution $y[x] = 0.106853\,e^{3x}$. You can also get the same solution directly from *Mathematica*:

In[4]:=
```
Clear[y,x];
sol = DSolve[{y'[x] == 3 y[x], y[1.71] == 18.06}, y[x], x]
```

Out[4]=
```
                          3 x
{{y[x] -> 0.106853 E     }}
```

Good.

■ B.2) The logistic differential equation $y'[x] = ry[x]\,(1 - y[x]/b)$ and how you get a formula for its solution

B.2.a) Given numbers r and b, you want a function $y[x]$ with the property that

$$y'[x] = r\,y[x]\left(1 - \frac{y[x]}{b}\right).$$

How do you find a formula for $y[x]$?

Answer: This one is harder than the others. Let's see what mighty *Mathematica* thinks about this one:

In[5]:=
```
Clear[y,x,r,b]; DSolve[y'[x] == r y[x] (1 - y[x]/b),y[x],x]
```

Out[5]=
```
                 r x
              b E
{{y[x] -> ---------------}}
            r x
           E    + b C[1]
```

Here $C[1]$ can be any constant. Call it k and check *Mathematica*'s solution:

In[6]:=
```
Clear[y,x,k,r,b]; y[x_] = b E^(r x)/(E^(r x) + b k);
Expand[y'[x]] == Expand[r y[x] (1 - y[x]/b)]
```

Out[6]=
```
True
```

Good. Later in the course, you'll learn how to pull this formula out of a hat with your own hands. In the meantime, everyone is happy with the machine solution because it checks out.

B.2.b) Here is *Mathematica*'s formula for the solution of

$$y'[x] = 2y[x]\left(1 - \frac{y[x]}{8}\right):$$

In[7]:=
```
Clear[x,y]
solution = DSolve[y'[x] == 2 y[x] (1 - y[x]/8),y[x],x]
```

Out[7]=
```
                2 x
               E
{{y[x] -> ---------------}}
            2 x
           E
           ---- + C[1]
            8
```

What data do you need to nail down what the value of the constant $C[1]$ is?

Answer: You know the formula for the solution is:

In[8]:=
```
y[x_] = y[x]/.solution[[1]]
```

Out[8]=

```
    2 x
   E
-------------
 2 x
E
---- + C[1]
8
```

So you have one unknown (namely $C[1]$) to determine. You need one data point on $y[x]$ to do this. For instance, if you know $y[0.85] = 5.38$, then you determine $C[1]$ by solving for $C[1]$ as follows:

In[9]:=

```
Solve[y[0.85] == 5.38,C[1]]
```

Out[9]=

```
{{C[1] -> 0.333219}}
```

This gives you the formula:

In[10]:=

```
y[x_] = y[x]/.C[1]->0.333219
```

Out[10]=

```
      2 x
     E
----------------
               2 x
              E
0.333219 + ----
              8
```

This is the same as:

In[11]:=

```
Clear[x,y]
solution = DSolve[{y'[x] == 2 y[x] (1 - y[x]/8), y[0.85]
  == 5.38},y[x],x]
```

Out[11]=

```
                      2 x
                     E
{{y[x] -> ---------------------}}
                           2 x
                          E
            0.333219 + ----
                          8
```

Hunky-dory.

■ B.3) Logistic growth is controlled growth

B.3.a.i) Here is a plot of the solution of the logistic differential equation

$$y'[x] = r\, y[x] \left(1 - \frac{y[x]}{b}\right)$$

with $r = 0.8$, $b = 30$, and $y[0] = 1.5$.

In[12]:=

```
Clear[x,y]; r = 0.8;b = 30;starter = 1.5;
logistic =
DSolve[{y'[x] == r y[x](1 - y[x]/b), y[0] == starter},y[x],x];
Clear[ylogistic]; ylogistic[x_] = y[x]/.logistic[[1]];
```

In[13]:=

```
logisticplot =
Plot[ylogistic[x],{x,0,8},
PlotStyle->{{DarkGreen,Thickness[0.01]}},
AxesLabel->{"x","ylogistic[x]"}];
```

This is the typical S curve most folks like to associate with logistic growth. Now take a look at a plot of the solution of the exponential differential equation $y'[x] = r\,y[x]$ using the same values of $y[0]$ and r as above:

In[14]:=

```
Clear[x,y,yexpon]
exponential = DSolve[{y'[x] == r y[x],y[0] == starter},y[x],x];
yexpon[x_] = y[x]/.exponential[[1]];
```

In[15]:=

```
exponplot =
Plot[yexpon[x],{x,0,8},
PlotStyle->{{Red,Thickness[0.01]}},
AxesLabel->{"x","yexpon[x]"}];
```

Typical exponential growth. Here are both plots together:

In[16]:=

```
both =
Show[exponplot,logisticplot,
PlotRange->{0,100},AxesLabel->{"x",""}];
```

They run nicely together for a while and then break apart.

> Look at the differential equations
>
> → exponential: $y'[x] = r\,y[x]$
>
> → logistic: $y'[x] = r\,y[x]\,(1 - y[x]/b)$
>
> and then try to explain why the last plot turned out the way it did.

Answer: Compare the differential equations

→ exponential: $y'[x] = r\,y[x]$

→ logistic: $y'[x] = r\,y[x]\,(1 - y[x]/b)$

To get the logistic differential equation, you take the exponential differential equation and tack on the extra factor $(1 - y[x]/b)$. Now remember that in both plots above, $y[0] = 1.5$ and $b = 30$ and look at:

In[17]:=
```
(1 - y[0]/b)/.y[0]->1.5
```

Out[17]=
```
0.95
```

This tells you that when x is close to 0, then

$$r\,y[x]\left(1 - \frac{y[x]}{b}\right) \qquad \text{and} \qquad r\,y[x]$$

are nearly the same thing. This means that solutions of

→ exponential: $y'[x] = r\,y[x]$

→ logistic: $y'[x] = r\,y[x]\,(1 - y[x]/b)$

move out from $\{0, 1.5\}$ with approximately the same instantaneous growth rate, and consequently they are forced to run very close together as x advances away from 0.

Confirm with a plot:

In[18]:=
```
Show[both,PlotRange->{{0,1.3},{0,5}},
AxesLabel->{"x",""}];
```

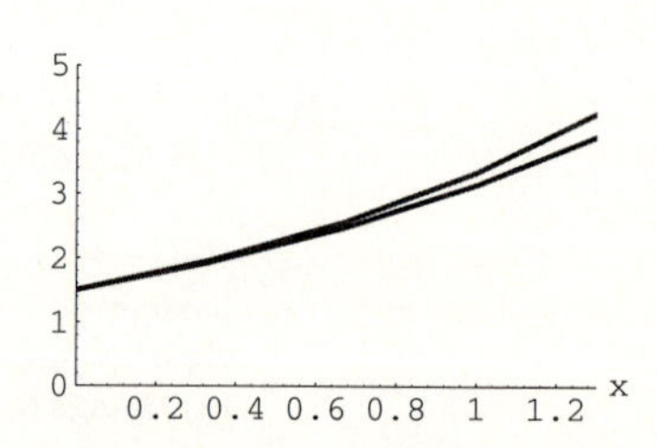

Just as the analysis predicted! As x advances from 0 to 1.2, the two solutions run together.

B.3.a.ii) What's the moral?

Answer: When you take the logistic differential equation

$$y'[x] = r\,y[x]\left(1 - \frac{y[x]}{b}\right)$$

with $y[0]$ small relative to b, then as x advances from 0, the solution of this differential equation runs close to the exponential solution of $y'[x] = r\,y[x]$.

In short, if $y[0]$ is small relative to b, then logistic growth mimics exponential growth at the start.

B.3.b) Here are plots of the solution of the logistic differential equation

$$y'[x] = r\,y[x]\left(1 - \frac{y[x]}{b}\right)$$

and the horizontal line $y = b$ for various choices of positive r, b, and $y[0]$ with $0 < y[0] < b$:

In[19]:=
```
r = 0.8; b = 30; starter = 1.5;
Clear[x,y]; logistic =
DSolve[{y'[x] == r y[x](1 - y[x]/b),
y[0] == starter},y[x],x];
Clear[ylogistic]; ylogistic[x_] =
y[x]/.logistic[[1]];
logisticplot = Plot[{ylogistic[x],b},{x,0,12},
PlotStyle->{{DarkGreen,Thickness[0.01]},Red},
PlotRange->All, AxesLabel->{"x","ylogistic[x]"}];
```

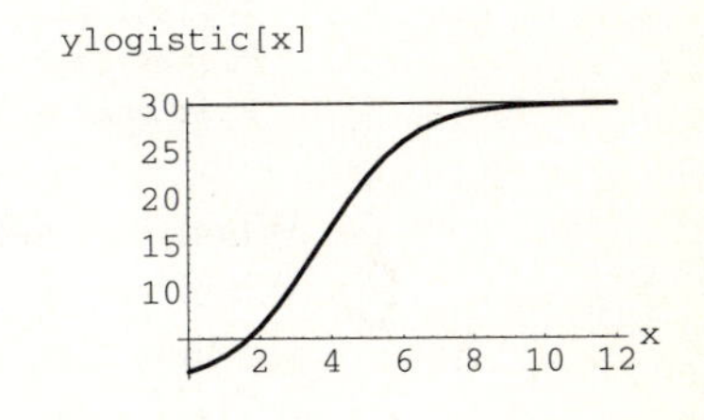

In[20]:=
```
r = 0.3; b = 50; starter = 5;
logistic =
DSolve[{y'[x] == r y[x](1 - y[x]/b),
y[0] == starter},y[x],x];
Clear[ylogistic]; ylogistic[x_] =
y[x]/.logistic[[1]];
logisticplot = Plot[{ylogistic[x],b},{x,0,50},
PlotStyle->{{DarkGreen,Thickness[0.01]},Red},
PlotRange->All, AxesLabel->{"x","ylogistic[x]"}];
```

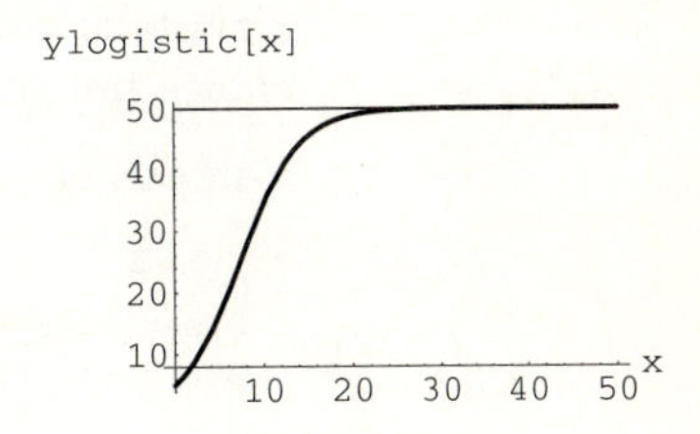

In[21]:=
```
r = 2; b = 200; starter = 30;
logistic =
DSolve[{y'[x] == r y[x](1 - y[x]/b),
y[0] == starter},y[x],x];
Clear[ylogistic]; ylogistic[x_] =
y[x]/.logistic[[1]];
logisticplot = Plot[{ylogistic[x],b},{x,0,20},
PlotStyle->{{DarkGreen,Thickness[0.01]},Red},
PlotRange->All, AxesLabel->{"x","ylogistic[x]"}];
```

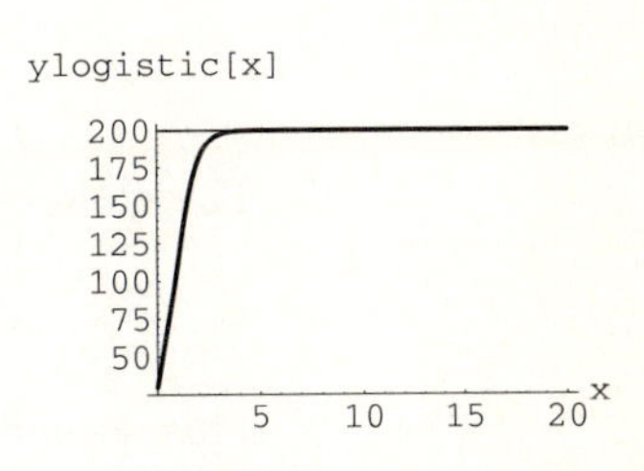

Evidently the solution of the logistic differential equation

$$y'[x] = r\,y[x]\left(1 - \frac{y[x]}{b}\right)$$

with r and $b > 0$ and $0 < y[0] < b$ goes into global scale with

$$\lim_{x\to\infty} y[x] = b.$$

> Explain why this is automatic.

Answer: To get one explanation, look at another sample case:

In[22]:=
```
r = 0.6; b = 10; starter = 1.1;
logistic = DSolve[{y'[x] == r y[x](1 - y[x]/b),
y[0] == starter},y[x],x];
Clear[ylogistic]; ylogistic[x_] = y[x]/.logistic[[1]];
logisticplot = Plot[{ylogistic[x],b},{x,0,25},
PlotStyle->{{DarkGreen,Thickness[0.01]},Red},
PlotRange->All, AxesLabel->{"x","ylogistic[x]"}];
```

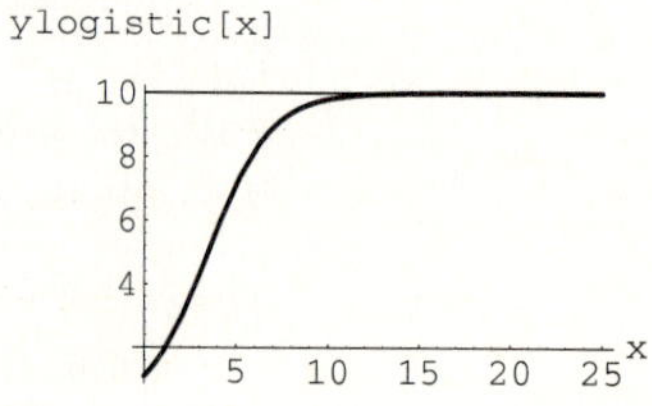

When $y[x]$ gets near the horizontal line $y = b$, then $y[x]$ snuggles in on this line. When you look at the logistic differential equation $y'[x] = r\,y[x]\,(1 - y[x]/b)$, you see that this behavior is dictated by the differential equation.

Reason: When $y[x]$ gets close to b, then $(1 - y[x]/b)$ gets close to 0.

So $y'[x] = r\,y[x]\,(1 - y[x]/b)$ gets close to 0.

This tells you that when $y[x]$ gets close to b, $y[x]$ cannot grow much and has no choice but to settle in along the horizontal line $y = b$.

You can get another explanation by running with cleared values of r, b, and $y[0]$:

In[23]:=
```
Clear[starter,b,r,y,x]
DSolve[{y'[x] == r y[x](1 - y[x]/b), y[0] == starter},y[x],x]
```

Out[23]=
```
                     r x
                  b E
{{y[x] -> ---------------------}}
                r x       b
          -1 + E    + -------
                      starter
```

The global scale of the solution is

$$\frac{b\,e^{rx}}{e^{rx}} = b.$$

This second explanation is short but it's a little disappointing because it doesn't reveal how you could see this directly by looking at the differential equation $y'[x] = r\,y[x]\,(1 - y[x]/b)$.

B.3.c)

> Some folks like to say that logistic growth is just controlled exponential growth. Comment on this.

Answer: Logistic growth is controlled because the solution of the logistic differential equation

$$y'[x] = r\,y[x]\left(1 - \frac{y[x]}{b}\right)$$

with r and $b > 0$ and $0 < y[0] < b$ must go into global scale with

$$\lim_{x \to \infty} y[x] = b.$$

Logistic growth is like exponential growth because if $y[0]$ is small relative to b, then as x advances from 0, $y[x]$ grows like an exponential function until it starts to move into its global scale.

Take another look:

In[24]:=

```
r = 0.15; b = 300; starter = 8.5;
Clear[x,y]; logistic =
DSolve[{y'[x] == r y[x](1 - y[x]/b),
y[0] == starter},y[x],x];
Clear[ylogistic]
ylogistic[x_] = y[x]/.logistic[[1]];
exponential = DSolve[{y'[x] == r y[x],
y[0] == starter},y[x],x];
Clear[yexpon]; yexpon[x_] = y[x]/.exponential[[1]];
Plot[{ylogistic[x],yexpon[x],b},{x,0,60},
PlotStyle-> {{DarkGreen,Thickness[0.01]},
{Red,Thickness[0.01]},Red},
AxesLabel->{"x",""},PlotRange->{0,b + 50},
PlotLabel->"Logistic versus exponential."];
```

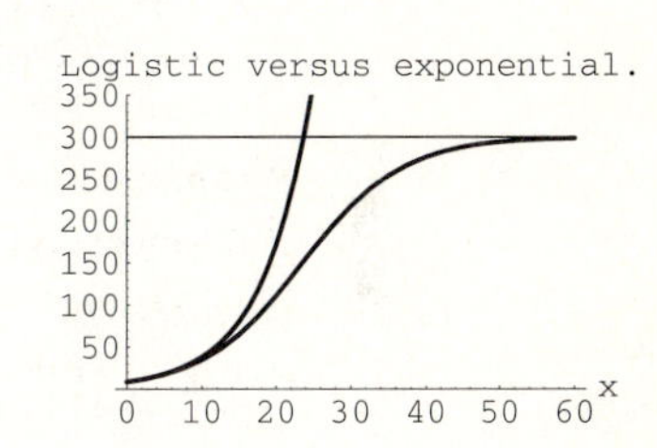

Beautiful.

■ B.4) The differential equation $y'[x] = r\,y[x] + b$

B.4.a) Given numbers r and b, you want a function $y[x]$ with the property that

$$y'[x] = r\,y[x] + b.$$

How do you find a formula for $y[x]$?

Answer: The differential equation is $y'[x] = r\,y[x] + b$. It seems reasonable to try: $y[x] = k[1]\,e^{rx} + k[2]$ where $k[1]$, $k[2]$ are constants:

In[25]:=

```
Clear[trialy,x,k,r,b]; trialy[x_] = k[1] E^(r x) + k[2]
```

Out[25]=

```
 r x
E    k[1] + k[2]
```

Plug trialy$[x]$ into the equation $y'[x] = r\,y[x] + b$:

In[26]:=
```
trialy'[x] == Expand[r trialy[x] + b]
```

Out[26]=
```
 r x               r x
E    r k[1] == b + E    r k[1] + r k[2]
```

This tells you to take $b + r\,k[2] = 0$:

In[27]:=
```
Solve[b + r k[2] == 0,k[2]]
```

Out[27]=
```
               b
{{k[2] -> -(-)}}
               r
```

To go for the solution $y[x]$, plug these into trialy$[x]$:

In[28]:=
```
Clear[y]; y[x_] = trialy[x]/.k[2]->-b/r
```

Out[28]=
```
  b      r x
-(-) + E    k[1]
  r
```

This means $k[1]$ can be any constant. Try it out to see that this $y[x]$ solves

$$y'[x] = r\,y[x] + b:$$

In[29]:=
```
y'[x] == Expand[ r y[x] + b]
```

Out[29]=
```
True
```

Nice going. You can also get the solution $y[x]$ directly from *Mathematica*:

In[30]:=
```
Clear[y,x,r,b]; DSolve[y'[x] == r y[x] + b,y[x],x]
```

Out[30]=
```
              b      r x
{{y[x] -> -(-) + E    C[1]}}
              r
```

This means $C[1]$ can be any constant. You can reconcile the solution

$$y[x] = k[1]\,e^{rx} - \frac{b}{r}$$

above with the *Mathematica* solution

$$y[x] = C[1]\,e^{rx} - \frac{b}{r}$$

by taking $k[1] = C[1]$.

B.4.b) Now that you know that $y[x] = k\,e^{2x} - 3/2$ solves the differential equation $y'[x] = 2\,y[x] + 3$, what data do you need to nail down what the value of k is?

Answer: You know $y[x] = k\,e^{2x} - 3/2$; so you have one unknown (namely k) to determine. You need one data point on $y[x]$ to do this. For instance, if you know $y[1.64]$ is to be equal to 9.38, then you solve for k as follows:

In[31]:=
```
Clear[k,x]; Solve[((k E^(2 x) - 3/2)/.x->1.64) == 9.38,k]
```

Out[31]=
```
{{k -> 0.409395}}
```

This gives you the solution:

In[32]:=
```
Clear[y]; y[x_] = 0.409395 E^(2 x) - 3/2
```

Out[32]=
```
  3             2 x
-(-) + 0.409395 E
  2
```

You can also get the same solution directly from *Mathematica*:

In[33]:=
```
Clear[y,x];
sol = DSolve[{y'[x] == 2 y[x] + 3, y[1.64] == 9.38},y[x],x]
```

Out[33]=
```
                3             2 x
{{y[x] -> -(-) + 0.409395 E    }}
                2
```

Not terribly hard.

Tutorials

■ T.1) Radioactive decay and carbon dating

You've probably seen this earlier in the course. But what was not talked about earlier is why radioactive decay is exponential decay. When you look at it from the point of view of a differential equation, an exponential differential equation pops up naturally.

Here you go:

Radioactive decay is the process in which a substance disintegrates while its mass is converting to radiation. Over a given time interval, each radioactive atom has the same chance for disintegration as any other.

The upshot:

If you have a given quantity P of radioactive atoms of one type, then you can expect twice as much decay if the supply is doubled to $2\,P$ and three times as much decay if the supply is tripled to $3\,P$.

In terms of instantaneous growth rate:

The rate of decay $P'[t]$ is proportional to the supply $P[t]$ at time t. In other words:

$$P'[t] = -a\,P[t]$$

for some constant of proportionality a. The negative sign is in here to emphasize that decay is happening. Most folks call the number a by the name "decay constant."

T.1.a.i) How is the notion of half life used to help to determine the decay constant?

Answer:

In[1]:=

```
Clear[P,t,a,Q]
solution = DSolve[{P'[t] == -a P[t], P[0] == Q},P[t],t]
```

Out[1]=

```
            Q
{{P[t] -> ----}}
           a t
          E
```

Fish out the solution:

In[2]:=

```
P[t_] = P[t]/.(solution[[1]])
```

Out[2]=

```
 Q
----
 a t
E
```

You get the half life by solving

In[3]:=

```
Solve[Q/2 == P[halflife],halflife]
```

Out[3]=

```
                   Log[2]
{{halflife -> ------}}
                     a
```

In decimals:

In[4]:=

```
N[Log[2]/a]
```

Out[4]=

```
0.693147
--------
   a
```

If you know the half life, then you see that

$$a = \frac{0.693147}{(\text{half life})}.$$

T.1.a.ii) The half life of radioactive carbon-14 $\left({}^{14}_{6}C\right)$ is about 5750 years. If measurements show that the radioactive carbon in a fossil has decayed by 30%, then how old is the fossil?

Answer: From part T.1.a.i) you know that $P[t] = Q\, e^{-at}$ where Q is the original amount of carbon-14 in the fossil when the bone was new. Also you know that

$$a = \frac{0.693147}{\text{halflife}} = \frac{0.693147}{5750}.$$

Because the half life is 5750 years.

So: Because the fossil has lost 30 percent of its radioactive carbon, you want to solve:

In[5]:=

```
a=0.693147/5750;
Solve[0.70 Q == P[t],t]
```

Out[5]=

```
{{t -> 2958.8}}
```

The fossil is about 3000 years old.

T.1.b) What are the biological principles behind carbon dating?

Answer: Living tissue contains two kinds of carbon. One kind, carbon-14 $\left({}^{14}_{6}C\right)$, is radioactive with a half life of about 5750 years; the other is not radioactive. In living tissue the ratio of the two is always constant, but when the tissue dies, the radioactive carbon begins to decay while the other carbon remains.

To date a fossil, scientists measure the amount of each kind of carbon present to determine how much of the radioactive carbon has decayed. This gives them what they need to date the fossil as above.

■ T.2) Socking money away

You deposit $\$P$ in a savings account bearing interest rate $100\,r$ percent per year compounded every instant. This means that if balance$[t]$ is the value in dollars of this untouched savings account at time t years after the deposit, then

$$\text{balance}[t] = P\, e^{rt}.$$

T.2.a) What differential equation does balance$[t]$ satisfy?

Answer:

$$\text{balance}'[t] = P\,e^{rt}\,r = \text{balance}[t]\,r = r\,\text{balance}[t];$$

so balance[t] solves the exponential differential equation

$$\text{balance}'[t] = r\,\text{balance}[t].$$

No big surprise.

T.2.b) At the time of your original deposit of $\$P$, you arrange with your bank to make a continuous deposit into your account at a rate of d dollars per year. This means that after t years, $t\,d$ additional dollars have been deposited. As the new money is deposited into the account, the new money begins to earn interest at the rate of $100\,r$ percent compounded every instant.

If balance[t] is the amount in the account t years after the arrangement was set up, then what differential equation does balance[t] satisfy and what is a formula for its solution?

Answer: The differential equation is

$$\text{balance}'[t] = r\,\text{balance}[t] + d \qquad \text{with balance}[0] = P.$$

(The derivative of the sum is the sum of the derivatives; the r balance[t] term comes from interest and the d term comes from your continuous deposit.) A formula for its solution is:

In[6]:=

```
Clear[balance,t,r,t,d,P,Derivative]
DSolve[{balance'[t] == r balance[t] + d, balance[0] == P},balance[t],t]
```

Out[6]=

```
                      r t     r t
              -d + d E    + E    P r
{{balance[t] -> -------------------}}
                        r
```

T.2.c) Plot balance[t] over the first 30 years for the case in which the original deposit $= \$5000$, $d = \$500$, $100\,r\% = 5.3\%$ compounded every instant.

On the same plot, show how much of balance[t] is in the account via interest payments and discuss the information contained in the plot.

Answer: It's a snap:

In[7]:=

```
P = 5000; d = 500; r = 0.053;
Clear[balance,t]; solution =
DSolve[{balance'[t] == r balance[t] + d, balance[0] == P},balance[t],t];
balance[t_] = balance[t]/.solution[[1]]
```

Out[7]=

```
                 0.053 t
-9433.96 + 14434. E
```

The part of balance[t] coming from interest is given by:

In[8]:=
```
Clear[interest];  interest[t_] = balance[t] - P - t d
```
Out[8]=
```
                      0.053 t
-14434. + 14434. E             - 500 t
```

Here comes the plot:

In[9]:=
```
Plot[{balance[t],interest[t]},{t,0,30},
PlotStyle->{{DarkGreen,Thickness[0.01]},
{Red,Thickness[0.01]}},
AxesLabel->{"t","$ balance"}];
```


Better than \$60,000 in 30 years. The lower curve is the interest curve. After a feeble start, the accrued interest eventually represents the biggest part of the nest egg.

You may be interested in changing the parameters r, d, and your original deposit P to see what develops in other situations. Do it.

T.2.d) You and your significant other have just had a beautiful bouncing baby. You can already tell that the kid is as smart as all get-out, and you decide to set up a bank account geared toward producing big-time college money in 18 years.

You can come up with \$10,000 for the account now and you can lock in an interest rate of $100\,r\% = 4.6\%$ compounded every instant.

> How do you set your yearly payment rate d (to result in \$$t\,d$ after t years) to be sure of having \$120,000 in your account by the end of 18 years?

Answer: Thanks to the magic of differential equations, coming up with the answer is cake.

In[10]:=
```
Clear[d]; P = 10000; r = 0.046;
Clear[balance,t]; solution =
DSolve[{balance'[t] == r balance[t] + d, balance[0] == P},balance[t],t];
balance[t_] = balance[t]/.solution[[1]]
```
Out[10]=
```
                                         0.046 t
-21.7391 d + (10000. + 21.7391 d) E
```

You want to see what d is needed to guarantee balance[18] = 120,000:

In[11]:=
```
Solve[balance[18] == 120000,d]
```

Out[11]=
```
{{d -> 3466.33}}
```

A bit less than $300 per month. If you have a good job, then this is a rather painless way of coming up with tuition, room, and board. Watch the account grow:

In[12]:=
```
d = 3466.33;
Plot[balance[t],{t,0,18},
PlotStyle->{{DarkGreen,Thickness[0.01]}},
AxesLabel->{"t","$ for the kid"}];
```

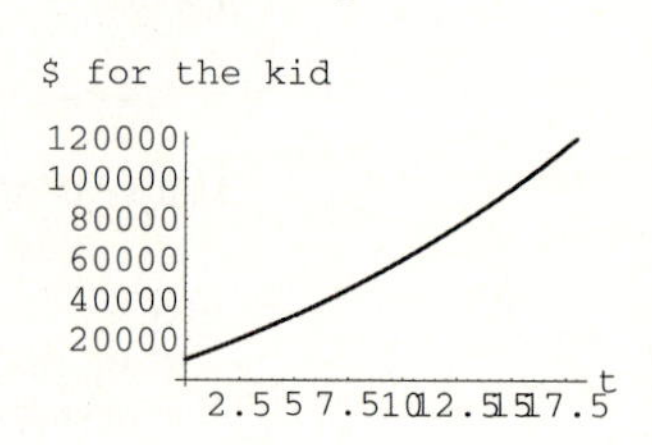

■ T.3) Wal-Mart: Exponential or logistic growth?

This problem appears only in the electronic version.

■ T.4) Pollution elimination

An artesian well supplies clear, pure water to a 2000-gallon cistern at a rate of 10 gallons per minute. It mixes with the old water and then overflows through a relief tube at a rate of 10 gallons per minute.

At a certain instant, a complete air-head foul-up spills 50 gallons of a polluting liquid into the cistern and the polluting liquid mixes thoroughly with the water instantly. As new water flows in and old flows out, the polluting liquid is continually diluted. You job is to find the time after which only 1/10 gallon of the pollutant is left in the cistern.

Agree that $P[t]$ stands for the number of gallons of the pollutant in the cistern at time t minutes after the spill.

The first goal here is to come up with a formula for $P[t]$. The strategy is to find a differential equation that $P[t]$ satisfies, to make some measurements to set the constants, and then come up with a complete formula for $P[t]$. This strategy is very common in making scientific measurements.

T.4.a) Find a differential equation that $P[t]$ satisfies, solve it, and obtain a reasonably accurate estimate of the time it will take for only 0.1 gallon of the pollutant to remain in the cistern.

Give a plot of $P[t]$ as a function of t and describe it.

Answer: Ten gallons of the mixed pollutant and fresh water leave per minute. If h is small and positive, then from time t to time $t + h$ (in minutes) the amount of pollutant $P[t] - P[t + h]$ leaving satisfies

$$\left(\frac{10}{2000}\right) P[t+h]\,h < P[t] - P[t+h] < \left(\frac{10}{2000}\right) P[t]\,h.$$

Here is the algebra that gives these inequalities:

The actual amount of pollutant that leaves during the time interval $[t, t+h]$ is

$$\text{actual} = P[t] - P[t+h].$$

Next note that $P[t] > P[t+h]$ because the mixture is continually being diluted. In the h minutes, $10\,h$ gallons of the mixture leave. So less than

$$\frac{10\,h\,P[t]}{2000} = \left(\frac{10}{2000}\right) P[t]\,h$$

gallons of the pollutant leave. But more than

$$\frac{10\,h\,P[t+h]}{2000} = \left(\frac{10}{2000}\right) P[t+h]\,h$$

gallons of the pollutant leave. So

$$\left(\frac{10}{2000}\right) P[t+h]\,h < (P[t] - P[t+h]) < \left(\frac{10}{2000}\right) P[t]\,h$$

as above.

Divide through by h:

$$\left(\frac{10}{2000}\right) P[t+h] < \frac{P[t] - P[t+h]}{h} < \left(\frac{10}{2000}\right) P[t].$$

Multiply through by -1:

$$-\left(\frac{10}{2000}\right) P[t+h] > \frac{P[t+h] - P[t]}{h} > -\left(\frac{10}{2000}\right) P[t].$$

That takes care of the inequalities. (Whew!)

Watching what happens as h closes in on 0, and remembering that

$$P'[t] = \lim_{h\to 0} \frac{P[t+h] - P[t]}{h},$$

you have no choice but to say

$$P'[t] = -\left(\frac{10}{2000}\right) P[t].$$

Now you are ready to go into calculating action:

In[13]:=

```
Clear[P,t,Derivative]
solution = DSolve[{P'[t] == -(10/2000) P[t],P[0] == 50},P[t],t]
```

Out[13]=

```
              50
{{P[t] -> ------}}
           t/200
          E
```

You put $P[0] == 50$ because when you started to measure time, there were 50 gallons of the pollutant in the cistern.

Fish out the solution:

In[14]:=
```
P[t_] = P[t]/.solution[[1]]
```

Out[14]=
```
 50
-----
 t/200
E
```

To find when there are only 0.1 gallons of the pollutant left, solve:

In[15]:=
```
N[Solve[P[t] == 0.1,t]]
```

Out[15]=
```
{{t -> 1242.92}}
```

About 1240 minutes.

In[16]:=
```
N[1243/60]
```

Out[16]=
```
20.7167
```

The better part of one day. Here is a plot of the amount of pollutant remaining in the cistern as a function of time measured in minutes:

In[17]:=
```
Plot[P[t],{t,0,2000},
AxesLabel->{"t","P[t]"},
PlotStyle->{{SpringGreen,Thickness[0.01]}},
PlotRange->All];
```

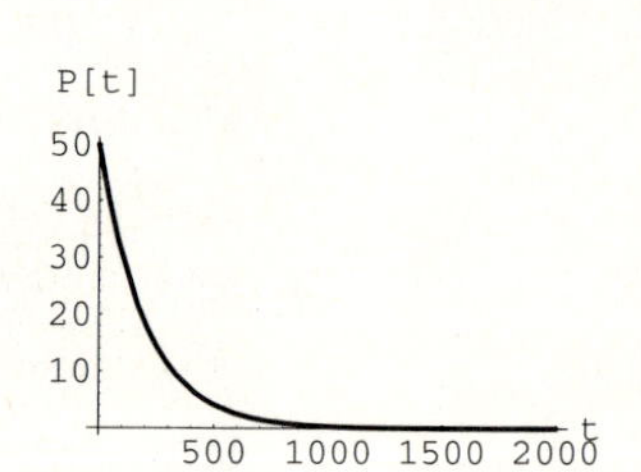

Exponential decay.

■ T.5) An irritating typo and how to recover from it

When you are trying to get a formula for the solution of a differential equation like

$$y'[x] = r\,y[x]\left(1 - \frac{y[x]}{b}\right)$$

with $y[0] = a$, you type and execute:

In[18]:=
```
Clear[r,a,b,x,y]
DSolve[{y'[x] == r y[x] (1 - y[x]/b), y[0] == a},y[x],x]
```

Out[18]=

```
                r x
              b E
  {{y[x] -> ──────────────}}
               b    r x
         -1 + - + E
               a
```

Fine and dandy. But sometime you might forget to put in the double equal signs. This might be what you actually typed and executed:

In[19]:=

```
Clear[r,a,b,x,y]
DSolve[{y'[x] = r y[x] (1 - y[x]/b), y[0] = a},y[x],x]
```

```
DSolve::deqn:
                                 y[x]
   Element r y[x] (1 - ────)
                                  b
     in the equation list is not an equation.
```

Out[19]=

```
                                y[x]
  DSolve[{r y[x] (1 - ────), a}, y[x], x]
                                 b
```

This drives *Mathematica* nuts. You look at what was typed and edit it to look correct and then execute:

In[20]:=

```
Clear[r,a,b,x,y]
DSolve[{y'[x] == r y[x] (1 - y[x]/b), y[0] == a},y[x],x]
```

```
DSolve::deqn:
   Element True in the equation list is not an equation.
```

Out[20]=

```
DSolve[{True, y[0] == a}, y[x], x]
```

It still fails.

> How do you fix this?

Answer: The confusion stems from when you typed and executed:

In[21]:=

```
y'[x] = r y[x] (1 - y[x]/b)
```

Out[21]=

```
                  y[x]
  r y[x] (1 - ────)
                   b
```

When you cleared y and x, *Mathematica* did not clear this formula for $y'[x]$. Check it out:

In[22]:=

```
Clear[y,x]; y'[x]
```

Out[22]=

```
                  y[x]
  r y[x] (1 - ────)
                   b
```

To clear this formula for $y'[x]$, you must clear *Mathematica*'s memory of specified derivatives. Try it:

In[23]:=

```
Clear[r,a,b,x,y,Derivative]
DSolve[{y'[x] == r y[x] (1 - y[x]/b), y[0] == a},y[x],x]
```

Out[23]=

```
                 r x
              b E
{{y[x] -> ──────────────}}
              b     r x
         -1 + - + E
              a
```

Now it works fine.

Give It a Try

Experience with the starred ($\star$) problems will be especially beneficial for understanding later lessons.

■ G.1) Quick calculations

G.1.a) When you start with $y'[x] = r\,y[x]$, give it one data point, and solve for $y[x]$, you always get something of the form $y[x] = k\,e^{rx}$ where k is a constant determined by the data point.

Try it with the data point $y[c] = d$ where c and d are constants:

In[1]:=

```
Clear[x,y,r,c,d]; DSolve[{y'[x] == r y[x],y[c] == d},y[x],x]
```

Out[1]=

```
                r (-c + x)
{{y[x] -> d E           }}
```

This gives you $y[x] = d\,e^{-rc+rx}$.

How do you reconcile this with the formula $y[x] = k\,e^{rx}$ above?

G.1.b) Find a formula for $y[x]$ if it is known that the instantaneous growth rate $y'[x]$ is proportional to $y[x]$ and it is known that $y[0] = 5$ and $y[5.7] = 10$.

Does the value of $y[x]$ double every 5.7 years?

G.1.c) When a condenser discharges electricity, the instantaneous rate of change of the voltage is proportional to the voltage in the condenser.

> Suppose you have a discharging condenser and the instantaneous rate of change of the voltage is one hundredth of the voltage (in volts per second). How many seconds does it take for the voltage to decrease by 90%?

G.1.d)

> A radium sample weighs 1 gram at time $t = 0$. At time $t = 10$ (years) it has diminished to 0.997 grams. How long will it take to diminish to 0.5 grams?

G.1.e) From page 1 of the *New York Times*, August 4, 1988: "Tritium is the basic fuel of hydrogen bombs and is used to increase the power for fission bombs. . . . In a basic atom bomb (or reactor), plutonium atoms are split, or fissioned, to release energy, but the fission can be promoted with a small amount of tritium because it has two extra atom-splitting neutrons. Plutonium is a relatively stable material, and its natural decay is not a major factor in bomb maintenance. Tritium, however, decays at a rate of 5.5 percent a year."

> What is the half life of tritium? Why is this an issue?

■ G.2) Data analysis*

G.2.a) Here are the United States population data $\{t, P[t]\}$ where t is the year since 1900 and $P[t]$ is the official census figure in millions for the corresponding year t. Here is a plot:

In[2]:=
```
data =
{{0,76.2},{10,92.2},{20,106},{30,123.2},
{40,132.2},{50,151.3},{60,179.3},
{70,203.3},{80,226.5},{90,248.7}};
dataplot = ListPlot[data,
AxesLabel->{"year","population"},
PlotStyle->{Red,PointSize[0.03]}];
```

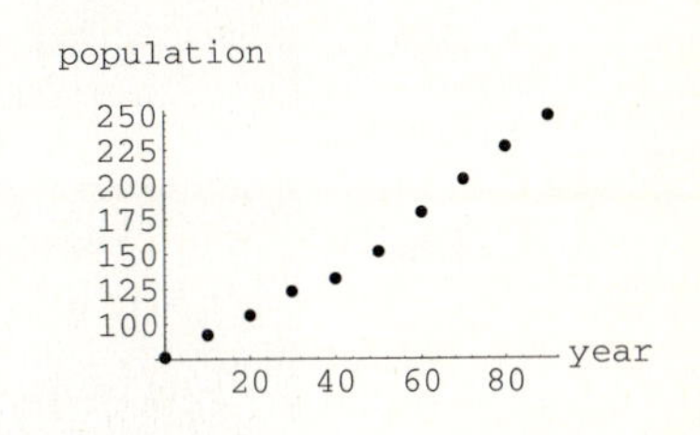

Here are the same data on semi-log paper:

In[3]:=
```
Clear[k]
logdata = Table[{data[[k,1]],
N[Log[data[[k,2]]]]},{k,1,Length[data]}];
ListPlot[logdata,
PlotStyle->{Red,PointSize[0.03]},
PlotLabel->"Semi-log paper plot"];
```

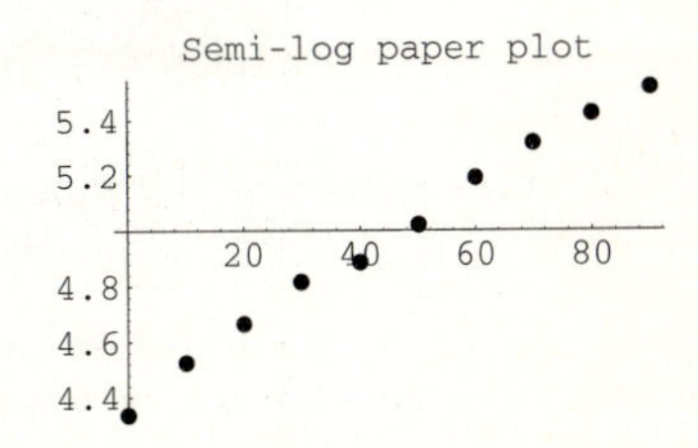

This shows that the later population figures lag behind exponential growth. Just for the heck of it, here is an exponential fit of the data:

In[4]:=

```
Clear[fitter,t]; fitter[t_] = E^Fit[logdata,{1,t},t]
```

Out[4]=

```
 4.38297 + 0.0130027 t
E
```

Try it out:

In[5]:=

```
fitplot = Plot[fitter[t],{t,0,90},
PlotStyle->{{Blue,Thickness[0.01]}},
DisplayFunction->Identity];
Show[dataplot,fitplot,
DisplayFunction->$DisplayFunction];
```

population

250
225
200
175
150
125
100

year

20 40 60 80

Pretty good fit. But take a look at what this exponential function predicts for the years 1900 to 1900 + 300 = 2200:

In[6]:=

```
fitplot = Plot[fitter[t],{t,0,300},
PlotStyle->{{Blue,Thickness[0.01]}},
DisplayFunction->Identity];
Show[dataplot,fitplot,PlotRange->All,
DisplayFunction->$DisplayFunction];
```

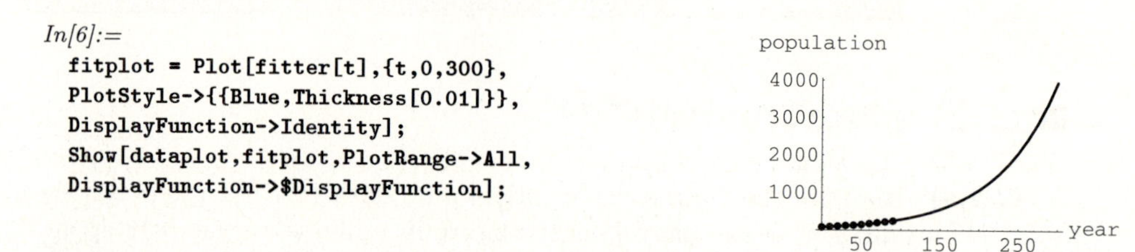

G.2.a.i) Do you think that this plot squares with reality?

G.2.a.ii) Use this exponential fit to get a good fit of the data with the solution of a logistic differential equation.

What does your logistic model predict the limiting population of the United States to be?

When does your logistic model predict that the U.S. population will settle into its limiting value?

G.2.b.i) For the most part, the cost of mailing a 1-ounce letter first class within the United States has increased steadily over time.

The following table gives figures in the form $\{t, \text{cost}[t]\}$ where t is the number of years between 1958 and $\text{cost}[t]$ is the cost in cents of mailing a 1-ounce letter first class within the United States at year $t + 1958$. This list is accurate through 1990.

In[7]:=
```
data = {{0, 4},{5, 5},{10, 6},{13, 8},{16, 10},
{17, 13},{20, 15},{23, 18},{24, 20},{27, 25}};
```

Take a look at the plot of these data:

In[8]:=
```
dataplot =
ListPlot[data,
PlotStyle->{Violet,PointSize[0.03]},
AxesLabel->{"Years\n since\n 1958","postcosts"}];
```

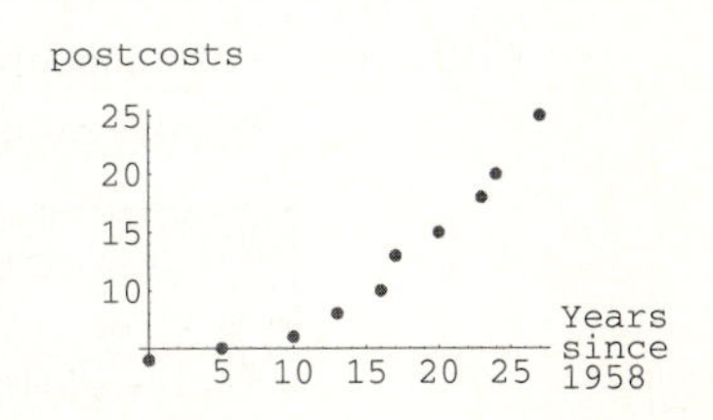

This looks approximately exponential. To get a better look, plot these data on log paper:

In[9]:=
```
Clear[j]; logdata =
Table[{data[[j,1]],Log[data[[j,2]]]},
{j,1,Length[data]}];
logdataplot =
ListPlot[logdata,
PlotStyle->{Violet,PointSize[0.02]}];
```

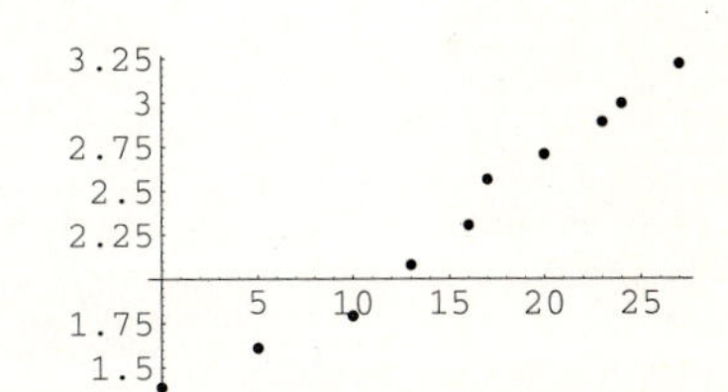

> Fit the data with a solution of the exponential differential equation $y'[t] = r\,y[t]$ or a solution of the logistic differential equation $y'[t] = r\,y[t]\,(1 - y[t]/b)$ as you judge appropriate.
>
> At what year does your model predict that the cost of mailing a 1-ounce first class letter will reach one dollar?

G.2.b.ii) In 1991, the cost of mailing a 1-ounce letter first class within the United States rose to 29 cents.

> How well does the function you got above predict this?
>
> Interpret the results.

G.2.c) In his book *Experimental Ecology in the Feeding of Fish*, V.S. Ivlev (English translation from the Russian by D. Scott, Yale, 1961) looks at the question of how much food fish eat in terms of the available food supply.

For carp eating denatured minnow roe, Ivlev made the following measurements $\{t, y[t]\}$ with t in mg/cm^2 and $y[t]$ in mg:

In[10]:=
```
data = {{1, 96.8},{2, 150.0},{3, 203.1},{4, 229.3},{5, 254.1},{6, 265.4},
{7, 264.8},{8, 281.9},{9, 292.0},{10, 291.3}};
```

where

t = concentration of food supply in weight of food per unit area of feeding and

$y[t]$ = average weight (in mg) of food eaten per fish per unit of time.

G.2.c.i) Ivlev measured that the fish would eat 296 mg of the food per unit of time when the food supply is much larger than the needs of the fish.

> How does this information reveal that an exponential model of these data is not appropriate? How does this information leave the door to a logistic model wide open?

G.2.c.ii)

> Determine whether these data are well modeled by a solution of the logistic equation
>
> $$y'[t] = r\, y[t]\left(1 - \frac{y[t]}{b}\right).$$
>
> Why does the information in part G.2.c.i) tell you that $b = 296$ is a good idea?

■ G.3) Logistic growth versus exponential growth★

Here is a formula for the exponential solution of $y'[x] = r\, y[x]$ with $y[0] = 2$ and $r = 0.6$.

In[11]:=

```
r = 0.6; starter = 2; Clear[y,x];
solution = DSolve[{y'[x] == r y[x], y[0] == starter},y[x],x];
Clear[yexpon]; yexpon[x_] = y[x]/.solution[[1]]
```

Out[11]=

```
   0.6 x
2 E
```

And a plot for $0 \leq x \leq 7$:

In[12]:=

```
left = 0; right = 7;
exponplot =
Plot[yexpon[x],{x,left,right},
PlotStyle->{{Red,Thickness[0.02]}},
AxesLabel->{"x","yexpon[x]"}];
```

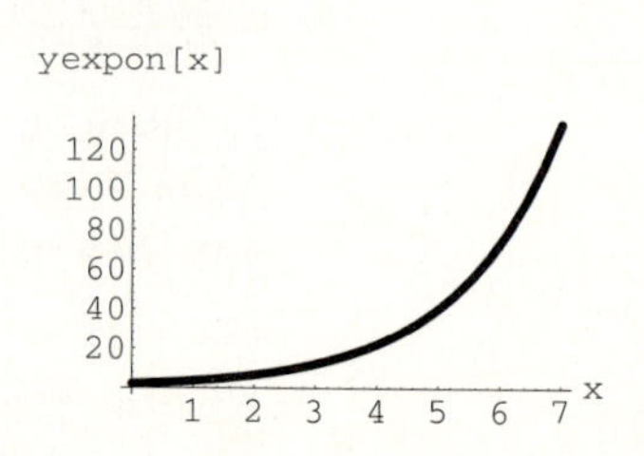

Here is a formula for the logistic solution of $y'[x] = r\, y[x]\,(1 - y[x]/b)$ with $y[0] = 2$ for $r = 0.6$, the same data as used above.

In[13]:=

```
r = 0.6; starter = 2; Clear[y,x,b];
solution = DSolve[{y'[x] == r y[x](1 - y[x]/b),y[0] == starter},y[x],x];
Clear[ylogistic]; ylogistic[x_,b_] = y[x]/.solution[[1]]
```

Out[13]=

```
     0.6 x
  b E
-------------
     b    0.6 x
-1 + - + E
     2
```

Here is a plot of what happens to ylogistic$[x, b]$ when you take $b = 10$ shown together with the exponential plotted above:

In[14]:=

```
b = 10;
logisticplot =
Plot[ylogistic[x,b],{x,left,right},
PlotStyle->Blue,DisplayFunction->Identity];
Show[exponplot,logisticplot,
PlotRange->{0,yexpon[right]},AxesLabel->{"x",""},
DisplayFunction->$DisplayFunction,
PlotLabel->b "= b"];
```

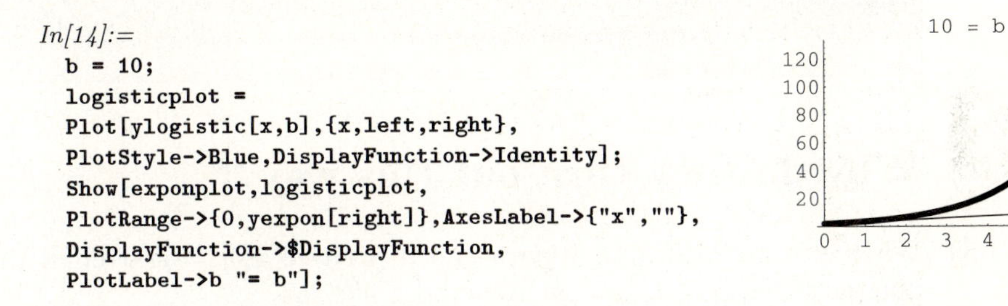

Here's what happens as you increase b:

In[15]:=

```
b = 100; logisticplot =
Plot[ylogistic[x,b],{x,left,right},
PlotStyle->Blue,DisplayFunction->Identity];
Show[exponplot,logisticplot,
PlotRange->{0,yexpon[right]},AxesLabel->{"x",""},
DisplayFunction->$DisplayFunction,
PlotLabel->b "= b"];
```

In[16]:=

```
b = 500;  logisticplot =
Plot[ylogistic[x,b],{x,left,right},
PlotStyle->Blue,DisplayFunction->Identity];
Show[exponplot,logisticplot,
PlotRange->{0,yexpon[right]},AxesLabel->{"x",""},
DisplayFunction->$DisplayFunction,
PlotLabel->b "= b"];
```

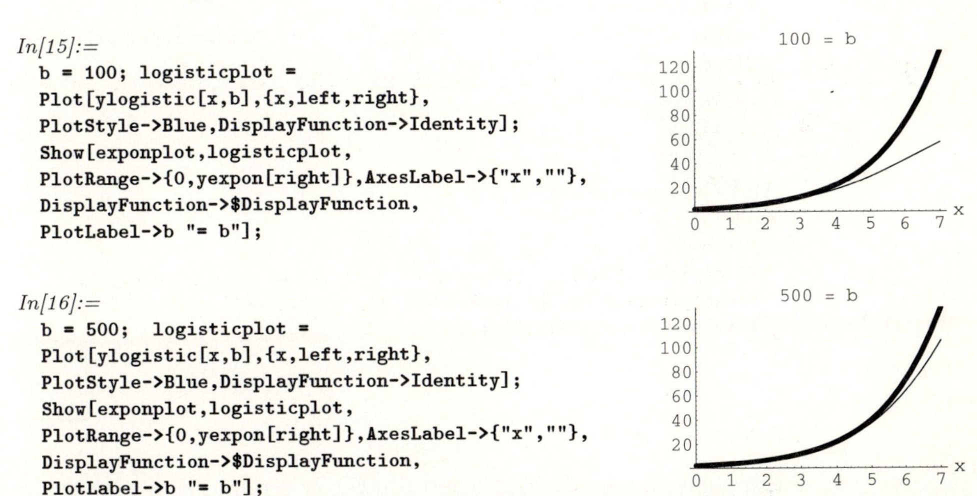

G.3.a) Come up with a number b so that the logistic plot shares ink with the exponential plot all the way from $x = 0$ to $x = 7$.

G.3.b) Still going with the same values of r and starter as used in the plots above, look at the differential equations $y'[x] = r\,y[x]$ with $y'[0] =$ starter and the solution of $y'[x] = r\,y[x]\,(1 - y[x]/b)$ with $y'[0]$ = starter with $0 <$ starter $< b$ and explain the statement:

No matter what positive b you go with, the logistic solution of the second differential equation has no choice but to run below the exponential solution of the first differential equation.

G.3.c) Discuss the statement:

No matter what numbers r and starter you go with, you can make the solution of $y'[x] = r\,y[x]$ with $y'[0]$ = starter and the solution of $y'[x] = r\,y[x]\,(1 - y[x]/b)$ with $y'[0]$ = starter plot out nearly the same by making b big enough.

■ G.4) Why do they turn out this way?★

G.4.a) When you go with a positive r and a positive $y[0]$ = starter in the exponential differential equation, you get something like this:

In[17]:=

```
r = 1.7; starter = 5; Clear[y,x];
solution = DSolve[{y'[x] == r y[x], y[0] == starter},y[x],x];
y[x_] = y[x]/.solution[[1]]
```

Out[17]=

```
   1.7 x
5 E
```

And the corresponding plot:

In[18]:=

```
Plot[y[x],{x,0,4},
PlotStyle->{{Blue,Thickness[0.01]}},
AxesLabel->{"x","y[x]"}];
```

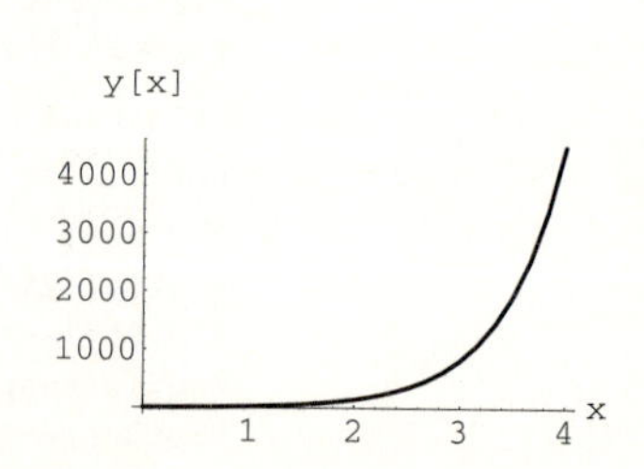

Exponential growth. But when you throw a minus sign against the same r and you keep everything else the same, you get:

In[19]:=

```
r = -1.7;  Clear[y,x];
solution = DSolve[{y'[x] == r y[x], y[0] == starter},y[x],x];
y[x_] = y[x]/.solution[[1]]
```

Out[19]=

```
  5
-----
 1.7 x
E
```

And the corresponding plot:

In[20]:=
```
Plot[y[x],{x,0,4},
PlotStyle->{{Blue,Thickness[0.01]}},
AxesLabel->{"x","y[x]"}];
```

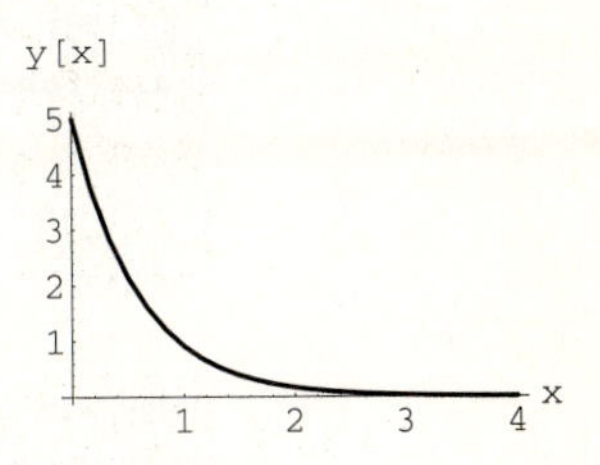

Exponential decay. All you did was change the sign of r.

Play with some more like these.

> Why do they turn out this way?

G.4.b) When you go with a positive r and a positive $y[0]$ = starter in the exponential differential equation, you get something like this:

In[21]:=
```
r = 2.5; starter = 4; Clear[y,x];
solution = DSolve[{y'[x] == r y[x],y[0] == starter},y[x],x];
y[x_] = y[x]/.solution[[1]]
```

Out[21]=
```
   2.5 x
4 E
```

And the corresponding plot:

In[22]:=
```
Plot[y[x],{x,0,4},
PlotStyle->{{Blue,Thickness[0.01]}},
AxesLabel->{"x","y[x]"}];
```

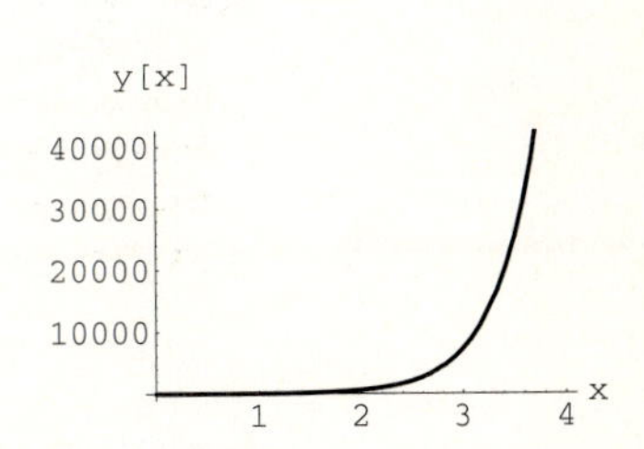

Exponential growth. But when you throw a minus sign against the same $y[0]$ = starter and you keep everything else the same, you get:

In[23]:=
```
starter = -4; Clear[y,x];
solution = DSolve[{y'[x] == r y[x],y[0] == starter},y[x],x];
y[x_] = y[x]/.solution[[1]]
```

Out[23]=
```
    2.5 x
-4 E
```

And the corresponding plot:

In[24]:=
```
Plot[y[x],{x,0,4},
PlotStyle->{{Blue,Thickness[0.01]}},
AxesLabel->{"x","y[x]"}];
```

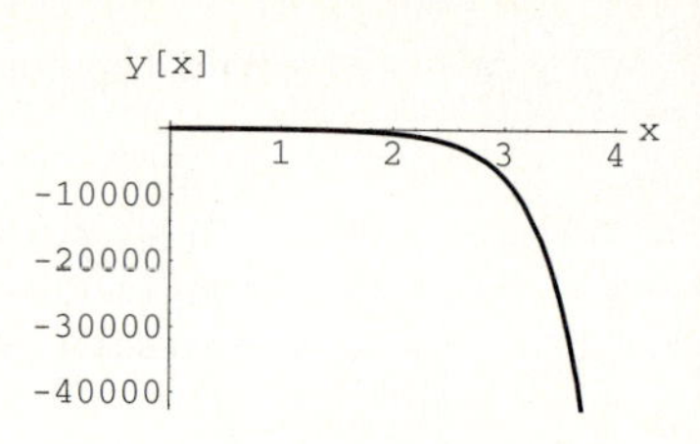

Exponential death. All you did was change the sign of $y[0]$ = starter. Play with some more like these.

> Why do they turn out this way?

G.4.c) When you go with a positive r and b in the logistic differential equation, and you put $y[0]$ = starter with $0 \leq$ starter $< b$, you get something like this:

In[25]:=
```
r = 1.7; b = 36; starter = b/8; Clear[y,x];
solution = DSolve[{y'[x] == r y[x] ( 1 - y[x]/b),y[0] == starter},y[x],x];
y[x_] = y[x]/.solution[[1]]
```

Out[25]=
```
          1.7 x
         E
---------------------------
                       1.7 x
0.194444 + 0.0277778 E
```

And the corresponding plot:

In[26]:=
```
Plot[{y[x],b},{x,0,4},
PlotStyle->{{Blue,Thickness[0.01]},Red},
AxesLabel->{"x","y[x]"}];
```

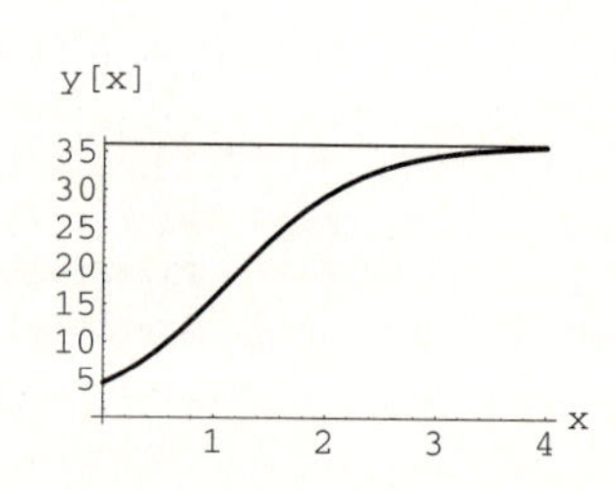

Logistic growth. But when you make $y[0]$ = starter $> b$ and keep everything else the same, then you get:

In[27]:=
```
starter = 2 b; Clear[y,x];
solution = DSolve[{y'[x] == r y[x] ( 1 - y[x]/b),y[0] == starter},y[x],x];
y[x_] = y[x]/.solution[[1]]
```

Out[27]=
```
           1.7 x
          E
----------------------------
                        1.7 x
-0.0138889 + 0.0277778 E
```

And the corresponding plot:

In[28]:=
```
Plot[{y[x],b},{x,0,4},
PlotStyle->{{Blue,Thickness[0.01]},Red},
AxesLabel->{"x","y[x]"}];
```

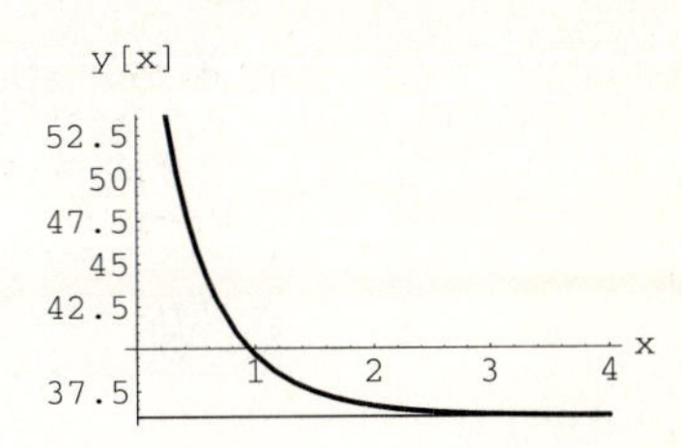

Logistic decay. Play with some more like these.

> Why do they turn out this way?

G.4.d) Here is a formula for the exponential solution of $y'[x] = r\,y[x]$ with $y[0] =$ starter for $r = 0.4$ and starter $= 2$:

In[29]:=
```
r = 0.4; starter = 2; Clear[y,x];
solution = DSolve[{y'[x] == r y[x],y[0] == starter},y[x],x];
Clear[yr]; yr[x_] = y[x]/.solution[[1]]
```

Out[29]=
```
  0.4 x
2 E
```

Here is a formula for the exponential solution of $y'[x] = r\,y[x]$ with $y[0] =$ starter with r increased from 0.4 to 0.6 and everything else the same:

In[30]:=
```
r = 0.6;  Clear[y,x];
solution = DSolve[{y'[x] == r y[x],y[0] == starter},y[x],x];
Clear[ynegr];  ybigr[x_] = y[x]/.solution[[1]]
```

Out[30]=
```
  0.6 x
2 E
```

And a plot of both for $0 \leq x \leq 10$:

In[31]:=
```
left = 0; right = 10;
exponplot =
Plot[{yr[x],ybigr[x]},{x,left,right},
PlotStyle->{{Blue,Thickness[0.01]},
{Red,Thickness[0.01]}},
AxesLabel->{"x",""}];
```

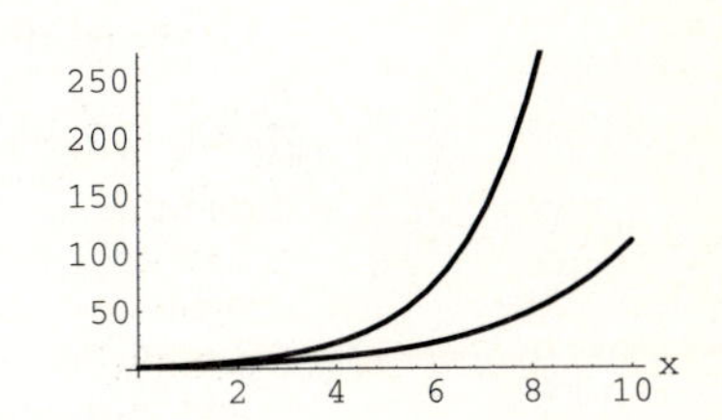

Play with some more like these. And look at this statement:

If you go with $r > s$ and a positive starter, then as x advances from 0 to ∞, the solution of

$$y'[x] = r\,y[x] \qquad \text{with } y[0] = \text{starter}$$

plots out above the solution of

$$y'[x] = s\,y[x] \qquad \text{with } y[0] = \text{starter}.$$

Why do they turn out this way?

G.4.e) Here is a formula for the exponential solution of $y'[x] = r\,y[x]$ with $y[0] = 2$ for $r = 0.6$:

In[32]:=
```
r = 0.6; starter = 2; Clear[y,x];
solution = DSolve[{y'[x] == r y[x],y[0] == starter},y[x],x];
Clear[yexpon]; yexpon[x_] = y[x]/.solution[[1]]
```

Out[32]=
```
   0.6 x
2 E
```

Here is a formula for the solution of $y'[x] = r\,y[x] + b$ with $y[0] = 2$ for $r = 0.6$; the same data as used above:

In[33]:=
```
Clear[y,x,b]
solution = DSolve[{y'[x] == r y[x] + b,y[0] == starter},y[x],x];
Clear[yother]
yother[x_,b_] = y[x]/.solution[[1]]
```

Out[33]=
```
                                     0.6 x
-1.66667 b + (2. + 1.66667 b) E
```

Here is a plot of what happens to yother$[x, b]$ and yexpon$[x]$ when you take $b = 0.8$.

In[34]:=
```
b = 0.8;
left = 0;
right = 6;
Plot[{yexpon[x],yother[x,b]},
{x,left,right},
PlotStyle->{{Red,Thickness[0.02]},{Blue}},
PlotLabel->b "= b"];
```

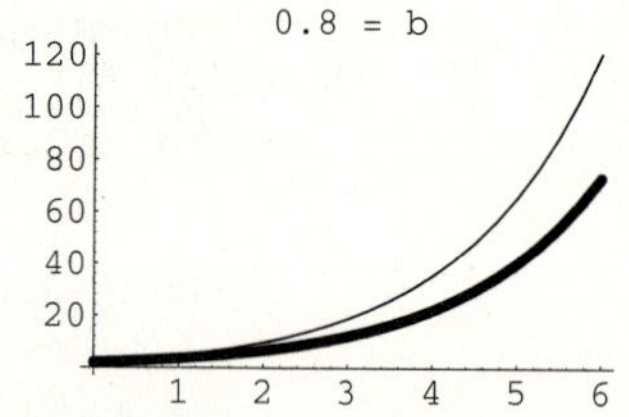

Here's what happens when you change b from 0.8 to -0.8 but keep everything else the same:

In[35]:=
```
b = -0.8;
left = 0;
right = 6;
Plot[{yexpon[x],yother[x,b]},
{x,left,right},
PlotStyle->{{Red,Thickness[0.02]},{Blue}},
PlotLabel->b "= b"];
```

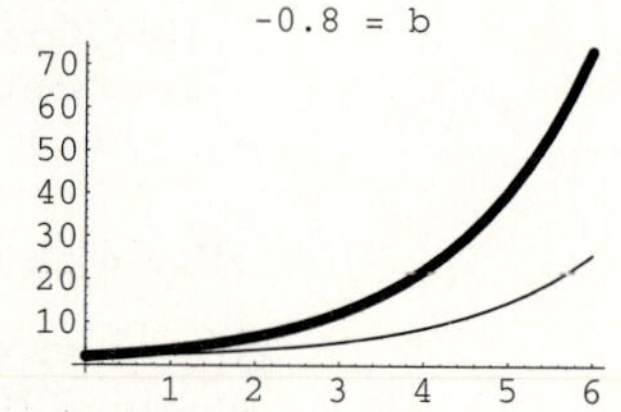

Experiment with other positive and negative values of b.

Now look at the statements:

→ Take a positive b. If you set r and starter, then solutions of $y'[x] = r\,y[x] + b$ with $y[0] =$ starter will always plot out on $[0, \infty)$ above solutions of $y'[x] = r\,y[x]$ with $y[0] =$ starter.

→ Take a negative b. If you set r and starter, then solutions of $y'[x] = r\,y[x] + b$ with $y[0] =$ starter will always plot out on $[0, \infty)$ below solutions of $y'[x] = r\,y[x]$ with $y[0] =$ starter.

Why do they turn out this way?

■ G.5) Other differential equations★

G.5.a) Here's *Mathematica*'s formula for a solution of the differential equation

$$y'[x] = \left(\frac{r}{x}\right) y[x]$$

with $y[1] = a$.

In[36]:=

```
Clear[y,x,a,r,Derivative]
DSolve[{y'[x] == (r/x) y[x], y[1] == a},y[x],x]
```

Out[36]=

```
                r
{{y[x] -> a x }}
```

Use the derivative formulas you know to explain the *Mathematica* output.

G.5.b) Here's *Mathematica*'s formula for a solution of the differential equation

$$y'[x] = \left(\frac{r}{x^2}\right) y[x]$$

with $y[1] = a$.

In[37]:=

```
Clear[y,x,a,r,Derivative]
DSolve[{y'[x] == (r/x^2) y[x], y[1] == a},y[x],x]
```

Out[37]=

```
                r - r/x
{{y[x] -> a E        }}
```

Use the derivative formulas you know to explain the *Mathematica* output.

G.5.c) Here's *Mathematica*'s formula for a solution of the differential equation

$$y'[x] = \sin[x]\, y[x]$$

with $y[0] = a$.

In[38]:=

```
Clear[y,x,a,r,Derivative]
DSolve[{y'[x] == Sin[x] y[x], y[0] == a},y[x],x]
```

Out[38]=

```
                          2
                2 Sin[x/2]
{{y[x] -> a E              }}
```

Use the derivative formulas you know to explain the *Mathematica* output.

G.5.d.i) Given a number r, to solve the differential equation $y'[x] = r$, you want a function $y[x]$ with the property that $y'[x] = r$ for all x's. Here's how *Mathematica* handles this differential equation:

In[39]:=

```
Clear[x,y,r]; DSolve[y'[x] == r,y[x],x]
```

Out[39]=

```
{{y[x] -> r x + C[1]}}
```

Here $C[1]$ can be any constant.

Explain why *Mathematica*'s formula is on target.

G.5.d.ii) Bubba is on a vacation in Virginia Beach. After a big night of partying, he sleeps in until 3:00 P.M., wakes up, and decides to moderate his beer consumption for the day. His plan for the day is to consume two 12-ounce light beers gradually each hour starting at 4:00 P.M. and calling it quits at 10:00 P.M.

Each light beer contains 11.3 grams of alcohol, and Bubba's liver eliminates 12 grams of alcohol per hour. From last night's festivities, Bubba's body fluids contain 5 grams of alcohol at 4:00 P.M.

Measuring t in hours with $t = 0$ corresponding to 4:00 P.M. and putting $y[t] =$ grams of alcohol in Bubba's body fluids for $0 \leq t \leq 6$, explain why $y[t]$ solves the differential equation $y'[t] = 22.6 - 12$ with $y[0] = 5$.

Get a formula for $y[t]$ and use it to plot the projected alcohol in Bubba's system for $0 \leq t \leq 6$.

■ G.6) Managing your money

G.6.a) You graduate and get one of those high-paying yuppie jobs everyone hopes to get. To announce your success, you decide to buy that new red BMW 325i convertible you've always had your heart set on. You go to the bank to talk about financing a \$25,000 loan for the BMW.

Here are the details:

→ The interest on the unpaid part of the loan is to be compounded every instant.

→ The payoff is to be continuous in the sense that a number d (the yearly rate) is to be set and then after t (whole or fractional) years, you are to pay $t\,d$ dollars. Thus after the first day you will have paid $\$d/365$, after the first month of the year you will have paid $\$d/12$, and after two years, you will have paid $\$2\,d$ dollars. The bank can transfer money from your account to their coffers every time they run their computers.

You can have the following options:

2-year loan at 4.9% per year interest compounded every instant,

3-year loan at 6.9% per year interest compounded every instant,

or

5-year loan at 10.9% per year interest compounded every instant.

> How much money $\$d$ would you have to pay each year under each of the options given? What is your total payout in each case?

G.6.b.i) Your Aunt Matilda passes away and leaves you with the tidy sum of \$50,000. You find a bank that will pay you 5.15% interest per year compounded every instant. You want to enjoy some of this money, but you don't want to blow it away too fast.

You decide to deposit the \$50,000 but to pull out r dollars per year so that after t years, you will have blown $r\,t$ dollars.

> If $P[t]$ is the amount in the account, then why does $P[t]$ solve the differential equation $P'[t] = 0.0515P[t] - r$?

G.6.b.ii) Continue with the setup in part G.6.b.i) above. After some thought, you decide:

→ r should be as big as possible.

→ But the balance in the account should never dip below your original \$50,000 inheritance.

> Estimate your r.

G.6.b.iii) Continue with the setup in part G.6.b.i) above.

After some more thought, you realize that you are young and can really have some fun with this bankroll. So you decide:

→ r should be as big as possible.

→ But between now and 10 years from now, the account should never dip below \$30,000.

Estimate your r.

■ G.7) Which animals grow faster after their birth than they are growing at the time of their birth?

This problem appears only in the electronic version.

■ G.8) Newton's law of cooling: How a differential equation can help you enjoy your favorite cooled beverage

You put a tepid liquid refreshment into a very large refrigerator. The refrigerator is kept at a constant temperature S degrees. Let $T[t]$ be the temperature of the liquid at time t minutes after the beverage is placed in the refrigerator. According to Newton's law of cooling, the rate of cooling $T'[t]$ is proportional to the difference $(T[t] - S)$; i.e.,

$$T'[t] = r\,(T[t] - S) = r\,T[t] - r\,S$$

where r is a proportionality constant.

G.8.a) The following questions may be of some practical importance on a hot July afternoon.

Suppose the refrigerator temperature S is held at 42 degrees (Fahrenheit), the original temperature of a bottle of a desirable beverage is 80 degrees, and the bottle of the beverage has cooled to 63 degrees after 10 minutes in the refrigerator.

Plot the temperature of the bottle of the beverage as a function of time (in minutes).

What will be the temperature of the bottle of the beverage after 30 minutes?

Approximately how many minutes will it take for the bottle of the beverage to cool to the refreshing temperature of 44 degrees?

■ G.9) Pressure altimeters

Calculus&*Mathematica* thanks State Climatologist Wayne M. Wendland of the State of Illinois Water Survey for his help in the preparation of this problem.

The first goal of this problem is to come up with a formula for atmospheric pressure $P[x]$ as a function of altitude x above sea level. The strategy is to find a differential equation that $P[x]$ satisfies, then make some measurements to set the constants, and finally come up with a complete formula for $P[x]$. This strategy is very common in making scientific measurements.

Imagine a 1-inch by 1-inch square marked out on the earth at sea level. Now imagine lifting this square perpendicularly to the earth's surface to form a square column of air.

The weight, P_0, of this column of air is the atmospheric pressure (in pounds per square inch) at sea level. P_0 is usually about 14.5 pounds per square inch, but this figure can vary with the weather and location.

Accordingly, if $w[x]$ is the weight of the air in the first x feet of the square column of air, then the atmospheric pressure $P[x]$ at altitude x above sea level is given by

$$P[x] = P_0 - w[x].$$

G.9.a.i) Agree that $d[x]$ stands for the weight of 1 cubic inch of air at altitude x feet above sea level.

Explain why $d[x]$ decreases as x increases and then explain why

$$d[x+h]\,h \le w[x+h] - w[x] \le d[x]\,h$$

for any positive h.

Next explain why

$$d[x+h] \le \frac{w[x+h] - w[x]}{h} \le d[x]$$

for any positive h.

G.9.a.ii) Use the last inequality to explain why $w'[x] = d[x]$.

G.9.a.iii) Take the derivatives of both sides of the equality $P[x] = P_0 - w[x]$ with respect to x and then explain why $P'[x] = -d[x]$.

G.9.a.iv) Boyle's Ideal Gas Law says that $d[x]$ is proportional to $P[x]$.

Use Boyle's Gas Law and the result of part G.9.a.iii) above to explain why $P'[x]$ is proportional to $P[x]$.

G.9.a.v) Explain the formula

$$P[x] = P_0 \, e^{-rx}$$

for some constant r and x in feet.

G.9.b.i) At 100 feet above sea level, the pressure $P[100]$ was recorded at 14.5 pounds per square inch. Then at a height of 6600 feet, the pressure $P[6600]$ was recorded at 12.3 pounds per square inch.

Use these readings to determine the values of r and P_0 in the formula

$$P[x] = P_0 \, e^{-rx}$$

for x in feet.

G.9.b.ii) Use the values of P_0 and r that you calculated in part G.9.b.i) to give the pressures at

(a) altitude $x = 0$ feet

(b) altitude $x = 10{,}000$ feet

(c) altitude $x = 25{,}000$ feet.

G.9.c) Each FAA-approved aircraft is equipped with a pressure altimeter. The pressure altimeter measures pressure and receives from the local weather station the information to set what amounts to the local values of P_0 and r.

To demonstrate how a pressure altimeter works, use the values of P_0 and r that you calculated in part G.9.b.i) to give the altitude x if

(i) the pressure is 11.8 pounds per square inch,

(ii) the pressure is 12.8 pounds per square inch.

LESSON 1.07

The Race Track Principle

Basics

■ B.1) The Race Track Principle

You've been using implicit versions of the Race Track Principle all along your Calculus&*Mathematica* odyssey. This problem gives you a chance to toss around this basic calculus idea.

B.1.a) One version of the Race Track Principle:

Horses: If two horses start a race at the same point, then the faster horse is always ahead.

Functions: If $f[a] = g[a]$ and $f'[x] \geq g'[x]$ for $x \geq a$, then $f[x] \geq g[x]$ for $x \geq a$.

This version of the Race Track Principle is good for explaining why one function plots out above another function.

B.1.a.i) Here is a plot of $f[x] = x$ and $g[x] = \sin[x]$ for $0 \leq x \leq 3$.

In[1]:=

```
Clear[f,g,x]
f[x_] = x;  g[x_] = Sin[x];
Plot[{f[x],g[x]},{x,0,3},
PlotStyle->{{Red,Thickness[0.01]},
Thickness[0.01]},AxesLabel->{"x",""}];
```

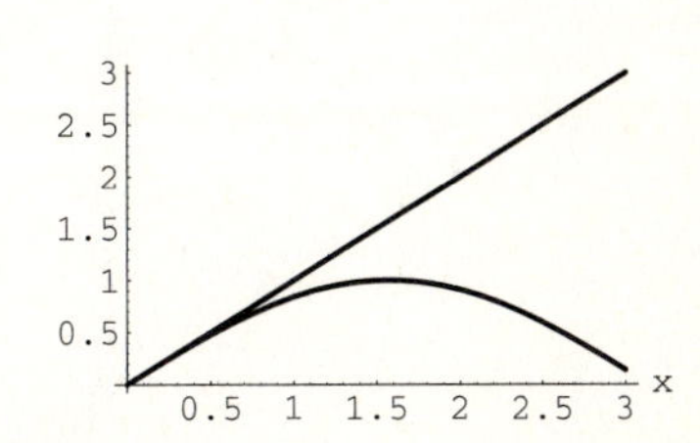

That's $f[x] = x$ sailing high above $g[x] = \sin[x]$.

> Use the Race Track Principle to explain why the plot turned out this way.

Answer: Look at $f[0]$ and $g[0]$:

In[2]:=
```
{f[0],g[0]}
```
Out[2]=
```
{0, 0}
```

The two functions start their race at the same point. Next look at $f'[x]$ and $g'[x]$ for $x \geq 0$:

In[3]:=
```
{f'[x],g'[x]}
```
Out[3]=
```
{1, Cos[x]}
```

Because $\cos[x]$ spends its miserable life oscillating between -1 and 1, you know that $1 \geq \cos[x]$ no matter what x is. Consequently, $f'[x] = 1 \geq \cos[x] = g'[x]$ for $x \geq 0$. In other words, $f[x] = x$ grows faster than $g[x] = \sin[x]$. The Race Track Principle tells you that $f[x] \geq g[x]$ for $x \geq 0$. And this explains the plot.

B.1.a.ii) Here's a plot of $f[x] = (1+x)^{3/2}$ and $g[x] = 1 + (3/2)\,x$ for $0 \leq x \leq 24$.

In[4]:=
```
Clear[f,g,x]
f[x_] = (1 + x)^(3/2);
g[x_] = 1 + (3/2) x;
Plot[{f[x],g[x]},{x,0,24},
PlotStyle->{{Red,Thickness[0.01]},
Thickness[0.01]},AxesLabel->{"x",""}];
```

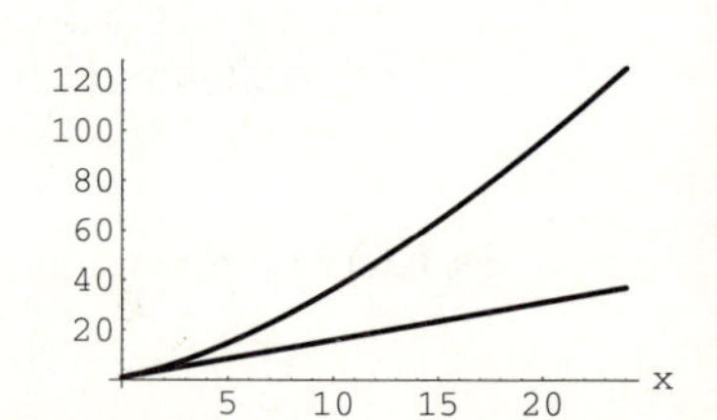

That's $f[x] = (1+x)^{3/2}$ sailing high above $g[x] = 1 + (3/2)\,x$.

> Use the Race Track Principle to explain why the plot turned out this way.

Answer: Look at $f[0]$ and $g[0]$:

In[5]:=
```
{f[0],g[0]}
```
Out[5]=
```
{1, 1}
```

The two functions start their race at the same point. Next look at $f'[x]$ and $g'[x]$ for $x \geq 0$:

In[6]:=
```
{f'[x],g'[x]}
```

Out[6]=
```
 3 Sqrt[1 + x]  3
{-------------, -}
       2        2
```

When $x \geq 0, \sqrt{1+x} \geq \sqrt{1+0} = 1$. Consequently, $f'[x] \geq g'[x]$ for $x \geq 0$. In other words, $f[x]$ grows faster than $g[x]$. The Race Track Principle tells you that $f[x] \geq g[x]$ for $x \geq 0$. And this explains the plot.

B.1.b) Another version of the Race Track Principle:

Horses: If two horses start at the same point and they run at nearly the same speed for the whole race, then they run very close together all the way and are likely to end the race in a photo finish.

Functions: If $f[a] = g[a]$ and $f'[x]$ is nearly the same as $g'[x]$ for $a \leq x \leq b$, then $f[x]$ is nearly the same as $g[x]$ for $a \leq x \leq b$.

This version of the Race Track Principle is good for explaining why one function plots nearly the same as another function. Take:

In[7]:=
```
Clear[f,g,x]; f[x_] = Sin[x]; g[x_] = x Cos[x]^(1/3);
```

Both functions start their race at $x = 0$ together:

In[8]:=
```
{f[0],g[0]}
```

Out[8]=
```
{0, 0}
```

Here's a plot of the instantaneous growth rates of both functions for $0 \leq x \leq 1.5$:

In[9]:=
```
growthplot =
Plot[{f'[x],g'[x]},{x,0,1.5},
PlotStyle->
{{Red,Thickness[0.01]},Thickness[0.01]},
AxesLabel->{"x",""},
PlotLabel->"f'[x] and g'[x]"];
```


> Use this plot and the Race Track Principle to predict a number b so that $f[x]$ and $g[x]$ are nearly the same for $0 \leq x \leq b$. Check yourself with a plot.

Answer: The two functions start their race with $f[0] = g[0]$ and the plot shows that the growth rates of $f[x]$ and $g[x]$ are nearly the same for $0 \leq x \leq 0.8$. The Race

Track Principle tells you that $f[x]$ and $g[x]$ are nearly the same for $0 \leq x \leq b = 0.8$. Confirm with a plot:

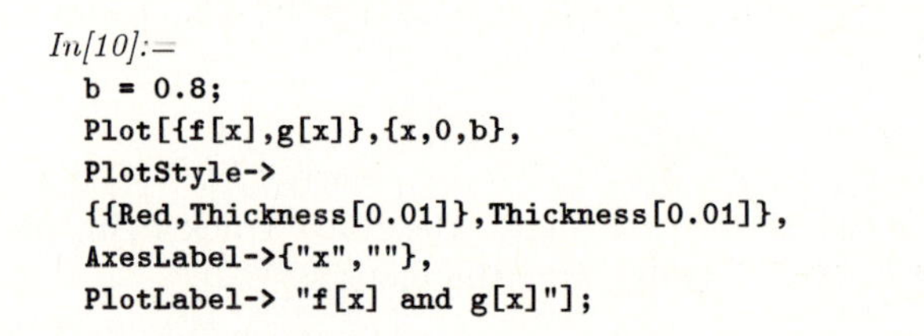

```
In[10]:=
b = 0.8;
Plot[{f[x],g[x]},{x,0,b},
PlotStyle->
{{Red,Thickness[0.01]},Thickness[0.01]},
AxesLabel->{"x",""},
PlotLabel-> "f[x] and g[x]"];
```

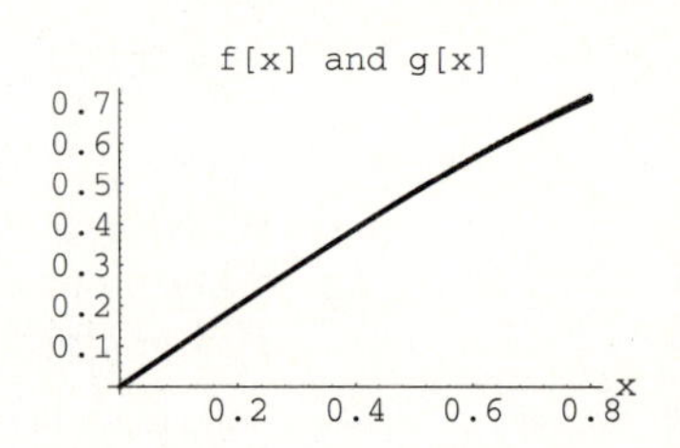

$f[x]$ and $g[x]$ are running together on this plot, just as the Race Track Principle predicted. Take a look on a longer interval:

```
In[11]:=
b = 1.5;
functionplot = Plot[{f[x],g[x]},{x,0,b},
PlotStyle->
{{Red,Thickness[0.01]},Thickness[0.01]},
AxesLabel->{"x",""},
PlotLabel->"f[x] and g[x]"];
```

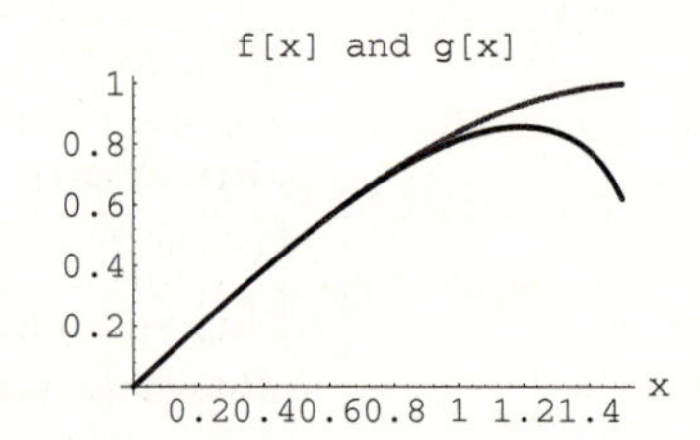

$f[x]$ and $g[x]$ begin to pull apart shortly after their instantaneous growth rates $f'[x]$ and $g'[x]$ begin to pull apart. This is in line with the Race Track Principle.

■ B.2) The Race Track Principle and differential equations

B.2.a) If $y[0] =$ starter is small relative to b, how does the Race Track Principle tell you that the solutions of $y'[x] = r\,y[x]$ and its companion logistic differential equation

$$y'[x] = r\,y[x]\left(1 - \frac{y[x]}{b}\right)$$

will run close together initially as x advances from 0?

Answer: Look at the equations again: $y'[x] = r\,y[x]$ and $y'[x] = r\,y[x]\,(1 - y[x]/b)$ both with $y[0] =$ starter. When you go with a $y[0] =$ starter that is small relative to b, then initially as x advances from 0, $y[x]/b$ is small and so $(1 - y[x]/b)$ is close to 1. The upshot: When you go with a $y[0] =$ starter that is small relative to b, then

$$y'[x] = r\,y[x] \qquad \text{and} \qquad y'[x] = r\,y[x]\left(1 - \frac{y[x]}{b}\right)$$

are nearly the same initially as x advances from 0. A version of the Race Track Principle tells you that the plots of the solutions to both equations share a lot of ink (provided you give them the same value at $x = 0$) initially as x advances from 0. Enough philosophy.

Check this out on an actual case:

In[12]:=

```
r = 0.091; b = 500; starter = 20;
Clear[x,y,Derivative]
logistic = DSolve[{y'[x] == r y[x](1 - y[x]/b),
y[0] == starter},y[x],x];
y[x_] = y[x]/.logistic[[1]];
logisticplot =  Plot[y[x],{x,0,4},
PlotStyle->{{Blue,Thickness[0.01]}},
DisplayFunction->Identity];
Clear[y]
exponential = DSolve[{y'[x] == r y[x],
y[0] == starter},y[x],x];
y[x_] = y[x]/.exponential[[1]];
exponplot = Plot[y[x],{x,0,4},
PlotStyle->{{Red,Thickness[0.01]}},
DisplayFunction->Identity];
Show[exponplot,logisticplot,AxesLabel->{"x",""},
DisplayFunction->$DisplayFunction];
```

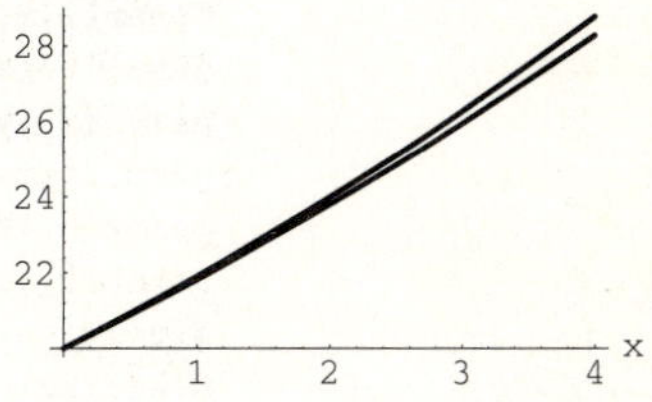

Just as the Race Track Principle predicted. Math happens.

■ B.3) Euler's method of faking the plot of the solution of a differential equation

Here's an innocent-looking differential equation: $y'[x] = y[x]\left(4\cos[x]^2 - y[x]\right)$ with $y[0] = \text{starter} = 1.7$. If you want to get a plot of the solution, you ask *Mathematica* for a formula for the solution:

In[13]:=

```
Clear[x,y,Derivative]; starter = 1.7;
DSolve[{y'[x] == y[x] (4 Cos[x]^2 - y[x]),
y[0] == starter},y[x],x]
```

```
Integrate::ilim: Integration limit 0 is not of the form {x,xmin,xmax}.

NIntegrate::vars:
  Integration range specification 0
    is not of the form {x, xmin, ..., xmax}.
```

Out[13]=

```
{{y[x] ->

                              2 x + Sin[2 x]
                             E
  ---------------------------------------------------------------}}
                                               2 x + Sin[2 x]
  0.588235 - 1. Integrate[1, 0] + Integrate[E               , x]
```

This made *Mathematica* ralph. Try as it might, *Mathematica* could not come up with a useful formula for the $y[x]$. But there is a way to fake the plot of the solution even when you can't get a formula for it. Here's some code to play with. Don't worry about the individual statements in the code right now.

In[14]:=

```
a = 0; b = 5; starter = 1.7; y[a] = starter; beginner = {a,y[a]};
Clear[x,y,next,jump,iterations,point,k]
jump[iterations_] := (b - a)/iterations;
next[{x_,y_},iterations_] := {x,y} +
jump[iterations]{1, y (4 Cos[x]^2 - y)};
point[0,iterations_] = beginner;
point[k_,iterations_] := point[k,iterations] =
N[next[point[k-1,iterations],iterations]];
Clear[Euler]
Euler[iterations_] := Show[Graphics[{Thickness[0.01],
Red,Line[Table[point[k,iterations],
{k,0,iterations}]]}],PlotRange->All,Axes->True,
AxesOrigin->beginner,AxesLabel->{"x",""},
PlotLabel->iterations " iterations",
DisplayFunction->Identity];
```

In[15]:=

```
primitive =
Show[Euler[10],
DisplayFunction->$DisplayFunction];
```

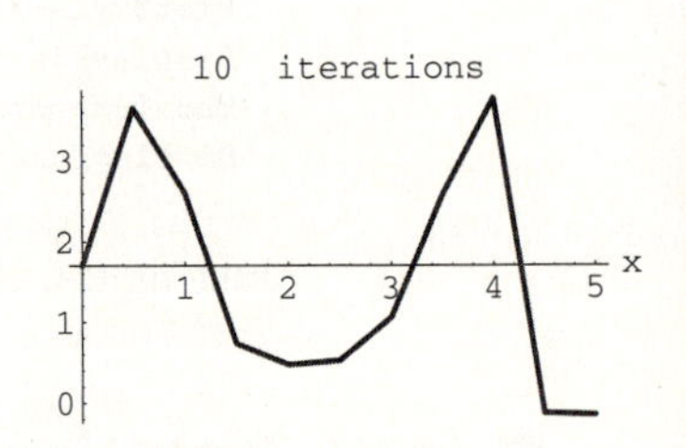

A primitive stick figure resulting from 10 iterations. Here's what you get with 20 iterations:

In[16]:=

```
better =
Show[Euler[20],
DisplayFunction->$DisplayFunction];
```

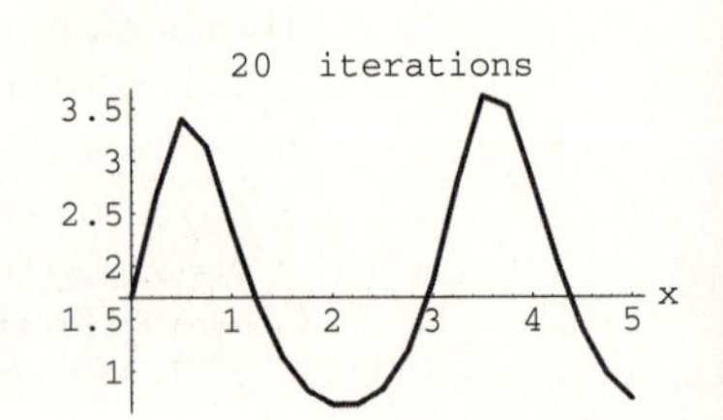

50 iterations:

In[17]:=

```
evenbetter =
Show[Euler[50],
DisplayFunction->$DisplayFunction];
```

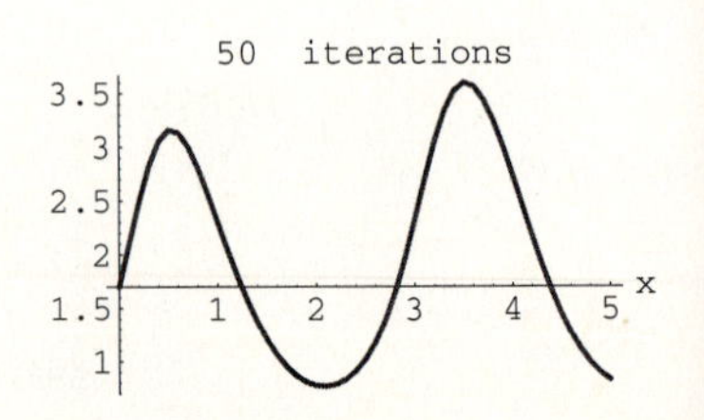

100 iterations:

In[18]:=

```
darngood =
Show[Euler[100],
DisplayFunction->$DisplayFunction];
```

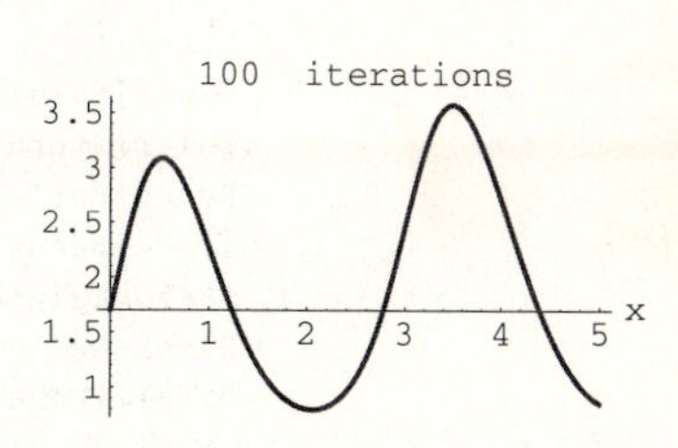

Notice that the last two plots are almost the same. This is your signal that they are both reasonably good fake plots of the solution of the differential equation. The more iterations you use, the better the fake.

B.3.a) Here is the differential equation $y'[x] = 0.3\,y[x]$ with $y[0] = 4$ and a formula for its solution:

In[19]:=

```
r = 0.3; starter = 4; Clear[x,y,Derivative];
solution = DSolve[{y'[x] == r y[x], y[0] == starter},y[x],x];
y[x_] = y[x]/.solution[[1]]
```

Out[19]=

```
   0.3 x
4 E
```

and a plot for $0 \leq x \leq 8$:

In[20]:=

```
realplot =
Plot[y[x],{x,0,8},
PlotStyle->{{Blue,Thickness[0.02]}},
PlotRange->All,AxesLabel->{"x",""}];
```

> Compare the Euler fake plots of the solution with the true plot. Discuss what you see.

Answer: Copy, paste, and edit the code used above to get a Euler faker of a plot of the solution of the differential equation $y'[x] = 0.3\,y[x]$ with $y[0] = 4$ and show the fake plot together with the plot of the true solution.

In[21]:=

```
a = 0; b = 8; starter = 4;
y[a] = starter; beginner = {a,y[a]};
Clear[x,y,next,jump,iterations,point,k]
jump[iterations_] := (b - a)/iterations;
next[{x_,y_},iterations_] := {x,y} + jump[iterations]{1, 0.3 y};
```

```
point[0,iterations_] = beginner;
point[k_,iterations_] := point[k,iterations] =
N[next[point[k-1,iterations],iterations]];
Clear[Euler]
Euler[iterations_] := Show[Graphics[{Thickness[0.01],
Red,Line[Table[point[k,iterations],
{k,0,iterations}]]}],PlotRange->All,Axes->True,
AxesOrigin->beginner,AxesLabel->{"x",""},
PlotLabel->iterations " iterations",
DisplayFunction->Identity];
```

In[22]:=

```
rough =
Show[Euler[10],realplot,
PlotLabel->"10 iterations",
DisplayFunction->$DisplayFunction];
```

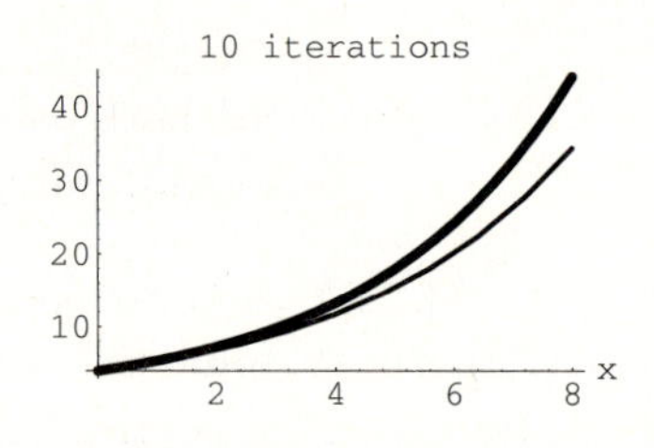

Step up the number of iterations to 100:

In[23]:=

```
evenbetter =
Show[Euler[100],realplot,
PlotLabel->"100 iterations",
DisplayFunction->$DisplayFunction];
```

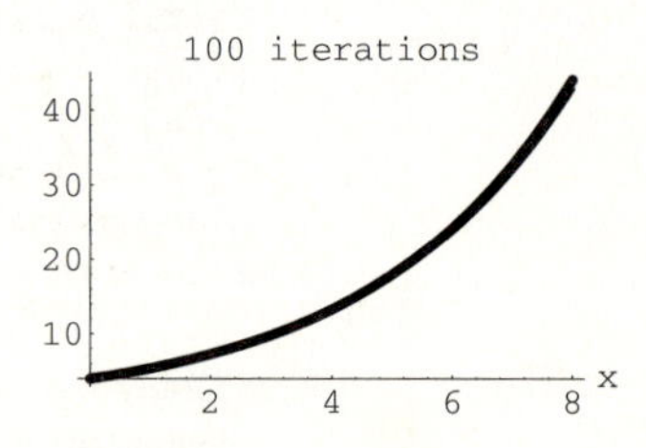

Right you are, governor. The more iterations you use, the better the Euler fake plot you get.

B.3.b) How does the Race Track Principle explain why Euler's method works?

Answer: Here is a differential equation $y'[x] = \sin[x]\, y[x]$ with $y[0] = 2$ and a formula for its solution:

In[24]:=

```
starter = 2; Clear[x,y,Derivative];
solution = DSolve[{y'[x] == Sin[x] y[x], y[0] == starter},y[x],x];
y[x_] = y[x]/.solution[[1]]
```

Out[24]=

```
                 2
    2 Sin[x/2]
 2 E
```

and a plot for $0 \leq x \leq 5$:

In[25]:=

```
realplot =
Plot[y[x],{x,0,5},
PlotStyle->{{Blue,Thickness[0.02]}},
PlotRange->All,AxesLabel->{"x",""}];
```

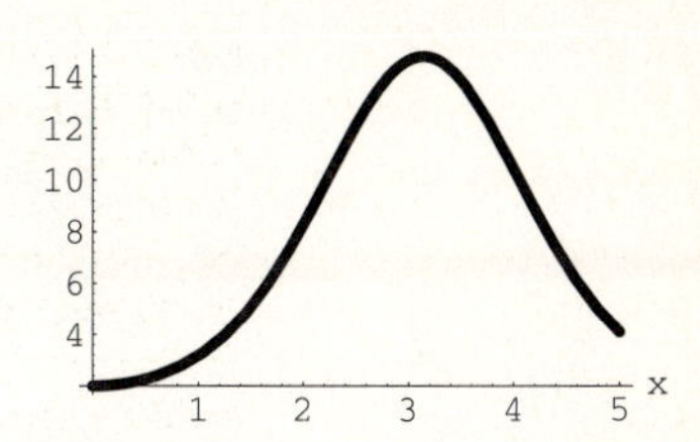

Now look at the Euler faker coming from eight iterations:

In[26]:=

```
a = 0; b = 5; y[a] = starter; beginner = {a,y[a]};
Clear[x,y,next,jump,iterations,point,k]
jump[iterations_] := (b - a)/iterations;
next[{x_,y_},iterations_] := {x,y} +
jump[iterations]{1, Sin[x] y};
point[0,iterations_] = beginner;
point[k_,iterations_] := point[k,iterations] =
N[next[point[k-1,iterations],iterations]];
Clear[Euler]
Euler[iterations_] := Show[Graphics[{Thickness[0.01],
Red,Line[Table[point[k,iterations],
{k,0,iterations}]]}],PlotRange->All,Axes->True,
AxesOrigin->beginner,AxesLabel->{"x",""},
PlotLabel->iterations " iterations",
DisplayFunction->Identity];
```

In[27]:=

```
veryrough =
Show[realplot,Euler[8],
PlotLabel->"8 iterations",
DisplayFunction->$DisplayFunction];
```

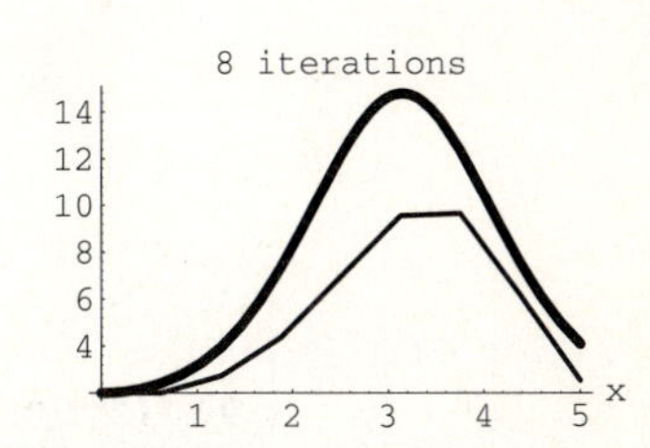

Not a very good fake, but this plot exposes what the faker does. You can see that the faker actually moves by generating a list of points and moving on straight line segments that connect the points. Here are the points:

In[28]:=

```
Clear[fakepoints,fakepointplot]
fakepoints[iterations_] :=
Table[point[k,iterations],{k,0,iterations}];
fakepointplot[iterations_] :=
ListPlot[fakepoints[iterations],PlotStyle->
{PointSize[0.03]},AxesOrigin->{0,0},
DisplayFunction->Identity];
```

In[29]:=
```
Show[realplot,fakepointplot[8]];
```

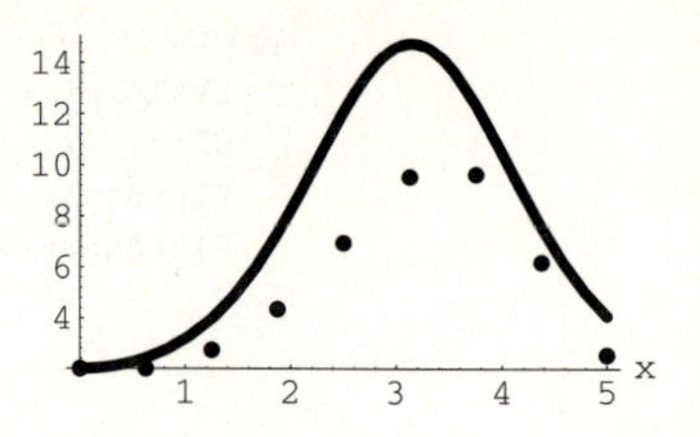

The faker connects these points with line segments:

In[30]:=
```
Show[realplot,Euler[8],fakepointplot[8],
PlotLabel->"8 iterations"];
```

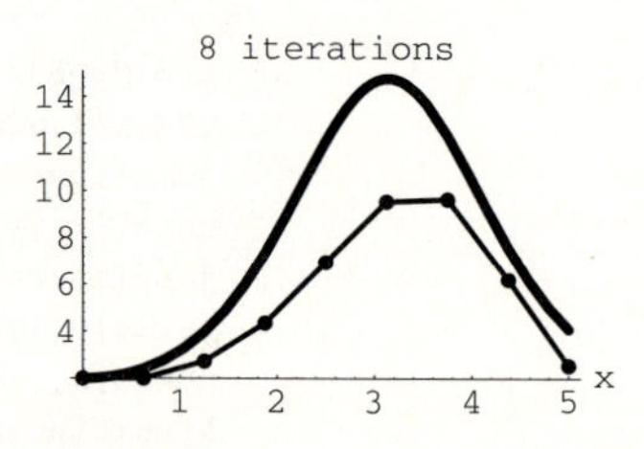

When you step up the iteration number, you shorten the jump between successive points:

In[31]:=
```
Show[realplot,Euler[20],fakepointplot[20],
PlotLabel->20 iterations];
```

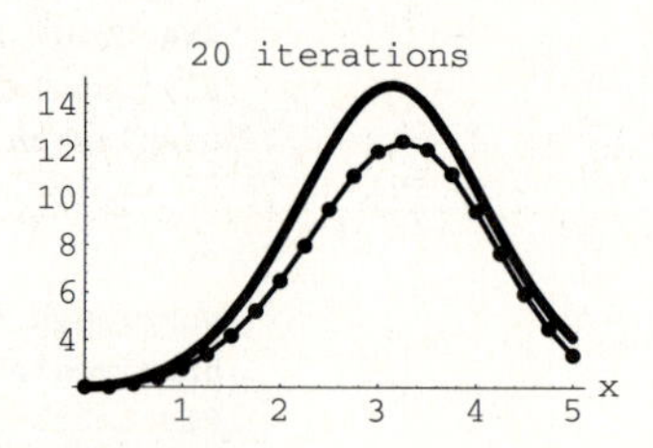

In[32]:=
```
Show[realplot,Euler[200],fakepointplot[200],
PlotLabel->200 iterations];
```

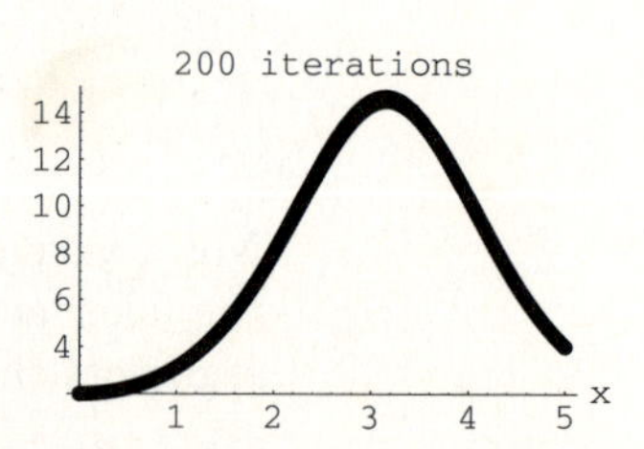

So far, so good. The real question that remains is: Why does the fake curve get better and better as you increase the number of iterations? To answer this, look at the code that generates the fake points:

```
next[{x,y}, iterations] = {x,y} + jump[iterations] {1,Sin[x] y};
point[k, iterations] = N[next[point[k-1], iterations], iterations]].
```

This code tells you that once the Euler faker has settled on a point $\{x, y\}$, then it moves off that point with the same growth rate as the solution of the differential equation $y'[x] = \sin[x]\, y[x]$ that passes through that point. So the big difference between the true plot and the fake plot is: As x advances, the actual solution $y[x]$

updates its growth rate $y'[x]$ instantaneously. But the fake waits for a while to update the growth rate, updating its growth rate only at each fake point. When you use a small jump, then you can expect the growth rate of the fake to be nearly the same as the actual growth rate of the function. So the version of the Race Track Principle that explains why the fake works is:

Horses: If two horses start at the same point and they run at nearly the same speed for the whole race, then they run very close together all the way and are likely to end the race in a photo finish.

Functions: If two functions start out together and their growth rates are nearly the same, then the two functions plot out nearly the same.

To get a small jump on a plotting interval $[a, b]$, you use a high iteration number because

$$\text{jump[iterations]} = \frac{b-a}{\text{iterations}}.$$

This is why high iteration numbers usually give better results than low iteration numbers.

B.3.c) Say a little bit about the great Swiss mathematician Leonhard Euler, 1707–1783, who originally came up with the idea of faking the plots of solutions of differential equations.

Answer: Euler (pronounced "Oiler") was certainly one of the greatest mathematical scientists of all time. In addition to his advanced work in geometry, he wrote the first accurate and complete treatise on calculus. Along the way, he introduced the symbols e, π, and $i = \text{Sqrt}[-1]$ $(= \sqrt{-1})$ and the abbreviations for sine, cosine, tangent, and the other trig functions. Even the notation $f[x]$ owes its life to the fertile mind of Euler. Over and above his mathematics, he had time to take up physics, astronomy, hydrodynamics, optics, electricity, magnetism, ship building, geographical maps, medicine, theology, and oriental languages. Even though he became blind 12 years before his death, nearly half his published works were written during the time he was blind. Not one to neglect the social side of life, Euler spent many years with friends in the famous court of Catherine I in St. Petersburg, Russia.

Tutorials

■ T.1) Using Euler's method to fake the plot of $f[x]$ given $f'[x]$ and one value of $f[x]$

Look at this plot of $f[x] = \sin[x]$:

In[1]:=
```
f[x_] = Sin[x];
a = 0; b = 2 Pi;
trueplot =
Plot[f[x],{x,a,b},
PlotStyle->{{Blue,Thickness[0.01]}},
PlotRange->{{a,b},{-1.5,1.5}}];
```

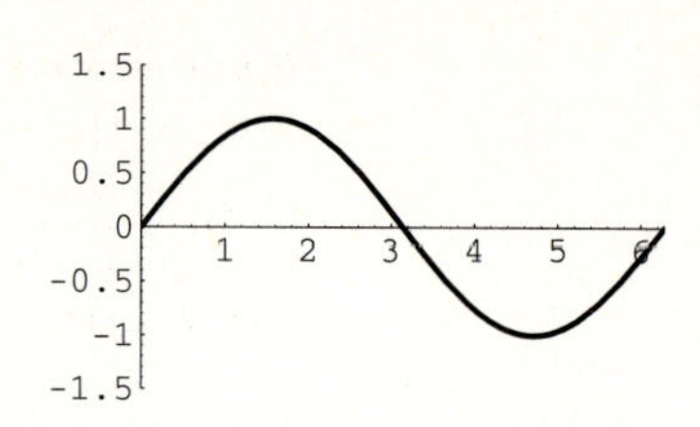

$f[x] = \sin[x]$ is the solution of the differential equation $f'[x] = \cos[x]$ with $f[0] = 0$:

In[2]:=
```
Clear[f,x]
DSolve[{f'[x] == Cos[x],f[0] == 0},f[x],x]
```

Out[2]=
```
{{f[x] -> Sin[x]}}
```

Here come some Euler fake plots of the solution to this differential equation shown with the true plots:

In[3]:=
```
a = 0; b = 2 Pi; starter = 0; y[a] = starter;
beginner = {a,y[a]}; Clear[x,y,next,jump,iterations,point,k]
jump[iterations_] := (b - a)/iterations;
next[{x_,y_},iterations_] := {x,y} + jump[iterations]{1, Cos[x]};
point[0,iterations_] = beginner;
point[k_,iterations_] := point[k,iterations] =
N[next[point[k-1,iterations],iterations]];
Clear[Euler]; Euler[iterations_] := Show[Graphics[{Thickness[0.01],
Red,Line[Table[point[k,iterations],
{k,0,iterations}]]}],PlotRange->All,Axes->True,
AxesOrigin->beginner,AxesLabel->{"x","Euler fake"},
PlotLabel->iterations " iterations", DisplayFunction->Identity];
```

In[4]:=

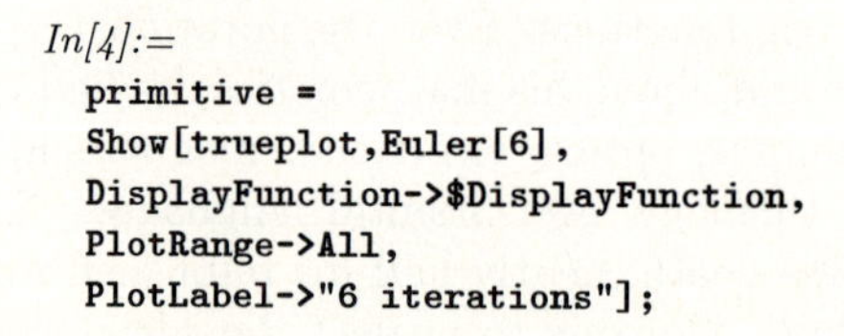

```
primitive =
Show[trueplot,Euler[6],
DisplayFunction->$DisplayFunction,
PlotRange->All,
PlotLabel->"6 iterations"];
```

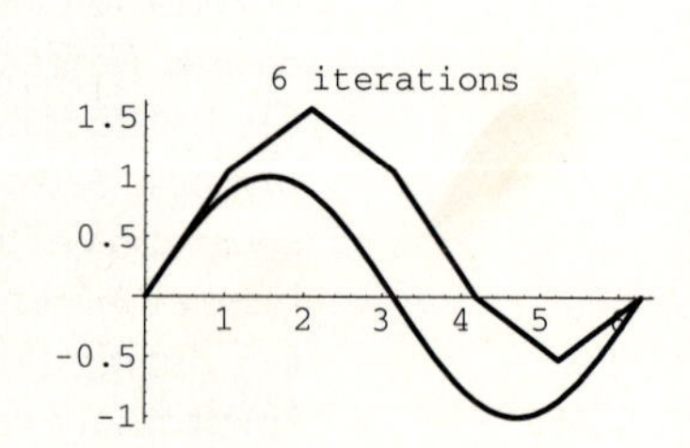

20 iterations:

In[5]:=
```
better =
Show[trueplot,Euler[20],
DisplayFunction->$DisplayFunction,
PlotRange->All,PlotLabel->"20 iterations"];
```

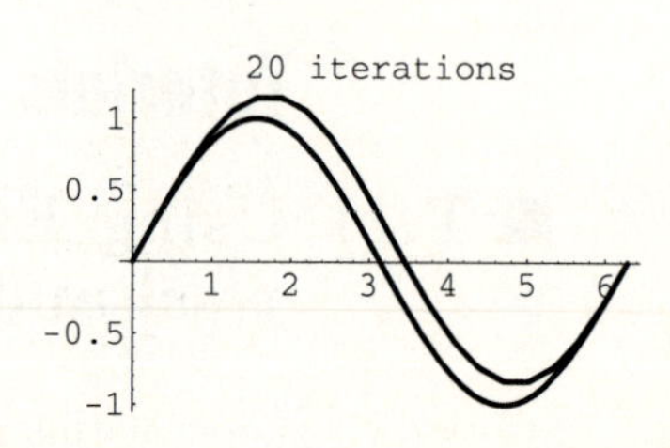

200 iterations:

In[6]:=
```
darngood =
Show[trueplot,Euler[200],
DisplayFunction->$DisplayFunction,
PlotRange->All,PlotLabel->"200 iterations"];
```

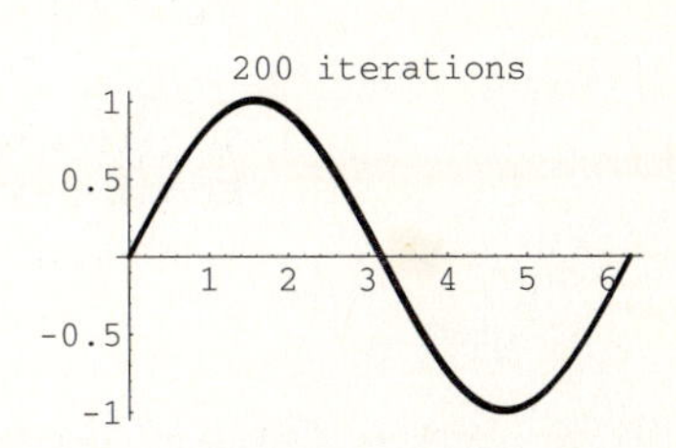

In principle, given a function $g[x]$ and numbers s and t, you can use Euler's method to fake a plot of the function $f[x]$ with $f'[x] = g[x]$ with $f[s] = t$.

T.1.a) Try as you might, you probably will not think of a formula for a function $f[x]$ with $f'[x] = e^{-x^2}$ and $f[0] = 1$. Use Euler's faker to come up with a plot of this function on the interval $[0, 4]$.

Answer: Cut, paste, and edit the code for Euler's faker.

In[7]:=
```
Clear[g,x]; g[x_] = E^(-x^2); a = 0; b = 4; starter = 1;
y[a] = starter; beginner = {a,y[a]};
Clear[x,y,next,jump,iterations,point,k]
jump[iterations_] := (b - a)/iterations;
next[{x_,y_},iterations_] := {x,y} + jump[iterations]{1, g[x]};
point[0,iterations_] = beginner;
point[k_,iterations_] := point[k,iterations] =
N[next[point[k-1,iterations],iterations]];
Clear[Euler]; Euler[iterations_] := Show[Graphics[{Thickness[0.01],
Red,Line[Table[point[k,iterations], {k,0,iterations}]]}],
PlotRange->All,Axes->True, AxesOrigin->beginner,AxesLabel->{"x",""},
DisplayFunction->Identity];
```

In[8]:=
```
firsttry =
Show[Euler[20],
DisplayFunction->$DisplayFunction,
PlotRange->All,PlotLabel->"20 iterations"];
```

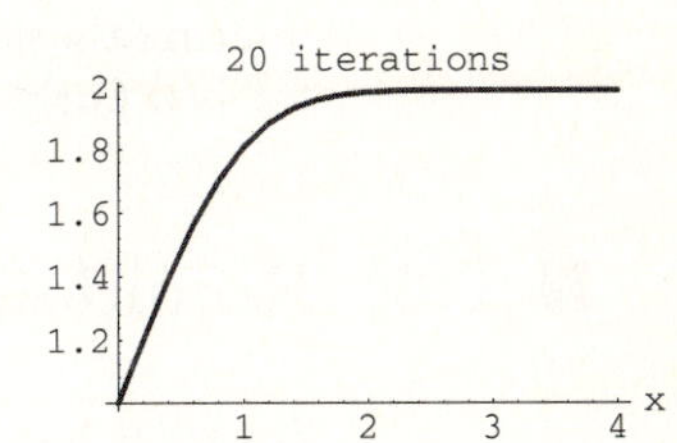

Try 100 iterations:

In[9]:=
```
secondtry =
Show[Euler[100],
DisplayFunction->$DisplayFunction,
PlotRange->All,PlotLabel->"100 iterations"];
```

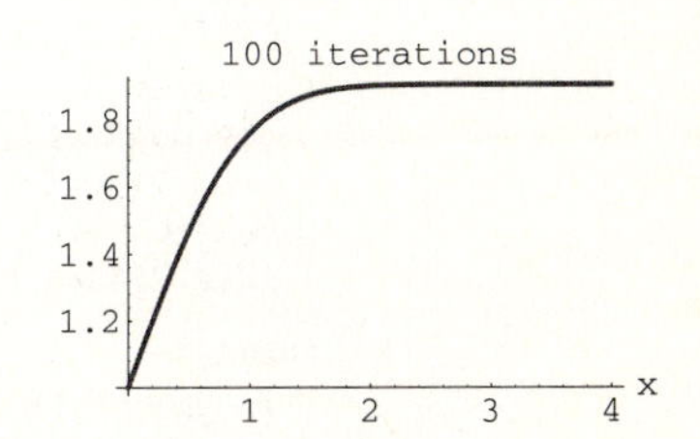

Compare:

In[10]:=
```
Show[firsttry,secondtry,
PlotLabel->"Comparison"];
```

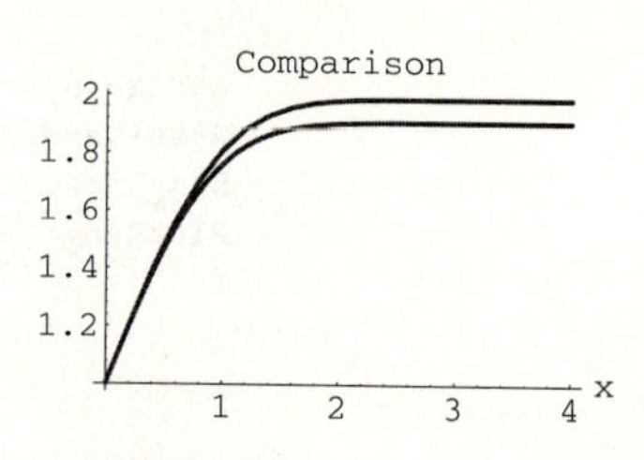

The plots are not the same. This is compelling evidence that the first try is probably not a trustworthy plot. Try again:

In[11]:=
```
thirdtry =
Show[Euler[200],
DisplayFunction->$DisplayFunction,
PlotRange->All,PlotLabel->"200 iterations"];
```

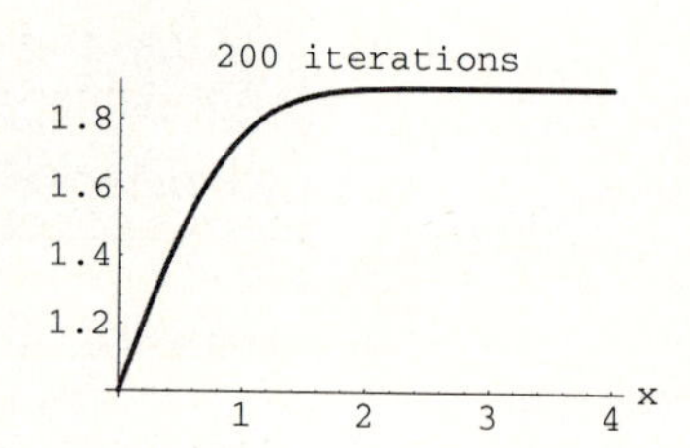

Compare:

In[12]:=
```
Show[secondtry,thirdtry,
PlotLabel->"Comparison"];
```

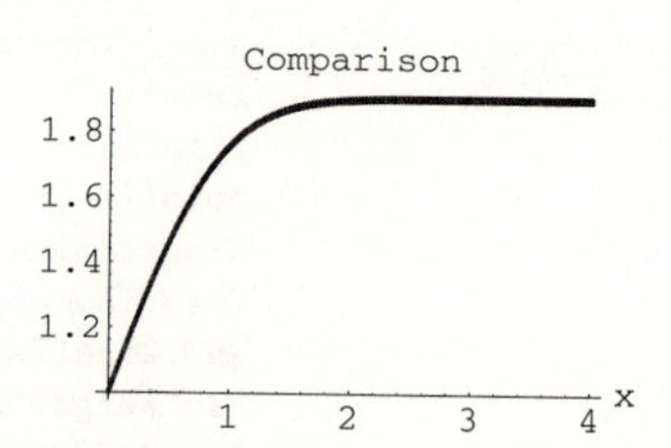

Sharing the same ink most of the way. This is fairly certain evidence that the third try is a fairly accurate plot of the function $f[x]$ with $f'[x] = e^{-x^2}$ and $f[0] = 1$.

■ T.2) Roundoff errors

See what happens when you feed four accurate decimals of e into e^2:

In[13]:=
```
N[E,6]
```

Out[13]=
```
2.71828
```

In[14]:=
```
{N[2.7183^2,5],N[E^2,5]}
```

Out[14]=
```
{7.3892, 7.3891}
```

You get three accurate decimals of e^2. See what happens when you feed eight accurate decimals of $10\,e$ into $(10\,e)^2$:

In[15]:=
```
N[10 E,15]
```
Out[15]=
```
27.1828182845904
```

In[16]:=
```
{N[(27.18281828)^2,12],N[(10 E)^2,12]}
```
Out[16]=
```
{738.905609644, 738.905609893}
```

You get six accurate decimals of $(10\,e)^2$. There is a way to use the Race Track Principle to help tell how many accurate decimals of a are needed to maintain a desired number of accurate decimals of $f[a]$. If this interests you, go on.

T.2.a.i) Here is a function and a plot of its derivative on the interval $a - 1 \leq x \leq a + 1$ for $a = \sqrt{2}$:

In[17]:=
```
Clear[x,f]; f[x_] = 3 E^(x/2) Cos[2 x^2];
a = Sqrt[2];
growthplot =
Plot[f'[x],{x,a - 1,a + 1},
AxesLabel->{"x","f'[x]"}];
```

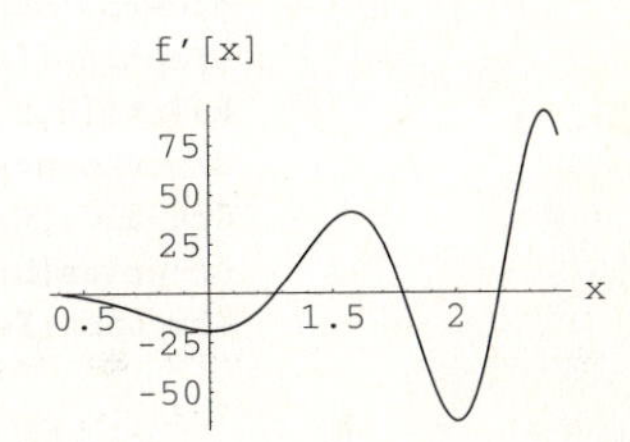

Read from the plot that $|f'[x]| < 100 = 10^2$ for all the x's with $a - 1 \leq x \leq a + 1$ where $a = \sqrt{2}$. Now plot on the same interval $f[x]$ and the lines that go through $\{a, f[a]\}$ with growth rates 10^2 and -10^2:

In[18]:=
```
p = 2; Clear[upline,downline]
upline[x_] = 10^p (x - a) + f[a];
downline[x_] = -10^p (x - a) + f[a];
lines = Plot[{f[x],upline[x],downline[x]}, {x,a - 1,a + 1},
PlotStyle->{GrayLevel[0.3],Red,Red}, DisplayFunction->Identity];
pointplot = Graphics[{Red,PointSize[0.03], Point[{a,f[a]}]}];
label = Graphics[Text["{a,f[a]}",{a,f[a]},{1,3}]];
```

In[19]:=
```
bowtie =
Show[lines,pointplot,label,
AxesLabel->{"x",""},
DisplayFunction->$DisplayFunction];
```

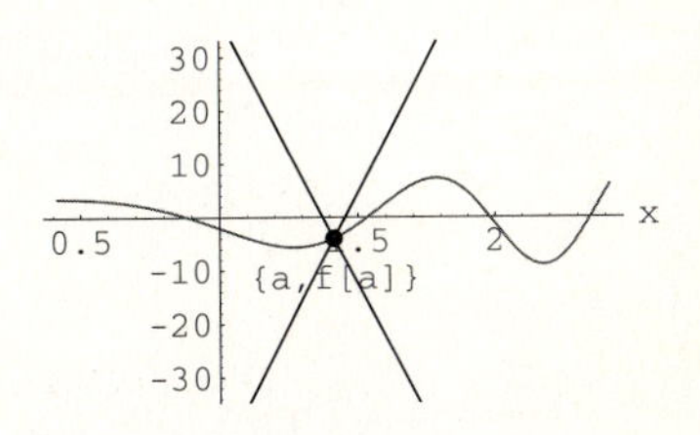

Why did the plot turn out this way?

Answer: One line has constant growth rate 10^2. The other line has constant growth rate -10^2. Because $|f'[x]| < 10^2$ for all the x's in this interval, the Race Track Principle tells you that $f[x]$ must plot out between the two lines. In short, $f[x]$ has no option except to be caught inside the bow tie.

T.2.a.ii) Continue to go with $f[x] = 3\,e^{x/2}\cos[2\,x^2]$ and $a = \sqrt{2}$. What does this plot tell you about the discrepancy between $f[x]$ and $f[a]$ for any other x with $a - 1 < x < a + 1$?

Answer: Look at an annotation of this plot:

In[20]:=
```
b = a + 0.3;
otherpointplot =
Graphics[{PointSize[0.03],
Point[{b,f[b]}]}];
discrepancy = Show[bowtie,otherpointplot,
Graphics[Line[{{a,f[a]},{b + 0.1,f[a]}}]],
Graphics[Line[{{b,f[a]},{b,f[b]}}]],
Graphics[Text["{x,f[x]}",{b,f[b]},{0,-1.5}]]];
```

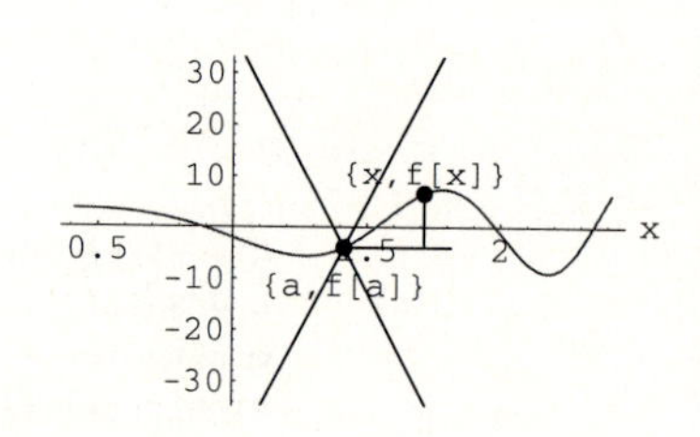

The discrepancy between $f[x]$ and $f[a]$ (which is the length of the little vertical line on the right) is smaller than 1/2 the length of the new vertical segment plotted below:

In[21]:=
```
bigdiscrepancy =
Show[Graphics[{Red,Thickness[0.01],
Line[{{b,downline[b]},{b,upline[b]}}]}],
discrepancy,Axes->True];
```

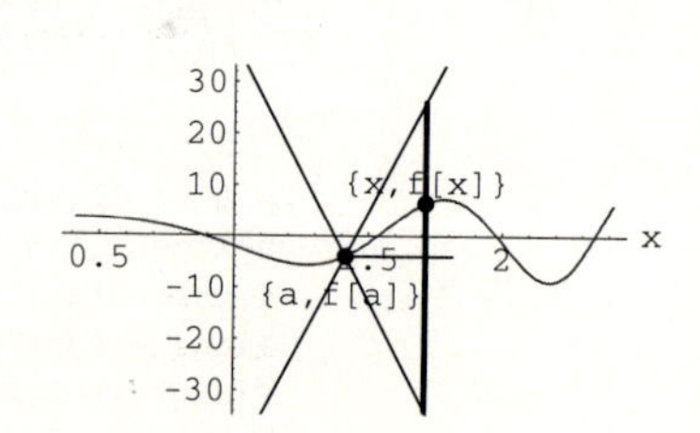

Half the length of the longer vertical line is

$$\frac{|\text{upline}[x] - \text{lowline}[x]|}{2} = \frac{|10^2\,(x-a) + f[a] - (-10^2\,(x-a) + f[a])\,|}{2}$$
$$= \frac{|210^2\,(x-a)\,|}{2}$$
$$= 10^2|x-a|.$$

This tells you that the discrepancy between $f[x]$ and $f[a]$ is smaller than $10^2|x-a|$.

T.2.a.iii)

> Continue to go with $f[x] = 3\,e^{x/2}\cos[2\,x^2]$ and $a = \sqrt{2}$. Now you know that if $a - 1 \leq x \leq a + 1$, then the discrepancy between $f[x]$ and $f[a]$ is no more than $10^2|x - a|$. How do you use this to tell how many accurate decimals (to the right of the decimal point) of $a = \sqrt{2}$ to feed into $f[a]$ to guarantee at least four accurate decimals of $f[a]$?

Answer: That's easy. The discrepancy between $f[x]$ and $f[a]$ is no more than $10^2|x-a|$. If $x = a$ to six accurate decimals (to the right of the decimal point), then $|x - a| \leq 10^{-6}$. So the discrepancy between $f[x]$ and $f[a]$ is less than $10^2 \times 10^{-6} = 10^{-4}$. The upshot: If x is a to six accurate decimals, then $f[x]$ is $f[a]$ to four accurate decimals. Try it:

In[22]:=
```
N[Sqrt[2],8]
```

Out[22]=
```
1.4142136
```

1.414213 is $a = \sqrt{2}$ to six accurate decimals; compare:

In[23]:=
```
{N[f[a],6],N[f[1.414213],6]}
```

Out[23]=
```
{-3.97699, -3.97701}
```

When you round these to four accurate decimals, the results are the same; just as the theory predicted.

T.2.b.i)

> How do you do this for other functions and other points?

Answer: You do it the same way; it works for the same reasons. Go with a couple of quickies:

In[24]:=
```
a = Sqrt[3]/2
```

Out[24]=
```
Sqrt[3]
-------
   2
```

In[25]:=
```
Clear[f,x]
f[x_] = E^(x^2)
```

Out[25]=
```
  2
 x
E
```

In[26]:=
```
Plot[f'[x],{x,a - 1,a + 1},
AxesLabel->{"x","f'[x]"}];
```

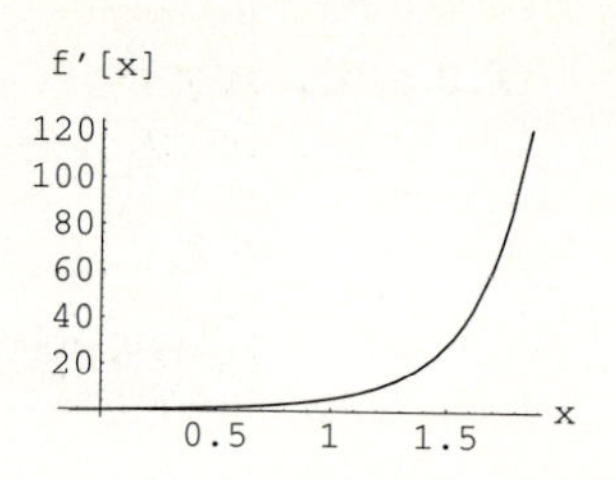

Read from the plot that $|f'[x]| < 1000 = 10^3$ for the x's in the interval. This tells you to feed $k+3$ accurate decimals of a into $f[a]$ to guarantee at least k accurate decimals of $f[a]$. For example: Twelve accurate decimals of a guarantee nine accurate decimals of $f[a]$:

In[27]:=
```
a
```

Out[27]=
```
Sqrt[3]
-------
   2
```

In[28]:=
```
N[a,13]
```

Out[28]=
```
0.8660254037844
```

In[29]:=
```
{N[f[a],12],N[f[0.866025403784],12]}
```

Out[29]=
```
{2.11700001661, 2.11700001661}
```

T.2.b.ii) Explain: No matter what point a you use, you can be assured that k accurate decimals of a guarantee k accurate decimals of $\sin[a]$.

Answer: Look at the derivative:

In[30]:=
```
Clear[f,x]
f[x_] = Sin[x];
f'[x]
```

Out[30]=
```
Cos[x]
```

You and everyone else know that $|f'[x]| = |\cos[x]| \leq 1 = 10^0$ no matter what x is because $\cos[x]$ spends its life oscillating between -1 and 1. As a result, you don't need any plot to be certain that $|f'[x]| \leq 10^0$ for the x's in the interval $[a-1, a+1]$. This tells you to feed $k+0 = k$ accurate decimals of a into $f[a]$ to guarantee at least k accurate decimals of $f[a]$. Done.

T.2.c) What can throw off these calculations?

Answer: These calculations are theoretical and do not allow for other rounding errors introduced as the calculations are processed by the machine.

Give It a Try

Experience with the starred ($\star$) problems will be especially beneficial for understanding later lessons.

■ G.1) Versions of the Race Track Principle★

G.1.a) Put $f[x] = (1 + x)^t$ and $g[x] = 1 + t\,x$ and look at:

In[1]:=
```
Clear[f,g,x,t]
f[x_] = (1 + x )^t;
g[x_] = 1 + t x;
{f[0],g[0]}
```

Out[1]=
```
{1, 1}
```

In[2]:=
```
{f'[x],g'[x]}
```

Out[2]=
```
               -1 + t
{t (1 + x)          , t}
```

> Use the Race Track Principle and what you see above to explain why $f[x] \geq g[x]$ for $x \geq 0$ provided $t \geq 1$.

G.1.b) Look at:

In[3]:=
```
Clear[f,g,x]; f[x_] = Sin[x]
```

Out[3]=
```
Sin[x]
```

In[4]:=
```
g[x_] = x (60 - 7 x^2)/(60 + 3 x^2)
```

Out[4]=
```
            2
x (60 - 7 x )
-------------
          2
 60 + 3 x
```

Both functions start their race at $x = 0$ together:

In[5]:=
```
{f[0],g[0]}
```
Out[5]=
```
{0, 0}
```

Here's a plot of the instantaneous growth rates of both functions for $0 \leq x \leq 3$:

In[6]:=
```
growthplot = Plot[{f'[x],g'[x]},{x,0,3},
PlotStyle->{Red,Thickness[0.01]},
AxesLabel->{"x",""},
PlotLabel->"f'[x] and g'[x]"];
```

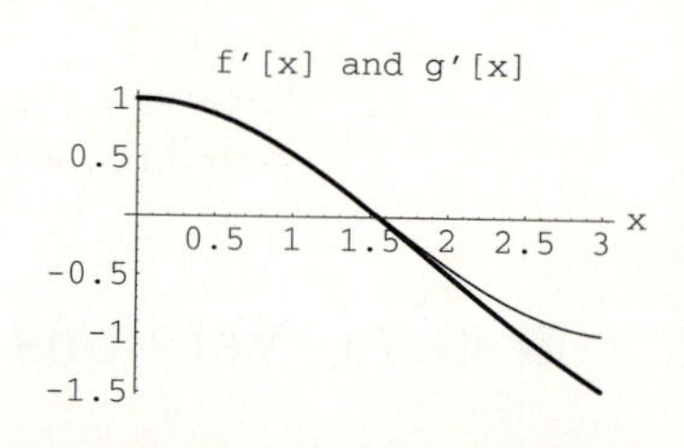

> Use this plot and the Race Track Principle to determine b so that $f[x]$ and $g[x]$ are nearly the same for $0 \leq x \leq b$. Check yourself with a plot.

G.1.c.i) Here's another version of the Race Track Principle:

Horses: If two horses end the race in a dead heat, then the faster horse was behind all the time except at the finish line.

Functions: If $f[b] = g[b]$ and $g'[x] \geq f'[x]$ for $a \leq x \leq b$, then $f[x]$________$g[x]$ for $a \leq x \leq b$.

> Fill in the blank above.

G.1.c.ii) Go with $f[x] = \sin[x]$ and $g[x] = x$ and look at:

In[7]:=
```
Clear[f,g,x]
f[x_] = Sin[x];
g[x_] = x;
{f[0],g[0]}
```
Out[7]=
```
{0, 0}
```
In[8]:=
```
{f'[x],g'[x]}
```
Out[8]=
```
{Cos[x], 1}
```

> Remembering that $\cos[x]$ spends all its life oscillating between -1 and 1, use the Race Track Principle and what you see above to explain why $f[x] \geq g[x]$ for $x \leq 0$.

G.1.d.i) Another version of the Race Track Principle:

Horses: If two horses run at exactly the same speed for the whole race, and they are tied at one point of the race, then they are tied throughout the race.

Functions: Suppose c is one point with $a \leq c \leq b$ and $f[c] = g[c]$ and $f'[x] = g'[x]$ for all x's with $a \leq x \leq b$; then $f[x]$________$g[x]$ for $a \leq x$.

> Fill in the blank above.

G.1.d.ii) Look at this:

In[9]:=
```
Clear[f,g,x]
f[x_] = Cos[x]^2 - Sin[x]^2
```
Out[9]=
```
      2         2
Cos[x]  - Sin[x]
```
In[10]:=
```
g[x_] = 1 - 2 Sin[x]^2
```
Out[10]=
```
              2
1 - 2 Sin[x]
```
In[11]:=
```
{f'[x],g'[x]}
```
Out[11]=
```
{-4 Cos[x] Sin[x], -4 Cos[x] Sin[x]}
```
In[12]:=
```
f[0] == g[0]
```
Out[12]=
```
True
```

> Explain why this gives a calculus explanation of the identity $f[x] = g[x]$ for all the x's.

■ G.2) Running Euler's faker★

G.2.a.i) Here's an innocent-looking differential equation: $y'[x] = y[x]\,(3 + 3\sin[x] - y[x])$ with $y[0] =$ starter $= 0.5$. If you want to get a plot of the solution, you ask *Mathematica* for a formula for the solution:

In[13]:=
```
Clear[x,y,Derivative]; starter = 0.5;
DSolve[{y'[x] == y[x] (3 + 3 Sin[x] - y[x]), y[0] == starter},y[x],x]
```

```
Integrate::ilim: Integration limit 0 is not of the form {x,xmin,xmax}.
NIntegrate::vars:
   Integration range specification 0
     is not of the form {x, xmin, ..., xmax}.
```

Out[13]=

```
            3 (x - Cos[x])
  {{y[x] -> E               /

                               3            -3                        3 (x - Cos[x])
       (0.0995741 - 0.0497871 E  Integrate[E  , 0] + Integrate[E                , x])}}
```

This made *Mathematica* barf. The output is useless. Try as it might, *Mathematica* could not come up with a useful formula for the true solution $y[x]$. But Calculus&*Mathematica* students like you are not easily dissuaded.

> Use Euler's method to come up with what you believe to be a reasonably accurate plot of the true solution for $0 \leq x \leq 7$.

G.2.a.ii) Here's a plot of $3 + 3\sin[x]$ for $0 \leq x \leq 7$:

In[14]:=

```
Clear[x]
auxillaryplot = Plot[3 + 3 Sin[x],{x,0,7},
PlotStyle->{{Thickness[0.01],VioletRed}}];
```

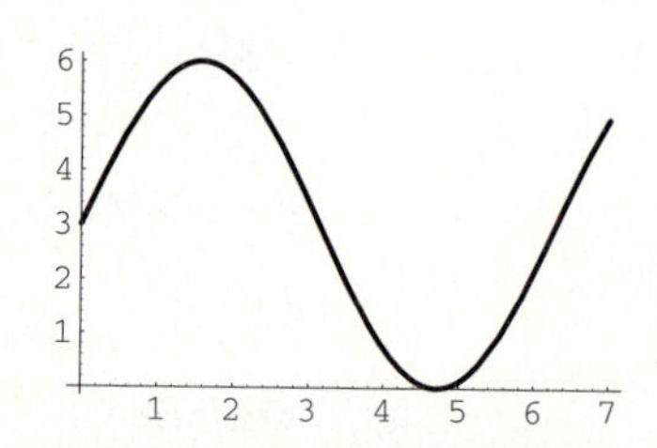

> Show this plot together with your accurate plot from part G.2.a.i). Then take another look at the differential equation $y'[x] = y[x]\,(3 + 3\sin[x] - y[x])$ and explain why it is automatic that: When the solution curve is below the $3 + 3\sin[x]$ curve, the solution curve is going up. When the solution curve is above the $3 + 3\sin[x]$ curve, the solution curve is going down.

G.2.b.i) Here's *Mathematica*'s attempt to find a formula for a function $f[x]$ with

$$f'[x] = e^{\sin[x]} - 1.5$$

and $f[0] = 1$.

In[15]:=

```
Clear[x,y,Derivative]; starter = 1;
DSolve[{y'[x] == E^(Sin[x]) - 1.5,
y[0] == starter},y[x],x]
```

```
Integrate::ilim: Integration limit 0 is not of the form {x,xmin,xmax}.
```

Out[15]=

```
                                              Sin[x]
{{y[x] -> 1 - 1.5 x - Integrate[1, 0] + Integrate[E      , x]}}
```

This made *Mathematica* spill its cookies again. The output is useless. Try as it might, *Mathematica* could not come up with a useful formula for $f[x]$.

> Use Euler's method to come up with what you believe to be a reasonably accurate plot of $f[x]$ for $0 \leq x \leq 8$.

G.2.b.ii) Here's a plot of $e^{\sin[x]} - 1.5$ for $0 \leq x \leq 8$:

In[16]:=

```
Clear[x]
auxillaryplot =
Plot[E^(Sin[x]) - 1.5,{x,0,8},
PlotStyle->{{Thickness[0.01],VioletRed}}];
```

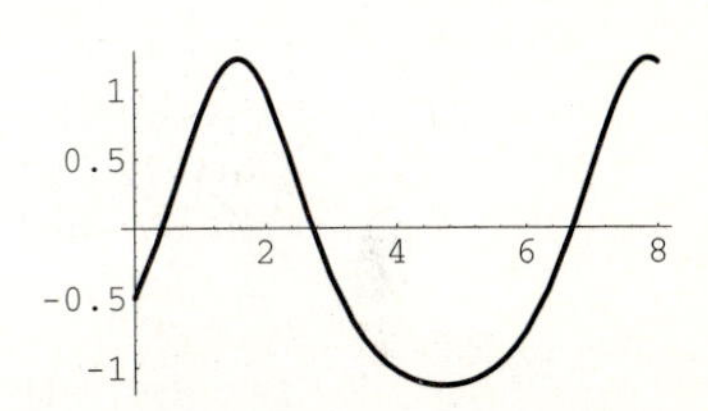

> Show this plot together with your accurate plot from part G.2.b.i). Then take another look at the specification $f'[x] = e^{\sin[x]} - 1.5$ and explain why it is automatic that: When the $e^{\sin[x]} - 1.5$ curve is positive, then the solution curve is going up. When the $e^{\sin[x]} - 1.5$ curve is negative, then the solution curve is going down.

■ G.3) The Race Track Principle and differential equations★

G.3.a.i)

> Here are three differential equations:
>
> $$y'[x] = 0.4\, y[x],$$
> $$y'[x] = 0.4\, y[x]\,(1 - y[x]/100) \text{ and}$$
> $$y'[x] = 0.4 \cos[x]\, y[x]$$
>
> with $y[0] = 1$ in all three. Use the DSolve instruction to get formulas for the solutions of each.

G.3.a.ii)

> Plot the solutions of all three on the same axes for $0 \leq x \leq 1$ and then answer the question: What is it about the three differential equations that guarantees that the plots of all three solutions will share lots of ink initially as x advances away from 0?

■ G.4) The error function Erf[x]

The error function, Erf[x], is the function with $\text{Erf}'[x] = (2/\sqrt{\pi}\,)\, e^{-x^2}$:

In[17]:=
```
Clear[x]
Erf'[x]
```

Out[17]=
```
     2
-----------
  2
 x
E   Sqrt[Pi]
```

and Erf[0] = 0:

In[18]:=
```
Erf[0]
```

Out[18]=
```
0
```

The derivative of Erf[x] is the famous bell-shaped curve:

In[19]:=
```
Clear[f,x]
f[x_] = Erf[x];
bell = Plot[f'[x],{x,-4,4},
PlotStyle->{{Blue,Thickness[0.01]}},
AxesLabel->{"x",""}];
```

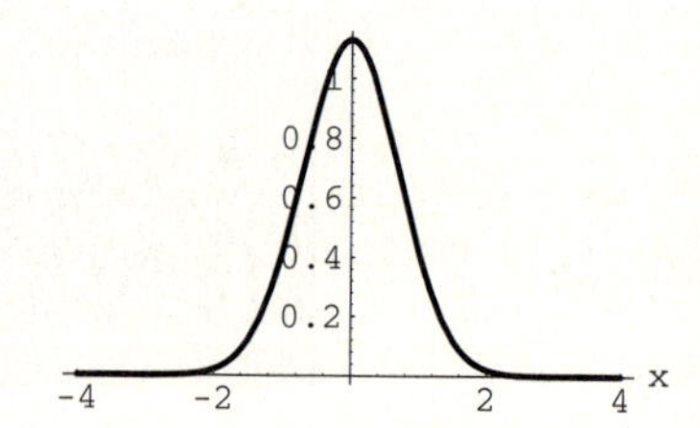

Here is a plot of Erf[x] itself:

In[20]:=
```
erfplot = Plot[Erf[x],{x,-4,4},
PlotStyle->{{GrayLevel[0.4],Thickness[0.01]}},
AxesLabel->{"x","Erf[x]"}];
```

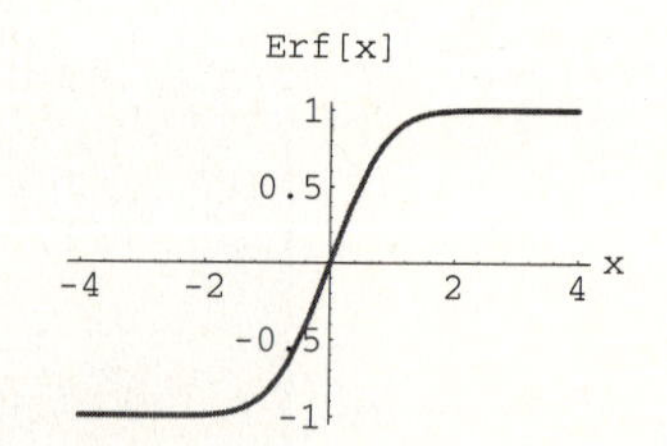

Here are Erf[x] and its derivative plotted together:

In[21]:=
```
Show[bell,erfplot];
```

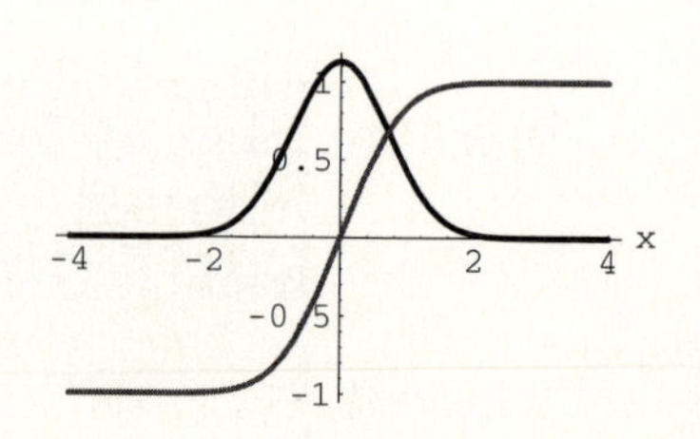

Looking good. Note the global scale behavior:

In[22]:=
```
Plot[{1,Erf[x],-1,},{x,-8,8},
PlotStyle->{{Red},
 {GrayLevel[0.4],Thickness[0.01]},{Red}},
AxesLabel->{"x","Erf[x]"}];
```

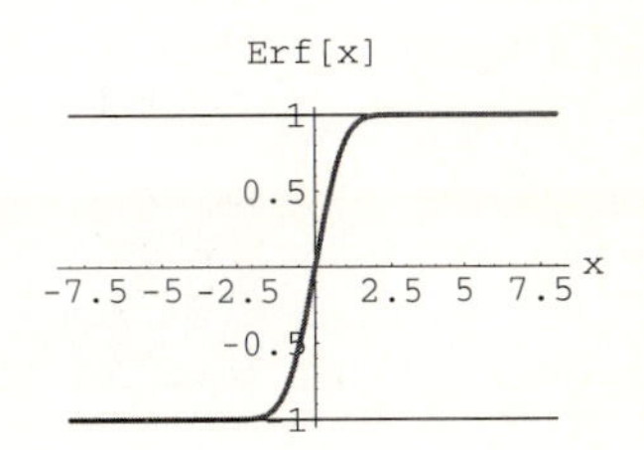

That's right;

$$\lim_{x\to\infty} \text{Erf}[x] = 1 \qquad \text{and} \qquad \lim_{x\to-\infty} \text{Erf}[x] = -1.$$

Giving a clean formula for Erf[x] in terms of other usual functions is not possible now or anytime in the future.

G.4.a) Take the information: $f'[x] = (2/\sqrt{\pi}\,)\,e^{-x^2}$ and $f[0] = 0$. Use Euler's method to give your own trustworthy fake plot of Erf[x] for $0 \le x \le 2.5$, and check your plot by showing your plot and the machine plot together.

G.4.b) Can you think of a way to use Euler's method to get a fake plot for $-2.5 \le x \le 0$?

■ G.5) Round off

G.5.a.i) Given $f[x] = x^2$: Use the derivative to try to predict how many accurate decimals of e guarantee eight accurate decimals of $f[e]$. Check yourself with a calculation.

G.5.a.ii) Given $f[x] = x^2$: Use the derivative to try to predict how many accurate decimals of $100\,e$ guarantee eight accurate decimals of $f[e]$. Check yourself with a calculation.

G.5.a.iii) Given $f[x] = \cos[x]$: Use the derivative to try to predict how many accurate decimals of $\pi/2$ guarantee eight accurate decimals of $f[\pi/2]$. Check yourself with a calculation.

G.5.a.iv) Given $f[x] = \sin[x]$:

In[23]:=
```
Clear[f,x]; f[x_] = Sin[x]
```
Out[23]=
```
Sin[x]
```

Predict how many accurate decimals of any old x guarantee eight accurate decimals of $\sin[x]$.

G.5.a.v) Given:

In[24]:=
```
Clear[f,x]; f[x_] =
x (166320 - 22260 x^2 + 551 x^4)/
            (15(11088 + 364 x^2 + 5 x^4))
```
Out[24]=
```
             2         4
x (166320 - 22260 x  + 551 x )
------------------------------
               2       4
  15 (11088 + 364 x  + 5 x )
```

Use the derivative to predict how many accurate decimals of π guarantee k accurate decimals of $f[\pi]$.

G.5.a.vi) For $f[x] = \mathrm{ArcTan}[x]$, use the derivative to try to predict how many accurate decimals of $1/\sqrt{3}$ guarantee seven accurate decimals of $f[1/\sqrt{3}\,]$.

G.5.b.i) Explain the statement: The bigger x is, the more sensitive e^x is to roundoff error.

G.5.b.ii) Explain: No matter what point x you use, you can be assured that k accurate decimals of x guarantee k accurate decimals of $\cos[x]$.

G.5.b.iii) Explain the statement: Roughly speaking, roundoff error is not much of a problem with $\sin[x]$ or $\cos[x]$, but roundoff error can become a big problem with e^x.

■ G.6) Calculating accurate values of $\log[x]$

This problem appears only in the electronic version.

G.7) Calculating accurate values of e^x

A method of calculating accurate values of e^x is essential for science because this is the most important function in mathematics or science. But e itself is not an easy number to work with. In fact at this stage, we have not seen how to slam out its decimals. *Mathematica* does a pretty good job:

In[25]:=
```
{N[E],N[E,9],N[E,12],N[E,35]}
```

Out[25]=
```
{2.71828, 2.71828183, 2.71828182846, 2.7182818284590452353602874713526625}
```

G.7.a.i) Use the Race Track Principle and the fact that $e^x \geq 1$ for $x \geq 0$ to explain why $e^x \geq 1 + x$ for $x \geq 0$.

G.7.a.ii) Use the Race Track Principle and the fact that $e^x \geq 1 + x$ for $x \geq 0$ to explain why $e^x \geq 1 + x + x^2/2$ for $x \geq 0$.

G.7.a.iii) Use the Race Track Principle and the fact that

$$e^x \geq 1 + x + \frac{x^2}{2}$$

for $x \geq 0$ to explain why

$$e^x \geq 1 + x + \frac{x^2}{2} + \frac{x^3}{3!}$$

for $x \geq 0$.

G.7.b) By now, a clear pattern emerges. Repeated applications of the Race Track Principle explain why

$$e^x \geq 1 + x + \frac{x^2}{2} + \frac{x^3}{3!} + \frac{x^4}{4!},$$

$$e^x \geq 1 + x + \frac{x^2}{2} + \frac{x^3}{3!} + \frac{x^4}{4!} + \frac{x^5}{5!},$$

$$e^x \geq 1 + x + \frac{x^2}{2} + \frac{x^3}{3!} + \frac{x^4}{4!} + \frac{x^5}{5!} + \frac{x^6}{6!},$$

etc. for $x \geq 0$. Consequently, if you set:

In[26]:=
```
Clear[under,n,x]
under[x_,n_] := 1 + Sum[(x^k)/k!,{k,1,n}]
```

then for each positive integer n, you know that $e^x \geq \text{under}[x, n]$ for all x's ≥ 0.

Evaluate the following cells; report what you think is happening.

In[27]:=
```
Clear[under,n,x]
under[x_,n_] := 1 + Sum[(x^k)/k!,{k,1,n}]
```

In[28]:=
```
N[{E,under[x,1]/.x->1},2]
```

Out[28]=
```
{2.7, 2.}
```

In[29]:=
```
N[{E,under[x,2]/.x->1},2]
```

Out[29]=
```
{2.7, 2.5}
```

In[30]:=
```
N[{E,under[x,3]/.x->1},3]
```

Out[30]=
```
{2.72, 2.67}
```

In[31]:=
```
N[{E,under[x,4]/.x->1},3]
```

Out[31]=
```
{2.72, 2.71}
```

In[32]:=
```
N[{E,under[x,5]/.x->1},4]
```

Out[32]=
```
{2.718, 2.717}
```

In[33]:=
```
N[{E,under[x,6]/.x->1},5]
```

Out[33]=
```
{2.7183, 2.7181}
```

G.7.c) Here are plots for e^x and under$[x, n]$ for $0 \leq x \leq 3$ and for bigger and bigger values of n.

In[34]:=
```
n = 2;
Plot[{E^x,under[x,n]},{x,0,3},
PlotStyle->
{{GrayLevel[0.5],Thickness[0.02]},{Red}},
PlotRange->{0,E^3},AxesLabel->{"x",""},
PlotLabel->n "= n"];
```

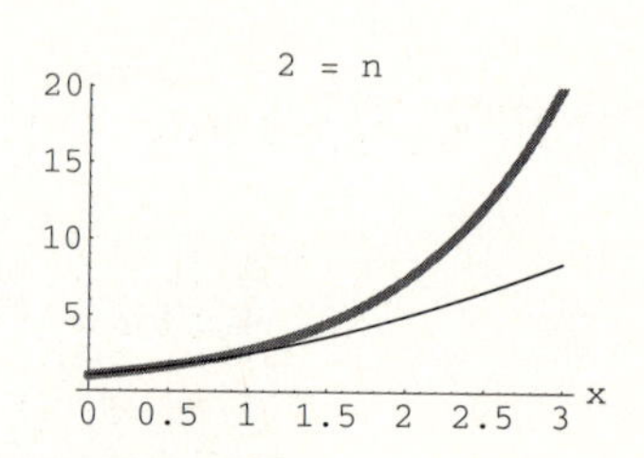

In[35]:=
```
n = 4;
Plot[{E^x,under[x,n]},{x,0,3},
PlotStyle->
{{GrayLevel[0.5],Thickness[0.02]},{Red}},
PlotRange->{0,E^3},AxesLabel->{"x",""},
PlotLabel->n "= n"];
```

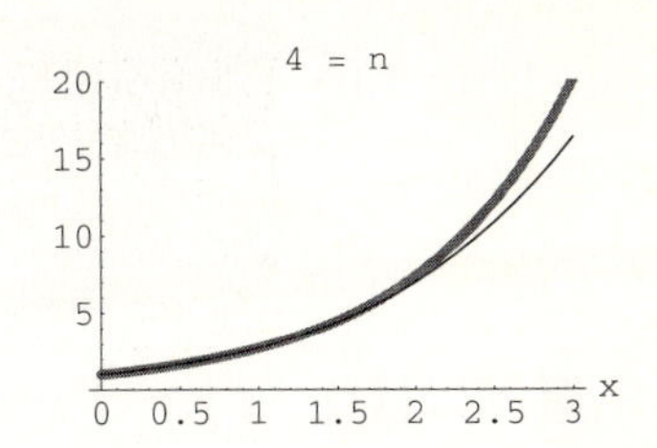

In[36]:=
```
n = 8;
Plot[{E^x,under[x,n]},{x,0,3},
PlotStyle->
{{GrayLevel[0.5],Thickness[0.02]},{Red}},
PlotRange->{0,E^3},AxesLabel->{"x",""},
PlotLabel->n "= n"];
```

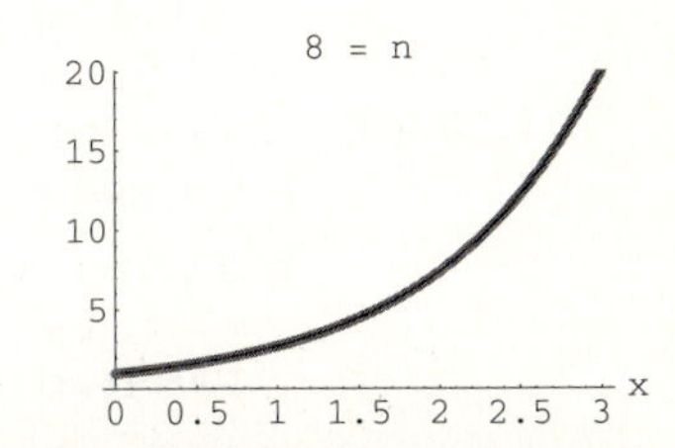

Describe what you see.

G.7.d.i) Find a positive integer n such that under$[x, n]$ is e^x to eight accurate decimals for all x's with $0 \leq x \leq 1$.

G.7.d.ii) What would you do to get even more accuracy?

G.7.d.iii) What's a half-decent way of calculating values of e^x on a calculator that can add, subtract, multiply, and divide but cannot calculate exponentials?

■ G.8) Euler's faker and the second derivative

This problem appears only in the electronic version.

■ G.9) Inequalities

G.9.a) Use the Race Track Principle to explain why the inequality

$$\log[x] \leq \frac{\sqrt{x} - 1}{\sqrt{x}}$$

is true for $x \geq 1$. Illustrate with a plot.

G.9.b) Fix a number $y \geq 0$ and keep it constant. Use the Race Track Principle to explain why $\sqrt{x} - \sqrt{y} \leq \sqrt{x - y}$ for $x \geq y$. Illustrate with a plot in the case $y = 10$.

■ G.10) The Law of the Mean

The Law of the Mean is sometimes called the Mean Value Theorem.

G.10.a.i) Here is a function $f[x]$ and a plot of its derivative on the interval $a \leq x \leq b$ for $a = 1$ and $b = 2$:

In[37]:=

```
a = 1; b = 2;
Clear[f,x]; f[x_] = (x^2 + x) E^(-x)
```

Out[37]=

```
     2
x + x
------
   x
  E
```

In[38]:=

```
growthplot =
Plot[f'[x],{x,a,b},
AxesLabel->{"x","f'[x]"}];
```

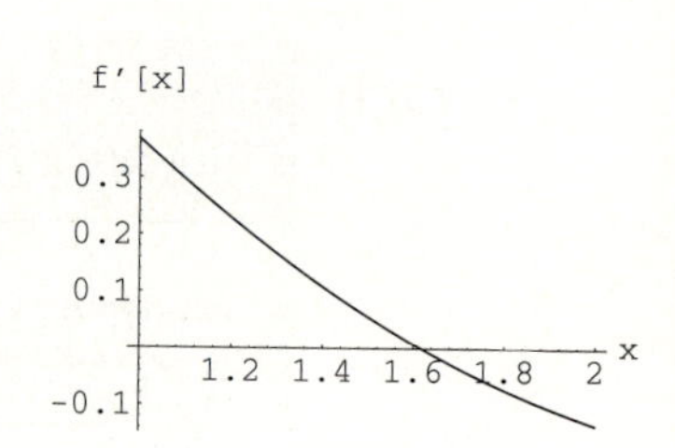

Read from the plot that

$$\text{mingrowth} = f'[2] \leq f'[x] \leq f'[1] = \text{maxgrowth}$$

for all the x's with $a \leq x \leq b$. Now plot on the same interval $f[x]$ and the lines that go through $\{a, f[a]\}$ with constant growth rates mingrowth and maxgrowth:

In[39]:=

```
mingrowth = f'[2]; maxgrowth = f'[1]; Clear[upline,downline]
upline[x_] = maxgrowth (x - a) + f[a];
downline[x_] =  mingrowth (x - a) + f[a];
lines = Plot[{f[x],upline[x],downline[x]}, {x,a,b},PlotStyle->{GrayLevel[0.4],
Red,Red},DisplayFunction->Identity];
pointplots = Graphics[{Red,PointSize[0.03],
Point[{a,f[a]}]},{Red,PointSize[0.03], Point[{b,f[b]}]}];
labels = Graphics[Text["{a,f[a]}",{a,f[a]},{-1,2}], Text["{b,f[b]}",{b,f[b]},{1,2}]];
```

In[40]:=

```
Show[lines,pointplots,labels,
AxesLabel->{"x",""},
DisplayFunction->$DisplayFunction];
```

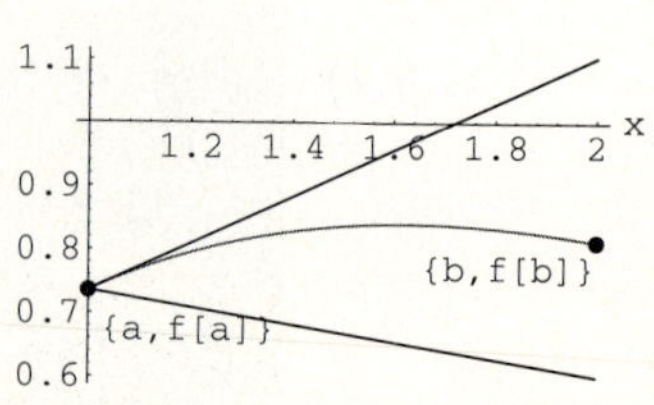

> Why did the plot turn out this way?

G.10.a.ii) Here is another function $f[x]$ and a plot of its derivative on the interval $a \le x \le b$ for $a = 1.7$ and $b = 3.5$:

In[41]:=
```
a = 1.7; b = 3.5; Clear[f,x]
f[x_] = -447 + 954 x - 720 x^2 + 260 x^3 - 45 x^4 + 3 x^5;
```

In[42]:=
```
growthplot =
Plot[f'[x],{x,a,b},
AxesLabel->{"x","f'[x]"}];
```

Read from the plot that

$$\text{mingrowth} = f'[2] \le f'[x] \le f'[3] = \text{maxgrowth}$$

for all the x's with $a \le x \le b$ where $a = 1.7$ and $b = 3.5$. Now plot on the same interval $f[x]$ and the lines that go through $\{a, f[a]\}$ with constant growth rates mingrowth and maxgrowth:

In[43]:=
```
mingrowth = f'[2]; maxgrowth = f'[3];
Clear[upline,downline]
upline[x_] = maxgrowth (x - a) + f[a];
downline[x_] =  mingrowth (x - a) + f[a];
lines = Plot[{f[x],upline[x],downline[x]},
{x,a,b},PlotStyle->{GrayLevel[0.4],
Red,Red},DisplayFunction->Identity];
pointplots = Graphics[{Red,PointSize[0.03],
Point[{a,f[a]}]},{Red,PointSize[0.03],
Point[{b,f[b]}]}];
labels = Graphics[Text["{a,f[a]}",{a,f[a]},{-1,2}],
Text["{b,f[b]}",{b,f[b]},{1,2}]];
```

In[44]:=
```
Show[lines,pointplots,labels,
AxesLabel->{"x",""},
DisplayFunction->$DisplayFunction];
```

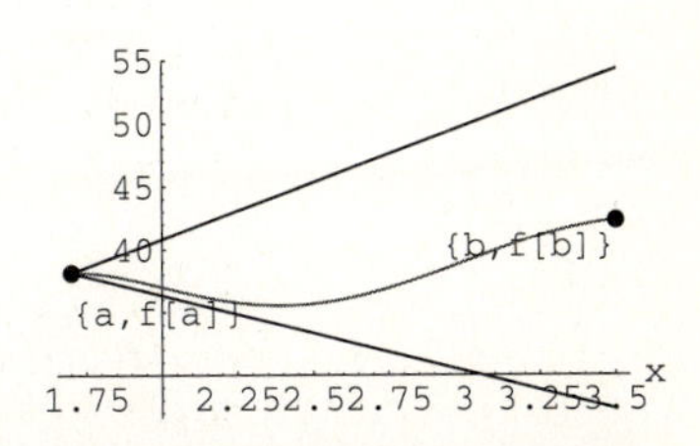

> Why did the plot turn out this way?

G.10.b.i)

Given $f[x]$, a and b with $a < b$. If you plot $f'[x]$ for $a \leq x \leq b$ and read off, as above, the maximum growth rate = maxgrowth and the minimum growth rate = mingrowth and then you put

$$\begin{aligned}\text{upline}[x] &= \text{maxgrowth}\,(x-a) + f[a]\\ \text{downline}[x] &= \text{mingrowth}\,(x-a) + f[a],\end{aligned}$$

why are you sure that $\text{downline}[b] \leq f[b] \leq \text{upline}[b]$?

G.10.b.ii)

The inequality $\text{downline}[b] \leq f[b] \leq \text{upline}[b]$ above is the same as

$$\text{mingrowth}\,(b-a) + f[a] \leq f[b] \leq \text{maxgrowth}\,(x-a) + f[a].$$

This is the same as

$$\text{mingrowth}\,(b-a) \leq f[b] - f[a] \leq \text{maxgrowth}\,(x-a)\,.$$

Explain why this tells you that there is a distinguished point c with $a \leq c \leq b$ such that

$$f[b] - f[a] = f'[c]\,(b-a)\,.$$

This is called the Law of the Mean or the Mean Value Theorem.

G.10.b.iii)

The Law of the Mean, as above, says that if you are given a and b with $a < b$, then there is a happy camper c with $a \leq c \leq b$ such that

$$f[b] - f[a] = f'[c]\,(b-a)\,.$$

In traditional courses, the Law of the Mean is proved using a quick trick, and then versions of the Race Track Principle are sometimes alluded to. Just for the fun of it, try to use the Law of the Mean to explain why it is that if $f'[x] > 0$ for $a \leq x \leq b$, then $f[b] > f[a]$.

LESSON 1.08

More Differential Equations

Basics

■ B.1) Euler's faker and *Mathematica*'s faker

There is only one new tool in this lesson, but quite a tool it is. Programmed into *Mathematica* is an instruction called NDSolve that will accomplish for you everything that Euler's method did for faking the plot of a solution of a differential equation. And there's an added plus: You don't have to screw around with setting the number of iterations.

Here's the NDSolve instruction working to get an excellent fake plot for $0 \leq x \leq 4$ of the differential equation

$$y'[x] = 3.2\sin[y[x]] - 0.3\,x^2$$

with $y[0] = 0.2$:

In[1]:=

```
a = 0; b = 4; starter = 0.2;
Clear[solution,x,y,fakey]
solution =
NDSolve[{y'[x] == 3.2 Sin[y[x]] - 0.3 x^2,
y[0] == starter},y[x],{x,a,b}];
fakey[x_] = y[x]/.solution[[1]];
masterfakeplot = Plot[fakey[x],{x,a,b},
PlotStyle->{{Blue,Thickness[0.01]}},
AxesLabel->{"x","y[x]"}, AxesOrigin->{a,starter},
PlotRange->All,PlotLabel->"Master forgery"];
```

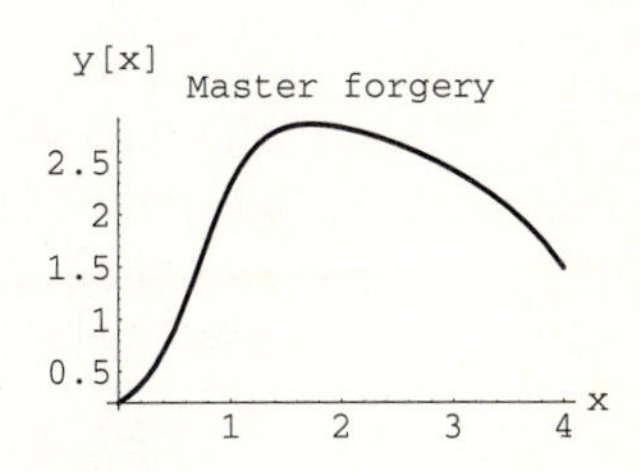

The NDSolve instruction is just waiting for you to use it.

B.1.a) Discuss the mathematical principles behind the NDSolve instruction.

Answer: The mathematical principles behind the NDSolve instruction are the same as the mathematical principles behind Euler's method. Except in the case of the NDSolve instruction, extra features typical of professionally written software are included. You'll hear more about these extra features later in Calculus&*Mathematica*.

B.1.b) Look at this:

In[2]:=

```
a = 0; b = 4; starter = 0.2; Clear[solution,x,y,fakey];
solution = NDSolve[{y'[x] == 3.2 Sin[y[x]] - 0.3 x^2,
y[0] == starter},y[x],{x,a,b}]; fakey[x_] = y[x]/.solution[[1]]
```

Out[2]=

```
InterpolatingFunction[{0., 4.}, <>][x]
```

What does the output mean?

Answer: This output reflects the fact that NDSolve first produces a bunch of points and then strings them together with an interpolating function—just as Euler's method does. The formula for this interpolating function is not available, but you can plot it:

In[3]:=

```
masterfakeplot=
Plot[fakey[x],{x,a,b},
PlotStyle->{{Blue,Thickness[0.01]}},
AxesLabel->{"x","y[x]"}, AxesOrigin->{a,starter},
PlotRange->All,PlotLabel->"Master forgery"];
```

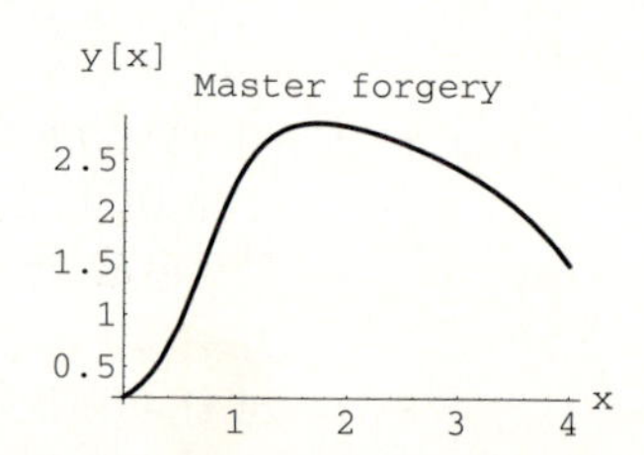

B.1.c) But isn't the actual formula always better than a fake?

Answer: It all depends on what you want to do. If what you want is a plot, then the fake should be as good as the actual formula. If you want to do theoretical analysis of the true solution, then the actual formula is what you want. The only trouble is that the actual formula is available only in special situations; the fake is nearly always available.

B.1.d) Does this mean that you should give up on Euler's method?

Answer: Hell no. For serious practical situations, you want something better, but as a theoretical tool to use to unlock more secrets of calculus, Euler's method is just the ticket. You'll see more about Euler's method from time to time in Calculus&*Mathematica*.

■ B.2) Simultaneous differential equations: The predator-prey model

This predator-prey model was orginally proposed and studied by Volterra in 1926 in an effort to explain the oscillatory levels of certain fish harvests in the Adriatic Sea.

Here's the idea: Two species coexist in a closed environment. One species, the predator, feeds on the other, the prey. There is always plenty of food for the prey, but the predators eat nothing but the prey. Put

$$\text{pred}[t] = \text{population of predators at time } t.$$

$$\text{prey}[t] = \text{population of prey at time } t.$$

It's reasonable to assume that there are positive constants a and b such that:

$$\text{prey}'[t] = a\,\text{prey}[t] - b\,\text{prey}[t]\,\text{pred}[t]$$

because the abundance of food for the prey allows the birth rate of the prey to be proportional to their current number, and the death rate of prey is proportional to both the current number of prey and the current number of predators.

It also makes some sense to assume that there are positive constants c and d such that:

$$\text{pred}'[t] = -c\,\text{pred}[t] + d\,\text{pred}[t]\,\text{prey}[t]$$

because it's reasonable to assume that the death rate of the predators is likely to be proportional to the current population of predators and that the birth rate of the predators is proportional to both the current number of the predators and the size of the food supply (the prey). This gives you two simultaneous differential equations

$$\text{prey}'[t] = a\,\text{prey}[t] - b\,\text{prey}[t]\,\text{pred}[t],$$
$$\text{pred}'[t] = -c\,\text{pred}[t] + d\,\text{pred}[t]\,\text{prey}[t].$$

Simultaneous differential equations are just the ticket when you want to see how one process interacts with another.

B.2.a) Examine what happens for the sample choice of proportionality constants $a = 0.7$, $b = 0.3$, $c = 0.44$, and $d = 0.08$ with the prey starting off with a population of four units and the predators starting out with a population of one unit.

Here the units could be thousands or millions so that a population of 0.35 can make sense.

Answer: Here is a plot of *Mathematica*'s faker of the prey population as a function of time t for the first 50 time units:

In[4]:=
```
endtime = 50; a = 0.7; b = 0.3; c = 0.44; d = 0.08;
Clear[pred,prey,t,Derivative]
approxsolutions =
NDSolve[{prey'[t] ==  a prey[t] - b prey[t] pred[t],
pred'[t] == -c pred[t] + d pred[t] prey[t],
prey[0] == 4, pred[0] == 1},
{prey[t],pred[t]},{t,0,endtime}];
Clear[fakepred,fakeprey]
fakepred[t_] = pred[t]/.approxsolutions[[1]];
fakeprey[t_] = prey[t]/.approxsolutions[[1]];
preyplot = Plot[fakeprey[t],{t,0,endtime},
PlotStyle->{{Blue,Thickness[0.01]}},
AxesLabel->{"t","prey"}];
```

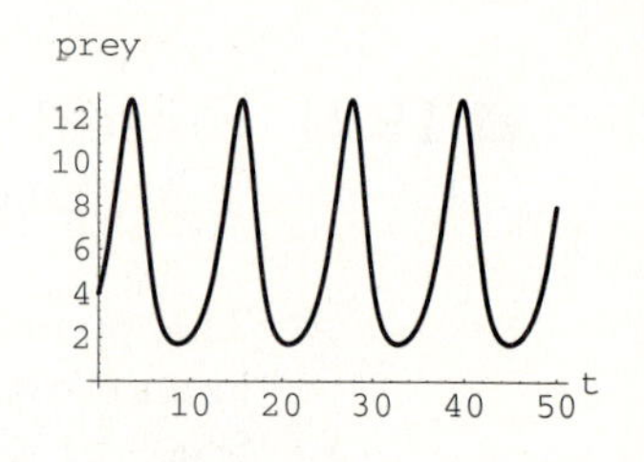

Hey! A cyclic pattern.

And the predator population as a function of time t:

In[5]:=
```
predatorplot =
Plot[fakepred[t],{t,0,endtime},
PlotStyle->{{Red,Thickness[0.005]}},
AxesLabel->{"t","predators"}];
```

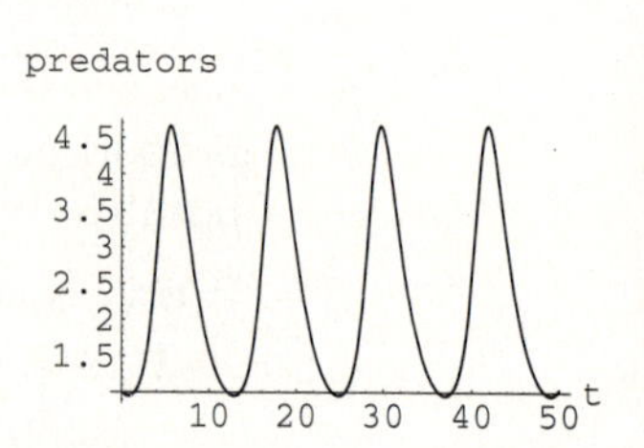

Another cyclic pattern. Here they are together: The plot of the prey population is thicker than the plot of the predator population.

In[6]:=
```
both =
Show[predatorplot,preyplot,AxesLabel->{"t",""}];
```

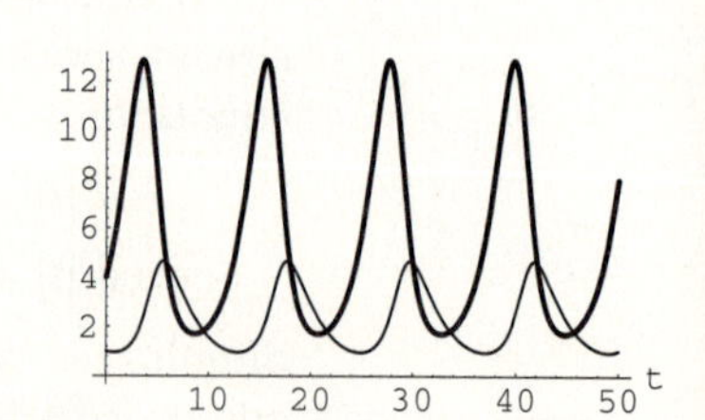

Look at those populations reacting with each other.

B.2.b) Look again at the plot and discuss the relationships between the curves.

Answer: Cyclic oscillations of both populations are evident as all get out. Feast your eyes on the vivid relationship between the crests and dips of the predator population and those of the prey population! The cycles are slightly out of sync. When you think about it, they should be slightly out of sync.

When there are enough prey to sustain strong predator growth, the predators begin to grow so much that they outstrip their food supply. This puts both populations into a decline until the point at which the predators are not a powerful menace; then the prey become numerous enough to support strong predator growth. The cycles go on and on.

B.2.c) Estimate the length of one cycle in time units.

Plot one cycle, two cycles, three cycles, and four cycles.

Answer: Put in horizontal lines shooting out from the vertical axis reflecting the information that pred[0] = 1 and prey[0] = 4.

In[7]:=

```
bothwithlines =
Show[both,Graphics[{Line[{{0,1},{50,1}}],
Line[{{0,4},{50,4}}]}]];
```

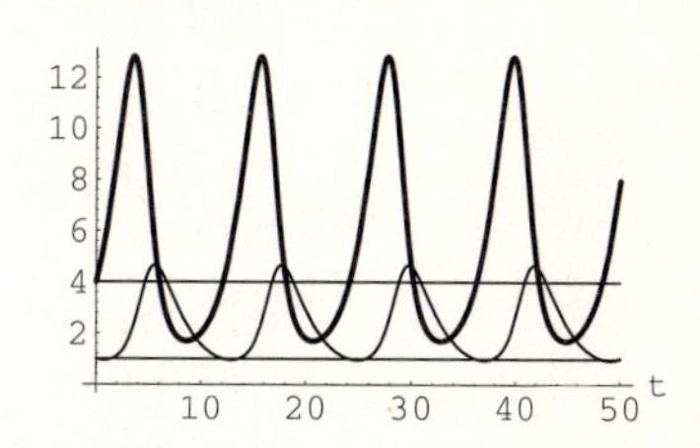

A closer look:

In[8]:=

```
Show[bothwithlines,PlotRange->{{0,15},Automatic}];
```

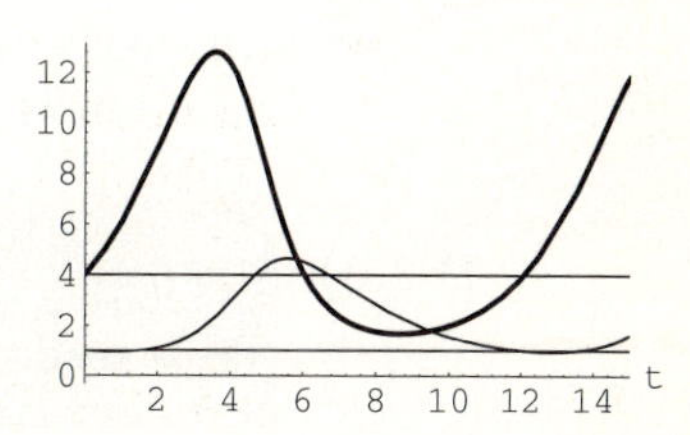

One cycle is about 12 time units by eyeball estimate. Use *Mathematica* to improve on this:

In[9]:=

```
FindRoot[fakeprey[t] == 4,{t,12}]
```

Out[9]=

```
{t -> 12.0777}
```

Check for the predators too:

In[10]:=

```
FindRoot[fakepred[t] == 1,{t,12}]
```

Out[10]=

```
{t -> 12.0777}
```

The eyeball was not so bad. A plot of one cycle:

In[11]:=
```
cycle = 12.0777;
Plot[{fakeprey[t],fakepred[t]},{t,0,cycle},
PlotStyle->{{Blue,Thickness[0.01]},
{Red,Thickness[0.01]}}, AxesLabel->{"t",""},
PlotRange->{{0,4 cycle},Automatic},
PlotLabel->"one cycle"];
```

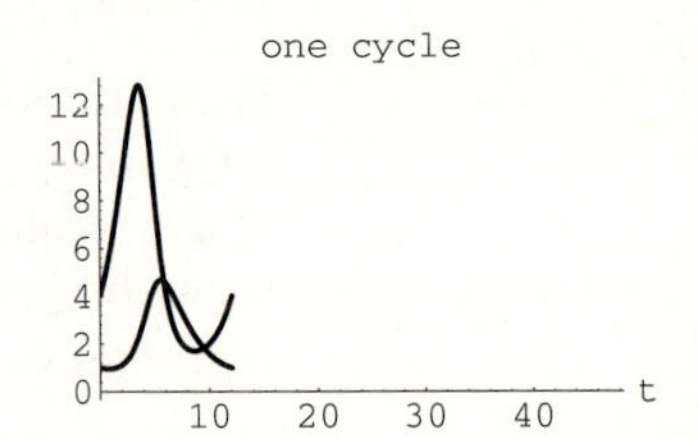

Two cycles:

In[12]:=
```
Plot[{fakeprey[t],fakepred[t]},{t,0,2 cycle},
PlotStyle->{{Blue,Thickness[0.01]},
{Red,Thickness[0.01]}}, AxesLabel->{"t",""},
PlotRange->{{0,4 cycle},Automatic},
PlotLabel->"two cycles"];
```

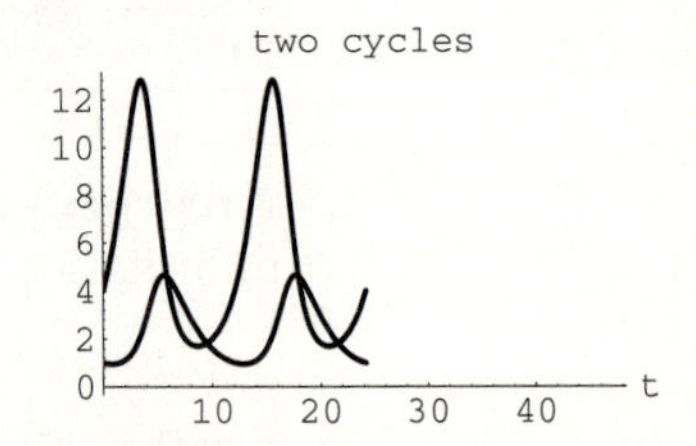

As William E. Boyce and Richard DiPrima point out in their book *Elementary Differential Equations* (Wiley, 1977), "Cyclic variations of predator and prey as [predicted above] are often observed in nature." In fact E.P. Odum (*Fundamentals of Ecology*, Saunders, 1953) was able to use the fur catch records of the Hudson Bay Company from 1850 to 1930 to estimate the Canadian lynx and snowshoe hare populations during these years. Odum's curves are not too different in character from the curves you saw above.

All mathematical ecologists agree that the predator-prey model is not an end in itself but is important as a tool in the quest for asking the right questions.

B.2.d) Why all this fooling around with fake plots?

Wouldn't actual formulas for pred[t] and prey[t] be more convenient?

Answer: Maybe you've seen this question before. It all depends on what you want to do. If what you want is a plot, then the fake should be as good as the actual formula. If you want to do theoretical analysis of the true solution, then the actual formula is what you want. The only trouble is that the actual formula is available only in special situations; the fake is always available. In the case of the predator-prey model, the formulas so treasured in old-fashioned math classes are not available. Try it:

In[13]:=
```
a = 0.7; b = 0.3; c = 0.44; d = 0.08;
Clear[pred,prey,t,Derivative]
DSolve[{prey'[t] ==  a prey[t] - b prey[t] pred[t],
pred'[t] == -c pred[t] + d pred[t] prey[t],
prey[0] == 4,pred[0] == 1}, {prey[t],pred[t]},t]
```

Out[13]=

```
DSolve[{prey'[t] == 0.7 prey[t] - 0.3 pred[t] prey[t],
        pred'[t] == -0.44 pred[t] + 0.08 pred[t] prey[t],
        prey[0] == 4, pred[0] == 1}, {prey[t], pred[t]}, t]
```

Mathematica failed to come up with formulas for pred$[t]$ and prey$[t]$. You can't chalk this up as a fault of *Mathematica*'s, because no scientist has ever succeeded in finding formulas for pred$[t]$ and prey$[t]$. For predator-prey, the fake is the only game in town.

B.2.e) Where do you go for further reading on predator-prey models?

Answer: Predator-prey models have been (and still are) under heavy study by biologists and mathematicians. For further reading, here are some good places to begin:

1. The next lesson on parametric plotting.
2. E. Batschelet, *Introduction to Mathematics for Life Scientists*, Springer-Verlag, 1979. (introductory and clear)
3. Peter Lax, Samuel Burstein, and Anneli Lax, *Calculus with Applications and Computing*, Springer-Verlag, 1976. (serious mathematics)
4. J. D. Murray, *Mathematical Biology*, Springer-Verlag, 1990. (serious mathematics and serious biology)

Few traditional calculus or differential equations courses attempt to touch this topic, and you know why.

B.2.f) What version of Euler's method underlies *Mathematica*'s fake plots above?

Answer: To fake a plot of the solution of $y'[x] = f[x, y[x]]$, Euler's method uses

```
next[{x,y}] = {x,y} + jump{1, f[x,y]}
point[k] = N[next[point[k-1]]]
```

To fake a plot of the solution of the simultaneous differential equations

```
x'[t] = f[t,y[t], x[t]]
y'[t] = g[t,x[t], y[t]]
```

Euler's method uses:

```
next[{t,x,y}] = {t,x,y} + jump{1,f[t,x,y], g[t,x,y]}
point[k] = point[k] = N[next[point[k-1]]]
```

The fake points for $x[t]$ as a function of t are fished out from the first two slots of point$[k]$, and the fake points for $y[t]$ as a function of t are fished out from the first and the third slots of point$[k]$. You could program it yourself, but why bother?

Tutorials

■ T.1) Using a differential equation to analyze Bubba's toot

Calculus&*Mathematica* thanks pharmacokineticist Professor Al Staubus of the College of Pharmacy at Ohio State University and Medical Doctor Jim Peterson of Urbana, Illinois for help on this problem.

T.1.a) You may remember this from Lesson 1.01. This time you'll see how to use a differential equation to analyze Bubba's sobriety (or lack thereof). Here are some facts:

→ A 12-ounce long neck beer contains 13.6 grams of alcohol.

→ The typical human liver can eliminate 12 grams of alcohol per hour.

→ In all states, if your body contains more than 1 gram of alcohol per liter of body fluids, then you are too drunk to drive legally.

→ In some states, if your body contains more than 0.8 grams of alcohol per liter of body fluids, then you are too drunk to drive legally.

→ Bubba weighs 187 pounds.

→ A college-age male in good shape (as Bubba is) weighing K kilograms has about $0.68\,K$ liters of fluids in his body.

To estimate the number of liters of fluids in Bubba's body, you use:

In[1]:=
```
Convert[187 PoundWeight,KilogramWeight]
```
Out[1]=
```
84.8217 KilogramWeight
```
In[2]:=
```
Bubbafluids = 0.68 (84.8217)
```
Out[2]=
```
57.6788
```

This is in liters.

Here is Bubba's plan for the big party: He plans to drink plenty of beer at the party. But he is also worried about his driver's license. So he comes up with the scheme: He'll drink beer gradually at the rate of one 12-ounce long neck every 10 minutes for the first hour; then he'll nurse one 12-ounce beer every half hour for the next two hours. After all this beer, he'll stop drinking and just sit around looking cool until he's sure his blood alcohol concentration is down to a safe, legal level for driving.

Here is the function of time (in hours) that measures Bubba's intake rate of alcohol in grams per hour and a plot for the first five hours:

In[3]:=
```
beer = 13.6; Clear[inrate,t];
inrate[t_] := 6 beer/;0 <= t < 1;
inrate[t_] := 2 beer/;1 <= t < 3;
inrate[t_] := 0/;3 <= t; Plot[inrate[t],{t,0,5},
PlotStyle-> {{Thickness[0.01],YellowBrown}},
PlotRange->All,AxesLabel->{"t","inrate[t]"}];
```

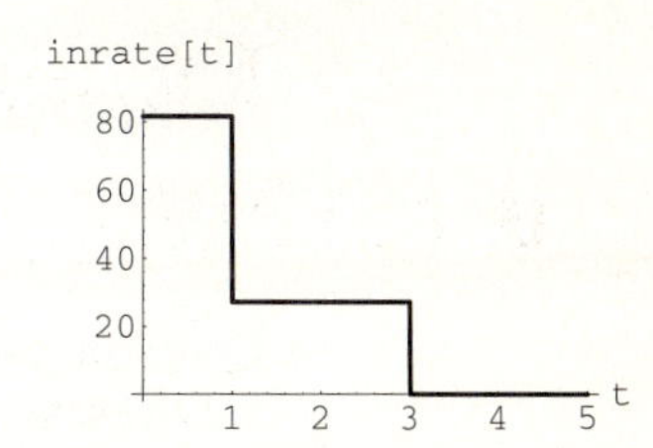

Assuming Bubba's liver is normal, you can say the function of time (in hours) that measures his elimination rate of alcohol in grams per hour is:

In[4]:=
```
Clear[outrate,t]; outrate[t_] = 12;
```

> Use the NDSolve instruction to help give a plot of the function that measures the alcohol (in grams) that will be found in Bubba's body anytime during the first nine hours. Then give a plot of the function that measures the concentration of alcohol in grams per liter in Bubba's bodily fluids.
>
> Estimate how long Bubba will have to stay at the party before he becomes a legal driver.

Answer: Call the function that measures the alcohol (in grams) that will be found in Bubba's body anytime during the first eight hours by the name alcohol$[t]$. Assuming Bubba has no residue from the night before, you know that at the start of the party ($t = 0$), alcohol$[0] = 0$. You also know that

$$\text{alcohol}'[t] = \text{inrate}[t] - \text{outrate}[t].$$

Here's the plot of alcohol$[t]$ for the first nine hours:

In[5]:=
```
beer = 13.6; Clear[inrate,outrate,t];
inrate[t_] := 6 beer/;0 <= t < 1;
inrate[t_] := 2 beer/;1 <= t < 3;
inrate[t_] := 0/;3 <= t; outrate[t_] = 12;
a = 0; b = 9; starter = 0;
Clear[solution,t,y,alcohol]
solution =
NDSolve[{y'[t] == inrate[t] - outrate[t],
y[0] == starter},y[t],{t,a,b}];
alcohol[t_] = y[t]/.solution[[1]];
Plot[alcohol[t],{t,a,b},
PlotStyle->{{YellowBrown,Thickness[0.01]}},
AxesLabel->{"t","alcohol[t]"},
AxesOrigin->{a,starter}, PlotRange->All];
```

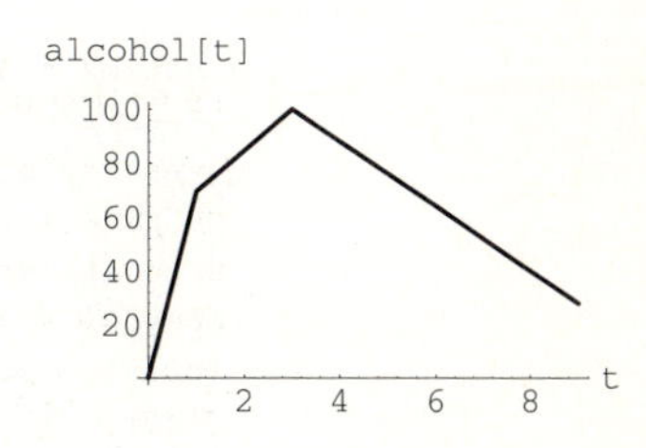

Call the function that measures the concentration of alcohol in grams per liter in Bubba's body fluids by the name concentration$[t]$. This function is given by:

In[6]:=

```
Clear[concentration]; concentration[t_] = alcohol[t]/Bubbafluids
```

Out[6]=

```
0.0173374 InterpolatingFunction[{0., 9.}, <>][t]
```

Here comes the plot showing concentration[t] together with the legal thresholds 1.0 grams per liter, 0.8 grams per liter, and the conservative threshold 0.5 grams per liter:

In[7]:=

```
Plot[{concentration[t],1.0,0.8,0.5},{t,a,b},
PlotStyle->{{YellowBrown,Thickness[0.01]},
Red,Red,Green},
AxesLabel->{"t","concentration[t]"},
PlotRange->All];
```

concentration[t]
1.75
1.5
1.25
1
0.75
0.5
0.25
2 4 6 8 t

In some states, Bubba can legally drive home about seven hours after the party begins. In other states, he must stay at the party for nearly eight hours. Just to be on the safe side, it might be a good idea for Bubba to stay at the party a full nine hours.

■ T.2) Analysis of the predator-prey model

T.2.a) Here is what happens in the predator-prey model for the sample choice of proportionality constants $a = 0.2$, $b = 0.05$, $c = 0.28$, and $d = 0.04$ with the prey starting off with a population of 20 units and the predators starting out with a population of two units.

In[8]:=

```
endtime = 50; a = 0.2; b = 0.05; c = 0.28; d = 0.04;
Clear[pred,prey,t,Derivative]
approxsolutions =
NDSolve[{prey'[t] ==  a prey[t] - b prey[t] pred[t],
pred'[t] == -c pred[t] + d pred[t] prey[t],
prey[0] == 20, pred[0] == 2},
{prey[t],pred[t]},{t,0,endtime}];
Clear[fakepred,fakeprey]
fakepred[t_] = pred[t]/.approxsolutions[[1]];
fakeprey[t_] = prey[t]/.approxsolutions[[1]];
predpreyplot =
Plot[{fakeprey[t],fakepred[t]},{t,0,endtime},
PlotStyle->{{Blue,Thickness[0.02]},
{Red,Thickness[0.01]}}, AxesLabel->{"t",""}];
```

20
15
10
5
10 20 30 40 50 t

The plot of the prey population is thicker than the plot of the predator population.

> Add the horizontal line that hits the vertical axis at a/b to this plot. Inspect the plot; then describe what you see and explain why you see it.

Answer: Here you go:

In[9]:=

```
Show[predpreyplot,
Graphics[Line[{{0,a/b},{endtime,a/b}}]]];
```

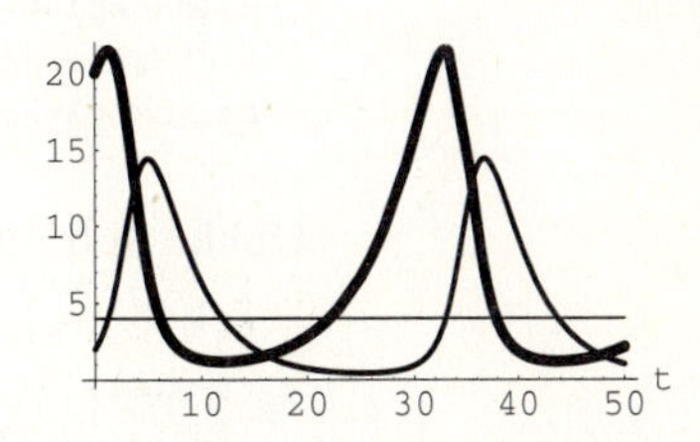

Interesting. When pred[t] $> a/b$, then prey[t] goes down. When pred[t] $< a/b$, then prey[t] goes up. And prey[t] takes its largest and smallest values when pred[t] $= a/b$.

To explain this phenomenon, look at one of the differential equations that went into this plot:

$$\begin{aligned}\text{prey}'[t] &= a\,\text{prey}[t] - b\,\text{prey}[t]\,\text{pred}[t] \\ &= \text{prey}[t]\,(a - b\,\text{pred}[t])\,.\end{aligned}$$

When pred[t] $> a/b$, then $(a - b\,\text{pred}[t]) < 0$. So when pred[$t$] $> a/b$, then

$$\text{prey}'[t] = \text{prey}[t]\,(a - b\,\text{pred}[t]) < 0.$$

That's why the prey curve goes down when pred[t] $> a/b$.

Similarly when pred[t] $< a/b$, then prey$'$[t] > 0. That's why the prey curve goes up when pred[t] $< a/b$.

When pred[t] $= a/b$, then prey$'[t] = \text{prey}[t]\,(a - b\,\text{pred}[t]) = 0$. That's why prey[$t$] takes its largest and smallest values when pred[t] $= a/b$. Amazing what a little calculus can do.

Give It a Try

Experience with the starred ($\star$) problems will be especially beneficial for understanding later lessons.

■ G.1) Variable interest rates★

If you borrow \$35,000 for 12 years and pay 8.9% interest compounded every instant, then the amount you owe at any time t (in years) after you borrow the \$35,000 plots out like this:

In[1]:=
```
Clear[balance,y,t]
starter = 35000;
solution =
DSolve[{y'[t] == 0.089 y[t],
y[0] == starter},y[t],t];
balance[t_] = y[t]/.solution[[1]];
balanceplot = Plot[balance[t],{t,0,12},
PlotStyle->{{DarkGreen,Thickness[0.01]}},
AxesLabel->{"years","balance owed"}];
```

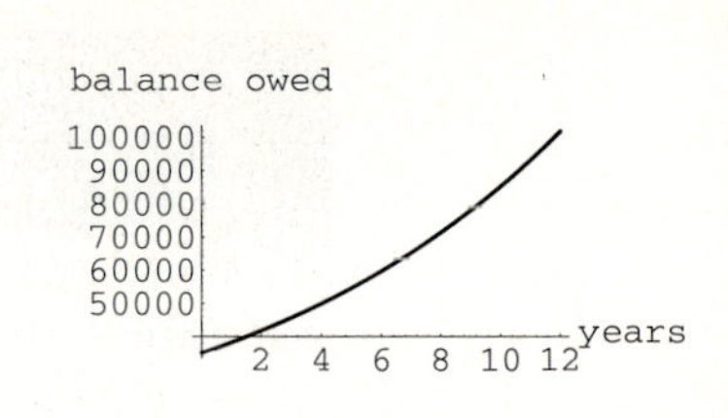

Ouch. A friend tells you about a deal a second bank is promoting. The deal is that you pay:

→ 9.9% interest for the first four years,

→ 8.9% interest for the second four years, and

→ 7.9% interest for the last four years.

In each case the interest is compounded every instant. Notice that the interest rates average out to 8.9% over the four years.

But your friend tells you that the second deal is better than the deal you got because the high rate (9.9%) is applied when the balance is small and the low rate (7.9%) is applied when the balance is high.

G.1.a) Set up a function $r[t]$ so that balance $2[t]$, the balance for the second scheme t years into the loan, solves the differential equation

$$\text{balance}\,2'[t] = r[t]\,\text{balance}\,2[t]$$

with balance $2[0] = 35{,}000$.

Use NDSolve to help get a plot of balance $2[t]$ for $0 \leq t \leq 12$.

Show this plot with the plot above and answer the questions:

Which is the better deal?

Why do you think the second bank is pushing this deal?

■ G.2) Drinking and driving

G.2.a) Melanie, a female student weighing 125 pounds, drives to the party with the idea of nursing three 12-ounce beers for the first hour and then nursing two 12-ounce beers each hour for the next three hours. If she starts with no alcohol in her system, how many grams of alcohol are there in each liter of her body fluids at the end of the four hours?

After the four hours, about how long must she go without drinking alcohol in order to reduce her body fluid alcohol concentration to a legal driving level of under 0.8 grams/liter?

Each 12-ounce beer Melanie is drinking contains 13.0 grams of alcohol.

■ G.3) Further analysis of the predator-prey model★

G.3.a.i) Here is what happens in the predator-prey model for the sample choice of proportionality constants $a = 0.15$, $b = 0.07$, $c = 0.36$, and $d = 0.04$ with the prey starting off with a population of 18 units and the predators starting out with a population of three units.

In[2]:=

```
endtime = 60;
a = 0.15; b = 0.07; c = 0.36; d = 0.04;
Clear[pred,prey,t,Derivative]
approxsolutions =
NDSolve[{prey'[t] ==  a prey[t] - b prey[t] pred[t],
pred'[t] == -c pred[t] + d pred[t] prey[t],
prey[0] == 18, pred[0] == 3},
{prey[t],pred[t]},{t,0,endtime}];
Clear[fakepred,fakeprey]
fakepred[t_] = pred[t]/.approxsolutions[[1]];
fakeprey[t_] = prey[t]/.approxsolutions[[1]];
predpreyplot =
Plot[{fakeprey[t],fakepred[t]},{t,0,endtime},
PlotStyle->{{Blue,Thickness[0.02]},
{Red,Thickness[0.01]}},PlotRange->All,
AxesLabel->{"t",""}];
```

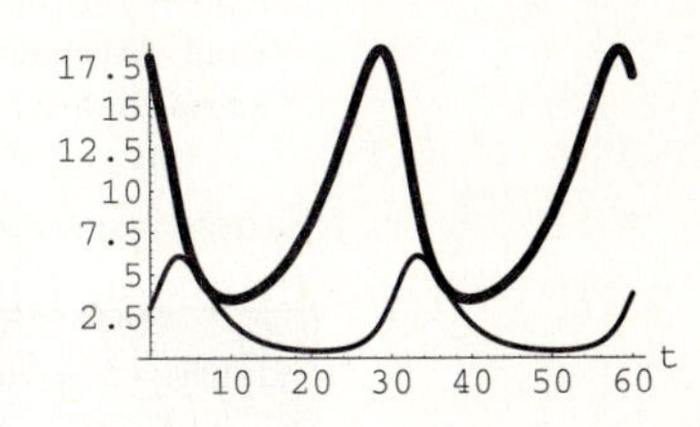

The plot of the prey population is thicker than the plot of the predator population.

Here is what you get when you add to the plot the horizontal line that hits the vertical axis at c/d:

In[3]:=

```
Show[predpreyplot,
Graphics[Line[{{0,c/d},{endtime,c/d}}]]];
```

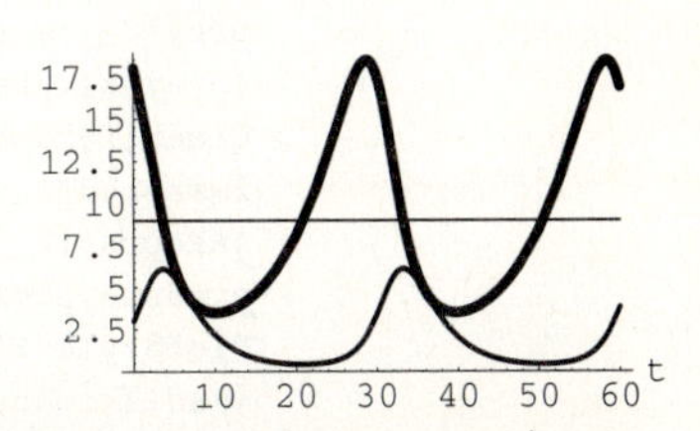

Inspect the plot; then describe what you see and explain why you see it.

G.3.a.ii) Here is what happens in the predator-prey model for the same proportionality constants $a = 0.15$, $b = 0.07$, $c = 0.36$, and $d = 0.04$ as in part G.3.a.i) but with the prey starting off with a population of c/d units and the predators starting out with a population of a/b units.

In[4]:=
```
endtime = 20; a = 0.15; b = 0.07; c = 0.36; d = 0.04;
Clear[pred,prey,t,Derivative]
approxsolutions =
NDSolve[{prey'[t] ==  a prey[t] - b prey[t] pred[t],
pred'[t] == -c pred[t] + d pred[t] prey[t],
prey[0] == c/d, pred[0] == a/b},
{prey[t],pred[t]},{t,0,endtime}];
Clear[fakepred,fakeprey]
fakepred[t_] = pred[t]/.approxsolutions[[1]];
fakeprey[t_] = prey[t]/.approxsolutions[[1]];
predpreyplot = Plot[{fakeprey[t],fakepred[t]},{t,0,endtime},
PlotStyle->{{Blue,Thickness[0.02]},
{Red,Thickness[0.01]}},PlotRange->All,
AxesLabel->{"t",""}];
```

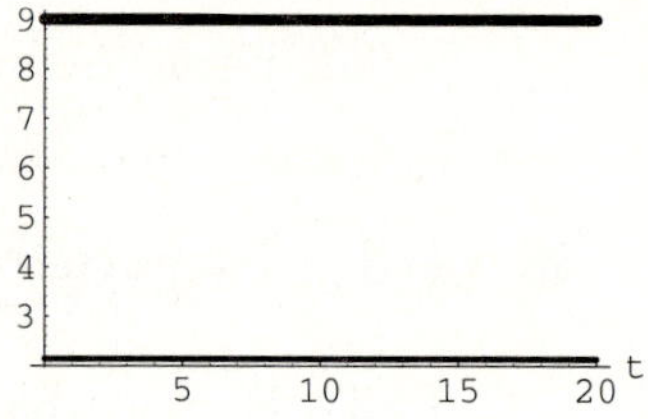

Fancy folks call this "equilibrium."

> Play with other choices of a, b, c, and d.
>
> Try to explain why this happened.

G.3.a.iii) Here is what happens in the predator-prey model for the same proportionality constants $a = 0.15$, $b = 0.07$, $c = 0.36$, and $d = 0.04$ as in part G.3.a.i) but with the prey starting off with a population of $c/d + 0.1$ units and the predators starting out with a population of $a/b + 0.1$ units.

In[5]:=
```
endtime = 60; a = 0.15; b = 0.07; c = 0.36; d = 0.04;
Clear[pred,prey,t,Derivative]
approxsolutions =
NDSolve[{prey'[t] ==  a prey[t] - b prey[t] pred[t],
pred'[t] == -c pred[t] + d pred[t] prey[t],
prey[0] == c/d + 0.1, pred[0] == a/b + 0.1},
{prey[t],pred[t]},{t,0,endtime}];
Clear[fakepred,fakeprey]
fakepred[t_] = pred[t]/.approxsolutions[[1]];
fakeprey[t_] = prey[t]/.approxsolutions[[1]];
predpreyplot = Plot[{fakeprey[t],fakepred[t]},{t,0,endtime},
PlotStyle->{{Blue,Thickness[0.02]},
{Red,Thickness[0.01]}},PlotRange->All,
AxesLabel->{"t",""}];
```

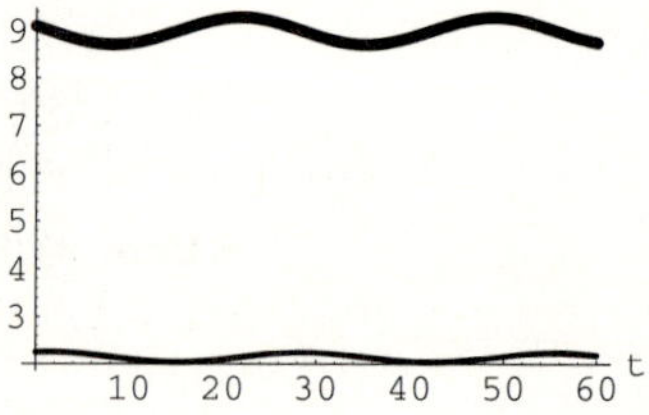

> Got any idea why this happened?

■ G.4) The drug equation★

Calculus&*Mathematica* thanks pharmacokineticist Professor Al Staubus, of the College of Pharmacy at Ohio State University, for lots of help on this problem.

Pharmacokineticists have determined that if a given human being has a given concentration of a given drug at time $t = 0$, then there are always positive constants A and k such that the amount in grams $y[t]$ of the drug in the body fluids is governed by the differential equation

$$y'[t] = \frac{-k\, y[t]}{A + y[t]}$$

with $y[0] =$ concentration at time $t = 0$.

Pharmacokineticists call this model the Michaelis-Menten equation and have used it since 1913. The original paper is Michaelis and Menten, "Die Kinetik der Invertinwirkung," *Biochem Z.*, 49(1913), 333–369. Before you get serious about this differential equation, there are two cases of big-time interest.

G.4.a.i) For drugs like cocaine, $y[0]$ is usually very small relative to A. In fact for cocaine, if $y[0]$ is big relative to A, then there is nothing to plot because a big $y[0]$ is lethal. Here is a sample with $y[0] = 0.0025$, $A = 6$, and $k = 3$:

In[6]:=
```
endtime = 10; A = 6; k = 3; starter = 0.0025;
Clear[t,y,Derivative]
solution = NDSolve[{y'[t] == -k y[t]/ (A + y[t]),
y[0] == starter},y[t],{t,0,endtime}];
Clear[coke]; coke[t_] = y[t]/.solution[[1]];
cokeplot = Plot[{coke[t]},{t,0,endtime},
PlotStyle->{{Blue,Thickness[0.015]}},
AxesLabel->{"t","cocaine"},PlotRange->All];
```


Holding the same constants, look at the differential equation again:

$$y'[t] = \frac{-k\, y[t]}{A + y[t]}.$$

Because $y[t]$ is very small all the time, you can expect the solution $y[t]$ to plot out about the same as the exponential solution of

$$y'[t] = \frac{-k\, y[t]}{A + 0} = \frac{-k}{A}\, y[t].$$

Check this out.

G.4.a.ii) Do a few more plots with $y[0]$ small relative to A and then explain the statement: For studying the decay of cocaine, most pharmacokineticists are happy to replace the basic model

$$y'[t] = \frac{-k\,y[t]}{A + y[t]}$$

with $y[0]$ = concentration at time $t = 0$ by the simpler exponential model

$$y'[t] = -\left(\frac{k}{A}\right) y[t]$$

with $y[0]$ = concentration at time $t = 0$.

G.4.a.iii) Explain why most folks say that cocaine concentration decays exponentially.

G.4.b.i) For drugs like alcohol, $y[0]$ is usually rather large relative to A. Here is a sample with $y[0] = 0.025$, $A = 0.005$, and $k = 0.01$:

In[7]:=

```
endtime = 10; A = 0.005; k = 0.01; starter = 0.025;
Clear[t,y,Derivative]
solution = NDSolve[{y'[t] == -k y[t]/ (A + y[t]),
y[0] == starter},y[t],{t,0,endtime}];
Clear[alcohol]
alcohol[t_] = y[t]/.solution[[1]];
sloshplot = Plot[{alcohol[t]},{t,0,endtime},
PlotStyle->{{Blue,Thickness[0.015]}},
AxesLabel->{"t","alcohol"}, PlotRange->All];
```

alcohol
0.025
0.02
0.015
0.01
0.005
2 4 6 8 10 t

Holding the same constants, look at the differential equation

$$y'[t] = \frac{-k\,y[t]}{A + y[t]}$$

again. Because $y[t]$ is large relative to A, you can expect

$$\frac{-k\,y[t]}{A + y[t]}$$

to be close to the global scale of $-k\,y/\,(A + y)$, which is $-k$.

As a result, you can expect the solution $y[t]$ to plot out about the same as the line solution of

$$y'[t] = -k$$

until $y[t]$ has decayed to the point at which it is very small.

Check this out.

G.4.b.ii) Do a few more with $y[0]$ large relative to A and then say why you think most pharmacokineticists are happy to replace the basic model

$$y'[t] = \frac{-k\,y[t]}{A + y[t]}$$

with $y[0] =$ concentration at time $t = 0$ by the simpler linear model

$$y'[t] = -k$$

with $y[0] =$ concentration at time $t = 0$.

G.4.b.iii) Explain why most folks say that alcohol concentration decays linearly until the alcohol concentration decays to the level at which no one cares.

■ G.5) War games

The first scientist who tried to apply calculus to warfare in the way you will see below was F. W. Lanchester, who worked in Britain during World War I. In his honor, fancy folks call all these models by the name Lanchesterian models.

G.5.a) The good guys and the bad guys are engaged in combat. Agree that good$[t]$ is the number of good guys still fighting t days after combat begins, and let bad$[t]$ be the number of bad guys still fighting t days after combat begins.

If neither side receives reinforcing troops, then why is it natural to propose a model

$$\text{good}'[t] = -\,a\,\text{bad}[t]$$
$$\text{bad}'[t] = -\,b\,\text{good}[t]$$

where a and b are positive numbers?

G.5.b.i) If the good guys bring in new troops at a rate of $f[t]$ new troops per day and the bad guys bring in new troops at a rate of $g[t]$ new troops per day, then why is it natural to propose a model

$$\text{good}'[t] = -\,a\,\text{bad}[t] + f[t]$$
$$\text{bad}'[t] = -\,b\,\text{good}[t] + g[t]$$

where a and b are positive numbers?

G.5.b.ii) Agree that the good guys win if bad$[t]$ becomes 0 before good$[t]$ becomes 0 and that the bad guys win if good$[t]$ becomes 0 before bad$[t]$ becomes 0.

The model

$$\text{good}'[t] = -0.5\,\text{bad}[t] + \sin[t]^2$$
$$\text{bad}'[t] = -0.5\,\text{good}[t] + \cos[t]^2$$

with good[0] = bad[0] = 2 suggests the situation in which the troops on both sides are equally effective and at the start ($t = 0$) they are of equal strength. The main difference is that the reinforcement rates are out of phase with the bad guys holding the initial advantage because

$$\cos[0]^2 = 1 > 0 = \sin[0]^2.$$

This suggests that the good guys lose.

> Check this out by plotting good[t] and bad[t] and showing both curves together.
>
> Determine how many days pass before the good guys are gone.

G.5.b.iii) The good guys find that they lose above, so they double their reinforcements. This gives the model:

$$\text{good}'[t] = -0.5\,\text{bad}[t] + 2\sin[t]^2$$
$$\text{bad}'[t] = -0.5\,\text{good}[t] + \cos[t]^2$$

with good[0] = bad[0] = 2.

> Who wins and how long does the battle last?

G.5.b.iv) This time the good guys have antiquated ammunition and heavy bureaucratic problems. As a result, their troops are only half as effective as the troops of the bad guys. This suggests the model:

$$\text{good}'[t] = -0.5\,\text{bad}[t] + 2\sin[t]^2$$
$$\text{bad}'[t] = -0.25\,\text{good}[t] + \cos[t]^2$$

with good[0] = bad[0] = 2.

> Who wins and how long does the battle last?

G.5.b.v) Again the good guys have antiquated ammunition and heavy bureaucratic problems. As a result, their troops are only half as effective as the troops of the bad guys, so they decide to find a positive number r as small as practical such that if

$$\text{good}'[t] = -0.5\,\text{bad}[t] + r\sin[t]^2$$
$$\text{bad}'[t] = -0.25\,\text{good}[t] + \cos[t]^2$$

with good[0] = bad[0] = 2, then the good guys do not lose during the first 15 days.

Estimate how small r can be.

G.5.c.i) Calculus&*Mathematica* is pleased to acknowledge the heavy influence of the book *Differential Equations and Their Applications* by Martin Braun, Springer-Verlag, 1978. If you like this stuff, you'll want to look at this book, paying special attention to section 4.3.

Iwo Jima is the largest of the volcano islands in the western Pacific Ocean. During World War II, it was a big Japanese air base. During February and March, 1945, American marines took it at great cost in a bloody assault. If you have been to Washington, D.C., you may have seen the Iwo Jima memorial statue which depicts U.S. Marines planting the Stars and Stripes on the beach during a very bloody stage of the battle.

The Americans kept careful records during their losses, and the records confirm that the Iwo Jima battle is correctly modeled by the ideas in the Lanchesterian model above. Here comes the nitty-gritty:

Agree that usa$[t]$ measures the number of active American troops and japan$[t]$ measures the number of active Japanese troops engaged in the battle t days after the battle began. For this battle, military scientists have arrived at the model

$$\text{usa}'[t] = -\,0.0544\,\text{japan}[t] + f[t]$$
$$\text{japan}'[t] = -\,0.0106\,\text{usa}[t] + g[t]$$

where $f[t]$ and $g[t]$ are the reinforcement rates.

The individual Japanese troop was five times more effective than the individual American troop because the Japanese held a defensive position.

At the beginning of the battle, usa$[0] = 5.4$ and japan$[0] = 2.15$ with units measured in tens of thousands. The reinforcement rate for Japan was

$$g[t] = 0.$$

There were no Japanese reinforcements.

In[8]:=
```
Clear[g,t]
g[t_] = 0;
```

The Americans piled in 5.4 (tens of thousands of) troops at the beginning of the battle ($t = 0$), 0 extra troops during the first and second day, 0.6 troops spread out over the third day, 0 troops during the fourth and fifth days, 1.3 troops spread out over the sixth day, and no other reinforcements. Remember that troops are measured in tens of thousands.

Here is a plot of $f[t]$:

In[9]:=
```
Clear[f,t]
f[t_] := 0.0/; t < 2;
f[t_] := 0.6/;2 <= t <= 3;
f[t_] := 0.0/;3 < t < 5;
f[t_] := 1.3/;5 <= t <= 6;
f[t_] :=  0.0/;6 < t;
Plot[f[t],{t,0,8},
PlotStyle->{{Thickness[0.015],Blue}},
AxesLabel->{"day","USA reinforce rate"},
PlotRange->All,AspectRatio->1/2];
```

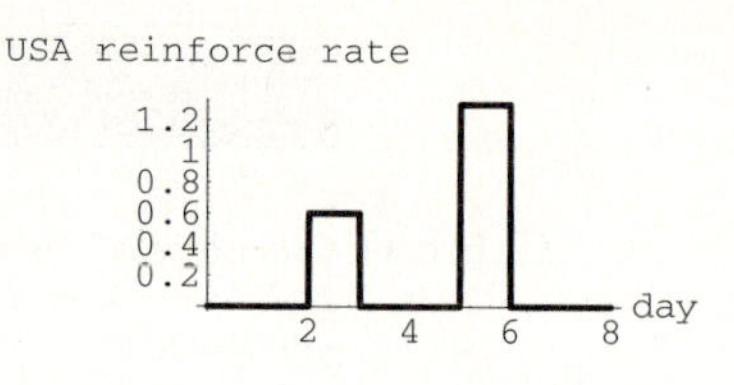

The Americans considered the island secure after 28 days of battle.

Replay the battle of Iwo Jima by giving plots of usa[t] and japan[t] for the first 28 days.

G.5.c.ii) In actuality, there were 5.28 (tens of thousands) American troops in the field on day 28. Compare this with the value predicted by the model.

G.5.c.iii) Estimate:

How many American troops were knocked out of action?

How many Japanese troops bit the dust?

G.5.d.i) Take the battle of Iwo Jima model

$$\text{usa}'[t] = -\,0.0544\,\text{japan}[t] + f[t]$$
$$\text{japan}'[t] = -\,0.0106\,\text{usa}[t] + g[t]$$

with usa[0] = 5.4 and japan[0] = 2.15 where $f[t]$ and $g[t]$ are the reinforcement rates.

Report what might have happened if the Americans had no reinforcements.

Would the Americans still have won in 28 days?

How long would it have taken to secure the island?

How many American troops would have been lost?

G.5.d.ii) Take the battle of Iwo Jima model

$$\text{usa}'[t] = -\,0.0544\,\text{japan}[t] + f[t]$$
$$\text{japan}'[t] = -\,0.0106\,\text{usa}[t] + g[t]$$

with usa[0] $= 5.4 - r$ and japan[0] $= 2.15$ where $f[t]$ and $g[t]$ are the reinforcement rates. But go with

In[10]:=

```
Clear[f,t]
f[t_] := 0.5 r/; t < 2;
f[t_] := 0.6/;2 <= t <= 3;
f[t_] := 0.0/;3 < t < 5;
f[t_] := 1.3/;5 <= t <= 6;
f[t_] :=  0.0/;6 < t;
```

and:

In[11]:=

```
Clear[g,t]; g[t_] = 0;
```

This means that you are withholding r ($0 \leq r \leq 5.4$) troops from the main assault force and are using them as reinforcements spread out over the first two days.

Is there a way to set r to improve the outcome from the American point of view?

G.5.d.iii) Put yourself in the role of the Japanese high command or the American high command or both and then try out possible scenarios that come to mind.

■ G.6) Logistic harvesting★

Take the logistic equation

$$y'[t] = a\, y[t] \left(1 - \frac{y[t]}{b}\right)$$

with $0 < a$ and $0 < b$ and start with $0 < y[0] < b$.

As t advances from 0, then you are guaranteed that $y[t]$ grows with some pep until $y[t]$ gets near b. Once $y[t]$ gets near b, then $y[t]$ settles into global scale with $\lim_{t\to\infty} y[t] = b$. Check it out:

In[12]:=

```
endtime = 52; a = 0.23; b = 1300; starter = 350;
Clear[t,y,Derivative]
solution = NDSolve[{y'[t] == a y[t] (1 - y[t]/b),
y[0] == starter},y[t],{t,0,endtime}];
Clear[fakey]; fakey[t_] = y[t]/.solution[[1]];
fakeplot = Plot[{fakey[t]},{t,0,endtime},
PlotStyle->{{Blue,Thickness[0.01]}},
DisplayFunction->Identity];
globalscale =  Graphics[{Red,Line[{{0,b},{endtime,b}}]}];
Show[fakeplot,globalscale, AxesLabel->{"t","y[t]"},
PlotRange->All, AspectRatio->1,
DisplayFunction->$DisplayFunction];
```

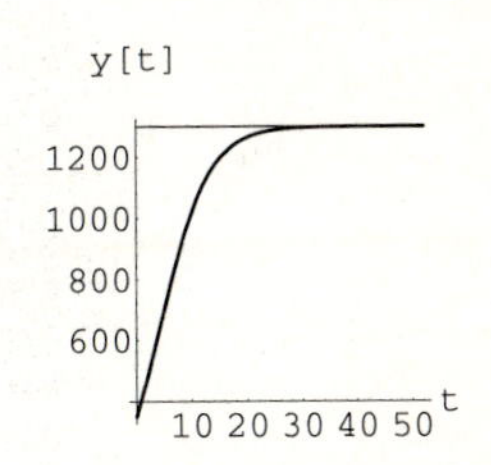

You get a reasonable interpretation of the logistic differential equation

$$y'[t] = a\,y[t]\left(1 - \frac{y[t]}{b}\right)$$

with $0 < y[0] < b$ by imagining that $y[t]$ is the number of catfish in a given lake on a catfish farm t weeks after the lake was stocked with $y[0]$ catfish. As time goes on, the catfish population increases until it reaches its steady-state population of b catfish.

But the catfish farmer doesn't grow catfish as pets; the farmer is in business to harvest catfish and to sell them so that hungry persons can fry them up and then wash them down with a couple of cold beers or iced teas.

G.6.a.i) Measure time in weeks and assume the farmer wants to harvest r fish per week and explain why

$$y'[t] = a\,y[t]\left(1 - \frac{y[t]}{b}\right) - r$$

lays the base for a reasonable model.

G.6.a.ii) The farmer opens the lake and lets the fish population increase to a level of 1000 fish. Then the farmer begins to harvest at a rate of $r = 100$ fish per week. Here's what happens over the next year:

This goes with specific data $a = 0.23$ and $b = 1300$. The reason that $b = 1300$ is that 1300 fish is the greatest number of fish the lake can support.

In[13]:=

```
endtime = 52; r = 100; a = 0.23; b = 1300;
starter = 1000;
Clear[t,y,Derivative]
solution = NDSolve[{y'[t] ==
a y[t] (1 - y[t]/b) - r,
y[0] == starter},y[t],{t,0,endtime}];
Clear[fakey]
fakey[t_] = y[t]/.solution[[1]];
fakeplot = Plot[{fakey[t]},{t,0,endtime},
PlotStyle->{{Blue,Thickness[0.015]}},
DisplayFunction->Identity];
withoutharvest =
Graphics[{Red,Line[{{0,b},{endtime,b}}]}];
Show[fakeplot,withoutharvest,
AxesLabel->{"t","y[t]"}, AspectRatio->1,
PlotRange->{-100,b},
DisplayFunction->$DisplayFunction];
```

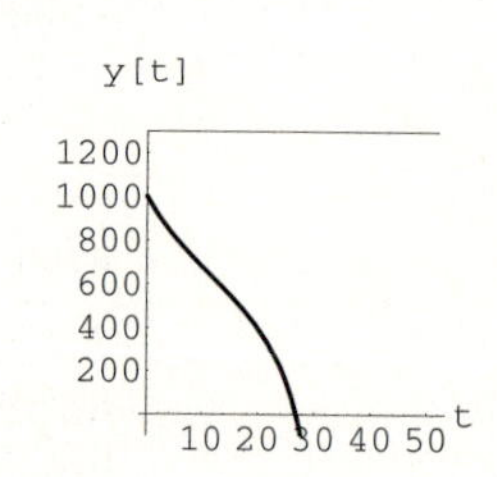

Don't abort. Disaster. In a little more than six months, the fish supply is gone. Here's what happens over a one-year period if the farmer lowers the weekly harvest from 100 to 50 fish:

In[14]:=

```
endtime = 52; r = 50; a = 0.23; b = 1300;
starter = 1000; Clear[t,y,Derivative]
solution =
NDSolve[{y'[t] == a y[t] (1 - y[t]/b) - r,
y[0] == starter},y[t],{t,0,endtime}];
Clear[fakey]; fakey[t_] = y[t]/.solution[[1]];
fakeplot = Plot[{fakey[t]},{t,0,endtime},
PlotStyle->{{Blue,Thickness[0.015]}},
DisplayFunction->Identity];
withoutharvest =  Graphics[{Red,Line[{{0,b},{endtime,b}}]}];
Show[fakeplot,withoutharvest,
AxesLabel->{"t","y[t]"}, AspectRatio->1,
PlotRange->{-100,b}, DisplayFunction->$DisplayFunction];
```

That's nice. The lake produces 50 fish per week and maintains a fish population of about 1000 fish at all times.

> Can the farmer harvest significantly more than 50 fish per week and avoid disaster?
>
> If so, then estimate how big the weekly harvest r can be before disaster sets in.

G.6.a.iii)

> Take your biggest possible weekly harvest estimate r from part G.1.a.ii) above and estimate how small $y[0] =$ starter can be before disaster sets in.

G.6.b.i) Look at these plots: The first is related to the case with $r = 100$ above, which ended up in a disaster.

In[15]:=

```
r = 100; a = 0.23; b = 1300;
Clear[f,y]
f[y_] = a y (1 - y/b) - r;
Plot[f[y],{y,0,b},AxesOrigin->{0,0},
AxesLabel->{"y",""}];
```

The second is related to the case with $r = 50$ above, which ended up with everything OK.

In[16]:=

```
r = 50; a = 0.23; b = 1300;
Clear[f,y]
f[y_] = a y (1 - y/b) - r;
Plot[f[y],{y,0,b},AxesOrigin->{0,0},
AxesLabel->{"y",""}];
```

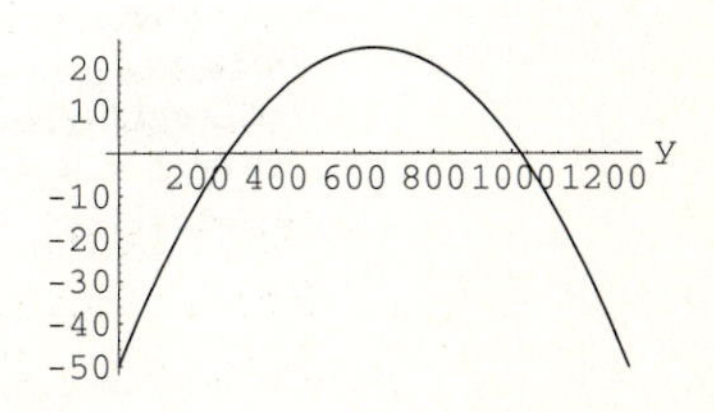

> Take your estimate of the largest r for which disaster is avoided and do the similar plot. Notice anything worth remarking on?

G.6.b.ii) Take any positive a, b, and r and look at:

In[17]:=
```
Clear[f,y,a,b,r]
f[y_] = a y (1 - y/b) - r;
Solve[f[y] == 0,y]
```

Out[17]=
```
             Sqrt[b] Sqrt[a b - 4 r]
        b + -----------------------
                   Sqrt[a]
{{y -> -------------------------------},
                     2

              Sqrt[b] Sqrt[a b - 4 r]
         b - -----------------------
                    Sqrt[a]
 {y -> -------------------------------}}
                      2
```

You can see that three cases pop up:

→ $f[y] = 0$ has two real roots; this happens when $a\,b - 4r > 0$.

→ $f[y] = 0$ has one real root; this happens when $a\,b - 4r = 0$.

→ $f[y] = 0$ has no real roots because $f[y] < 0$ for all y's; this happens when $a\,b - 4\,r < 0$.

This is just like high school algebra except that this is algebra with a purpose.

> Explain the following statements about the harvest model
>
> $$y'[t] = a\,y[t]\left(1 - \frac{y[t]}{b}\right) - r$$
>
> for any given positive a, b, and r (not just the a, b, and r given above):

→ If a, b, and r are set so that $a\,b - 4\,r < 0$, then, no matter what $y[0]$ is, when the farmer begins the harvest, disaster is inevitable.

→ If a, b, and r are set so that $a\,b - 4\,r > 0$, then harvesting can continue successfully provided $y[0]$ is bigger than at least one of the following numbers:

In[18]:=
```
Clear[f,y,a,b,r]
f[y_] = a y (1 - y/b) - r;
Solve[f[y] == 0,y]
```

Out[18]=

```
                    Sqrt[b] Sqrt[a b - 4 r]
                b + -----------------------
                            Sqrt[a]
        {{y -> -----------------------------},
                              2

                     Sqrt[b] Sqrt[a b - 4 r]
                 b - -----------------------
                             Sqrt[a]
         {y -> -----------------------------}}
                               2
```

→ Given a and b, you can get the biggest safe harvest rate r by setting r so that

$$a\,b - 4r = 0;$$

in other words $r = a\,b/4$ is the biggest harvest rate the farmer can get without disaster for a given positive a and b. And to take advantage of this number, the farmer should begin the harvest with $y[0] = \text{starter} > b/2$.

→ Given a and b, if the farmer sets the harvest rate at $r = a\,b/4$ but begins the harvest with $y[0] = \text{starter} < b/2$, then disaster is inevitable.

→ Given a and b, the risk-adverse farmer should begin the harvest with $y[0] =$ starter comfortably larger than $b/2$ and should set the harvest rate r so that r is comfortably smaller than $a\,b/4$.

■ G.7) The logistic predator-prey model

G.7.a) The basic predator-prey model is:

$$\text{prey}'[t] = a\,\text{prey}[t] - b\,\text{prey}[t]\,\text{pred}[t],$$
$$\text{pred}'[t] = -c\,\text{pred}[t] + d\,\text{pred}[t]\,\text{prey}[t]$$

where a, b, c, and d are given positive constants. In the absence of the predator, this gives you $\text{prey}'[t] = a\,\text{prey}[t]$ for a positive constant a.

> Explain why this allows the prey population to grow without bound in the absence of the predator.

G.7.b.i) As a response to the problem of allowing the prey to grow without bound, some scientists have proposed the logistic predator-prey model:

$$\text{prey}'[t] = a\,\text{prey}[t]\left(1 - \frac{\text{prey}[t]}{v}\right) - b\,\text{prey}[t]\,\text{pred}[t],$$
$$\text{pred}'[t] = -c\,\text{pred}[t] + d\,\text{pred}[t]\,\text{prey}[t]$$

where a, b, c, d, and v are positive constants.

> Why are you fairly certain that in the absence of the predator, prey[t] can never exceed v?
>
> Why are you even more certain that in the presence of the predator, prey[t] can never exceed v?

G.7.b.ii) Plot for the first 50 time units what happens for the sample choice of constants $a = 0.8$, $b = 0.20$, $c = 0.30$, $d = 0.04$, and $v = 10$ with the prey starting off with a population of three units and the predators starting out with a population of two units. These are the same data as in the plots at the beginning.

> Are the populations cyclic?

G.7.b.iii) Plot for the first 50 time units what happens for the same data but with $v = 10$ replaced by $v = 50$.

> Are the populations cyclic?

G.7.b.iv) Plot for the first 50 time units what happens for the same data but with $v = 50$ replaced by $v = 500$.

> Are the populations cyclic?

G.7.b.v) Plot for the first 50 time units what happens for the same data but with $v = 500$ replaced by $v = 1000$.

> Are the populations cyclic?

G.7.b.vi) When you keep all data but v the same, as you increase v, the more the plots resulting from the logistic predator-prey model resemble the plots resulting from the basic predator-prey model.

> Why is this totally natural?

■ G.8) Epidemics★

This model was developed by W. O. Kermack and A. G. McKendrick and was studied in their papers titled "Contributions to the mathematical theory of epidemics," *Proceedings of the Royal*

Society, A 115, 700–721 (1927); 138, 55–83 (1932); 141, 94–122 (1933). It forms the base for many epidemic models used today.

The SIR model deals with disease in a closed population. This model is geared at diseases like certain strains of flu which confer immunity to those who contract the disease and recover. Put

$$\begin{aligned}\text{Sus}[t] &= \text{the number of susceptible people at time } t \text{ after measurements begin;}\\ \text{Inf}[t] &= \text{the number of infected people at time } t \text{ after measurements begin;}\\ \text{Recov}[t] &= \text{the number of recovered people at time } t \text{ after measurements begin.}\end{aligned}$$

The model says that there are positive constants a and b such that

$$\begin{aligned}\text{Sus}'[t] &= -a\,\text{Sus}[t]\,\text{Inf}[t]\\ \text{Recov}'[t] &= b\,\text{Inf}[t].\end{aligned}$$

This says that the number of susceptibles decreases at a rate jointly proportional to the number of susceptibles and the number of infecteds. And this says that the recovery rate is proportional to the number of infecteds.

You also know that

$$\text{Sus}[t] + \text{Inf}[t] + \text{Recov}[t] = \text{total size of the population under study.}$$

For the purposes of this problem, assume that the total population stays constant. This tells you

$$\text{Sus}'[t] + \text{Inf}'[t] + \text{Recov}'[t] = 0;$$

so

$$\begin{aligned}\text{Inf}'[t] &= -\text{Sus}'[t] - \text{Recov}'[t]\\ &= a\,\text{Sus}[t]\,\text{Inf}[t] - b\,\text{Inf}[t].\end{aligned}$$

This gives you the full SIR epidemic model:

$$\begin{aligned}\text{Sus}'[t] &= -a\,\text{Sus}[t]\,\text{Inf}[t],\\ \text{Inf}'[t] &= a\,\text{Sus}[t]\,\text{Inf}[t] - b\,\text{Inf}[t]\\ \text{Recov}'[t] &= b\,\text{Inf}[t].\end{aligned}$$

with Sus[0] = number of susceptibles, Inf[0] = number of infecteds, and Recov[0] = number of recovered at the time the measurements begin.

G.8.a) This information was secured from the book *Mathematical Biology* by J. D. Murray (Springer-Verlag, 1990). For a lot more on the modeling of epidemics, see this book.

The March 4, 1978, issue of the *British Medical Journal* reported all the details about a flu epidemic that swept through a boarding school in early 1978. Scientists were delighted that the actual data fell in line with those predicted by the SIR model.

The model for this epidemic is:

$$\begin{aligned}\text{Sus}'[t] &= -a\,\text{Sus}[t]\,\text{Inf}[t],\\ \text{Inf}'[t] &= a\,\text{Sus}[t]\,\text{Inf}[t] - b\,\text{Inf}[t].\\ \text{Recov}'[t] &= b\,\text{Inf}[t]\end{aligned}$$

with Sus[0] = 762, Inf[0] = 1, Recov[0] = 0, $a = 0.00218$, and $b = 0.44036$.

One snotty-nosed little kid caused all the trouble.

Replay this epidemic by plotting Sus[t] and Inf[t] for $0 \leq t \leq 14$ with t measured in days.

Comment on its severity.

G.8.b.i) Go back to the basic SIR model:

$$\begin{aligned}\text{Sus}'[t] &= -\,a\,\text{Sus}[t]\,\text{Inf}[t],\\ \text{Inf}'[t] &= a\,\text{Sus}[t]\,\text{Inf}[t] - b\,\text{Inf}[t],\\ \text{Recov}'[t] &= b\,\text{Inf}[t].\end{aligned}$$

Why is it common sense that Sus′[t] < 0; so that Sus[t] goes down as time goes up?

G.8.b.ii) On the basis of part G.8.b.i) above, you know that Sus[0] $\geq$ Sus[t] for $t \geq 0$. Now look at the key ingredient of the model:

$$\begin{aligned}\text{Inf}'[t] &= a\,\text{Sus}[t]\,\text{Inf}[t] - b\,\text{Inf}[t]\\ &= (a\,\text{Sus}[t] - b)\ \text{Inf}[t].\end{aligned}$$

Explain the statements:

→ If Sus[0] $< b/a$, then the model predicts no epidemic because in this case Inf[t] goes down as t goes up.

→ If Sus[0] $> b/a$, then the model predicts that the disease will spread and the infected population will increase until $S[t]$ gets small enough that Sus[t] $< b/a$.

In other words, if Sus[0] $> b/a$, then an epidemic is in the cards.

G.8.b.iii) Take the boarding school data and show the plots of Sus[t] and Inf[t] together with the horizontal line through $\{0, b/a\}$ on the vertical axis.

Does this plot confirm the second statement in part G.8.b.ii)?

If you want to see how you can modify the SIR epidemic model to get a basic model for AIDS epidemics, go to page 626 of the book by Murray mentioned above. If you want to see how you can modify the SIR epidemic model to get a basic model for gonorrhea, go to page 620 of the same book.

G.9) Hints of chaos

Take the logistic equation

$$y'[t] = a\,y[t]\,(1 - y[t])$$

with limiting value 1 for $y[t]$ in its global scale. The value of $y[t]$ for a given t can be regarded as the fraction of the potential population that is alive at time t.

Mitchell Fiegenbaum, one of the central characters in the new field of chaos, wondered what happens when you take this differential equation and turn it into a function

$$f[x] = a\,x\,(1 - x)\,.$$

Now say that if x is the fraction of the potential population living during generation 0, then

→ $f[x]$ is the fraction of the potential population living during the first generation;

→ $f[f[x]]$ is the fraction of the potential population living during the second generation;

→ $f[f[f[x]]]$ is the fraction of the potential population living during the third generation, etc.

Here is what happens for the first three generations for the case $a = 2$ and when you start with 0.3 of the potential population alive in generation 0:

In[19]:=

```
Clear[x,a,f]
f[x_] := a x ( 1 - x)
a = 2.0; start = 0.3;
{start,f[start],f[f[start]],f[f[f[start]]]}
```

Out[19]=

```
{0.3, 0.42, 0.4872, 0.499672}
```

You can get the same results from:

In[20]:=

```
{Nest[f,start,0],Nest[f,start,1], Nest[f,start,2],Nest[f,start,3]}
```

Out[20]=

```
{0.3, 0.42, 0.4872, 0.499672}
```

Fancy folks say that $\text{Nest}[f, x, 3] = f[f[f[x]]]$ iterates $f[x]$ three times.

Now you can see what happens through the first 10 generations:

In[21]:=
```
Clear[k]
Table[Nest[f,start,k],{k,0,10}]
```

Out[21]=
```
{0.3, 0.42, 0.4872, 0.499672, 0.5, 0.5, 0.5, 0.5, 0.5, 0.5, 0.5}
```

In the long term, the population settles in at 0.5 of its potential. Here is a plot exhibiting what happens after five generations.

In[22]:=
```
Plot[Nest[f,x,5],{x,0,1},
PlotStyle->{{Thickness[0.015],Red}},
PlotRange->All,
AxesLabel->{"x","fraction of potential  alive"},
PlotLabel->"generation 5"];
```

fraction of potential alive
generation 5
0.5
0.4
0.3
0.2
0.1
x
0.2 0.4 0.6 0.8 1

This indicates that unless you start off in generation 0 with x close to 0 or 1, then after five generations the population will be about one half of its potential. Here is what happens after five generations when you use $a = 3.8$ instead of $a = 2$:

In[23]:=
```
Clear[x,a,f]
f[x_] := a x ( 1 - x)
a = 3.8;
Plot[Nest[f,x,5],{x,0,1},
PlotStyle->{{Thickness[0.015],Red}},
PlotRange->All,
AxesLabel->{"x","fraction of potential alive"},
PlotLabel->"generation 5"];
```

fraction of potential alive
generation 5
0.8
0.6
0.4
0.2
x
0.2 0.4 0.6 0.8 1

The fraction of the potential population after five generations depends shakily on the fraction x of the potential population in generation 0. Check out what happens in generation 10:

In[24]:=
```
Clear[x,a,f]
f[x_] := a x ( 1 - x)
a = 3.8;
Plot[Nest[f,x,10],{x,0,1},
PlotStyle->{{Thickness[0.015],Red}},
PlotRange->All,
AxesLabel->{"x","fraction of potential alive"},
PlotLabel->"generation 10"];
```

fraction of potential alive
generation 10
0.8
0.6
0.4
0.2
x
0.2 0.4 0.6 0.8 1

Total chaos. Reason: Nearby starting fractions x give rise to really different tenth-generation fractions of potential population. For example:

In[25]:=

```
{Nest[f,0.39,10],Nest[f,0.40,10],Nest[f,0.41,10]}
```

Out[25]=

```
{0.685426, 0.390778, 0.927035}
```

Investigate what happens through the first twenty generations starting out with $x = 0.40$ and with $x = 0.41$:

In[26]:=

```
gen = 20;
Clear[k]
pointfour =
Interpolation[Table[{k,Nest[f,0.4,k]},{k,0,gen}],
InterpolationOrder->1][k];
pointfourone =
Interpolation[Table[{k,Nest[f,0.41,k]},{k,0,gen}],
InterpolationOrder->1][k];
Plot[{pointfour,pointfourone},{k,0,gen},
PlotStyle->{{Red,Thickness[0.02]},
{Blue,Thickness[0.01]}},
AxesLabel->{"generation","pop potential"}];
```

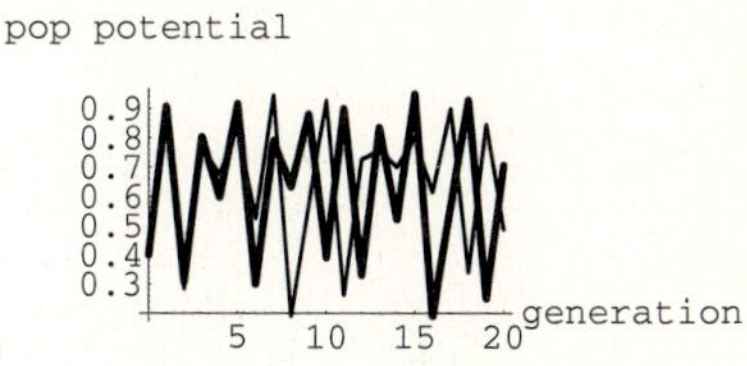

Follow the thinner curve to see what happens generation by generation when you start with $x = 0.4$; follow the other curve to see what happens when you start with $x = 0.41$. Play by trying two other x's that are close to each other.

The two plots start out close together, but after several generations the plots break apart. This is really amazing because the same process produced both curves! This is what chaos is all about.

All this chaos gives you one warning: For some processes, small changes at the start can result in wildly different long-term outcomes.

Maybe there is a butterfly flying in Argentina whose wing motion can be the root cause of a tornado in Missouri six months from now.

When, in the late 1970s, scientists became aware that this type of chaotic behavior could happen even with functions as simple as this one, a firestorm swept through math and physics. Heavy rollers like Feigenbaum threw themselves into the study of chaos, and now a theory is beginning to form.

For exciting reading on chaos, see James Glieck, *Chaos, Making a New Science*, Viking, 1987. For advanced mathematics on chaos, see Robert Devaney, *An Introduction to Chaotic Dynamical Systems*, Addison-Wesley, 1987.

You are invited to participate by engaging in some of the early research experiments that went on in the 1970's:

G.9.a) Go with

$$f[x] = a\,x\,(1 - x)\,.$$

When you use $a = 2.0$ as at the beginning, no chaotic behavior was suggested. But when you use $a = 3.8$, terrible chaos emerged.

Your problem is to use experimental plots to get an estimate of the number A such that if you set $a < A$, then no chaos seems to show up, but when you set $a > A$, chaotic behavior does seem to show up.

LESSON 1.09

Parametric Plotting

Basics

■ B.1) Parametric plots in two dimensions: Circular parameters

A handy way to plot the circle

$$x^2 + y^2 = 9$$

is to write

$$x[t] = 3\cos[t] \qquad \text{and} \qquad y[t] = 3\sin[t]$$

and then to plot the points

$$\{x[t], y[t]\} = \{3\cos[t], 3\sin[t]\}$$

as t advances from 0 to 2π.

In[1]:=

```
Clear[x,y,t]; {x[t_],y[t_]} = {3 Cos[t], 3 Sin[t]};
circle =
ParametricPlot[{x[t],y[t]},{t,0,2 Pi},
PlotStyle->{{Blue,Thickness[0.015]}},
AspectRatio->Automatic,
PlotRange->{{-4,4},{-4,4}}];
```

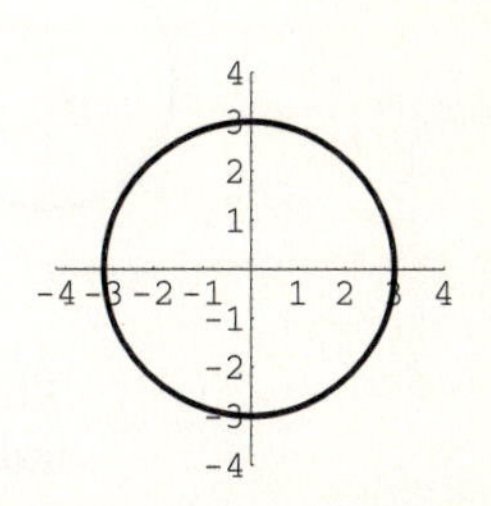

In this way, the original variables x and y are plotted by means of a third variable (t in this case) called a *parameter*. A parameter is an auxiliary variable that plays a back-room role.

B.1.a) When you plot the circle of radius 3 centered at $\{0,0\}$ parametrically through the parametric formula

$$\{x[t], y[t]\} = \{3\cos[t], 3\sin[t]\},$$

what is the physical meaning of the parameter t?

Answer: Look at this plot showing the circle and the point $\{x[t], y[t]\}$ you get with $t = \pi/4$:

In[2]:=
```
Clear[ray,t]
ray[t_] = Graphics[
Line[{{0,0},{x[t],y[t]}}],
{Red,PointSize[0.05],Point[{x[t],y[t]}]}];
Show[circle,ray[Pi/4]];
```

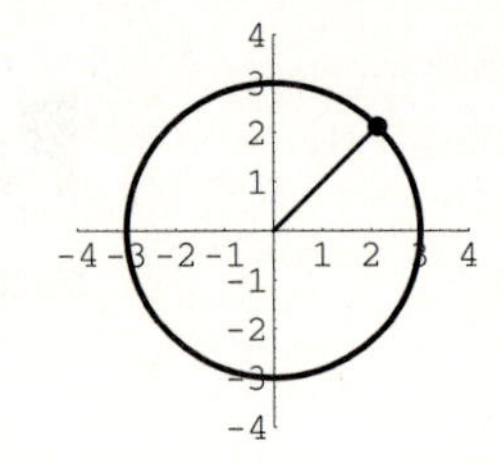

Now look at this:

In[3]:=
```
Table[Show[circle,ray[t],
Graphics[Text[t,1.4 {x[t],y[t]}]],
AspectRatio->Automatic,Ticks->None,
PlotRange->{{-5,5},{-5,5}}],
{t,0,2 Pi,Pi/4}];
```

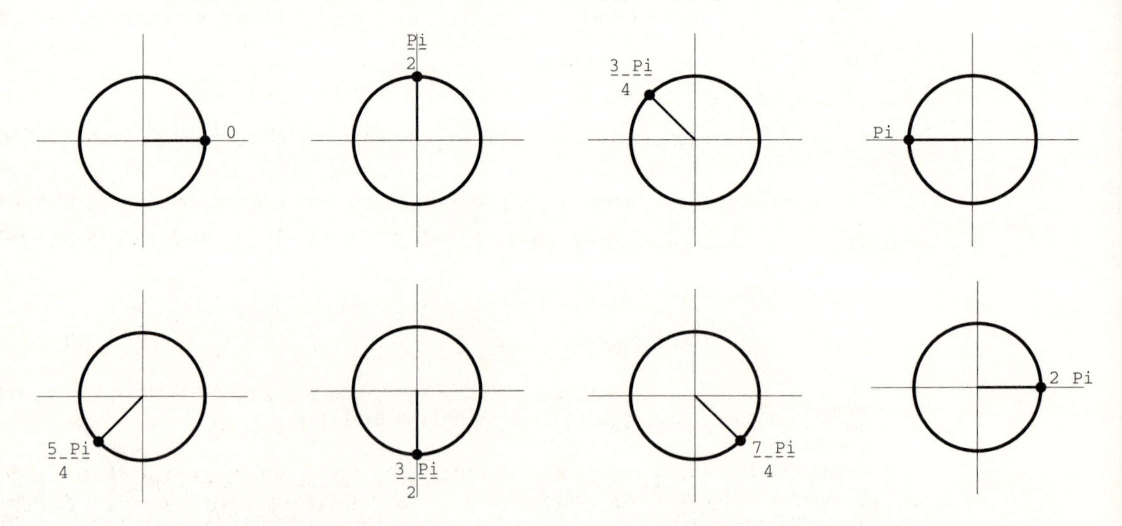

The labels on the graphs give the values of t that make $\{x[t], y[t]\}$ plot out at the indicated point.

B.1.b) Parameters sometimes give you plotting freedom normal plotting does not allow. Some curves are best described via parametric equations and are difficult to describe in the usual $y = f[x]$ terms.

To see what this means, plot the spiral

$$x = x[t] = t\sin[t], \qquad y = y[t] = t\cos[t].$$

Answer:

In[4]:=
```
Clear[x,y,t]
x[t_] = t Cos[t];
y[t_] = t Sin[t];
ParametricPlot[{x[t],y[t]},{t,0,8 Pi},
AxesLabel->{"x","y"},
AspectRatio->Automatic,
PlotStyle->{{Thickness[0.01],Red}}];
```

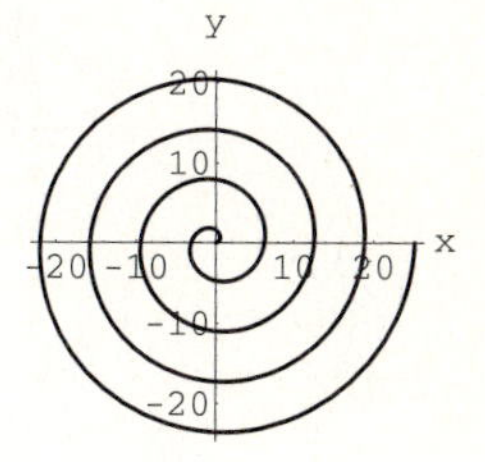

Extend the values of the parameter t to see more of the spiral:

In[5]:=
```
ParametricPlot[{x[t],y[t]},{t,0,16 Pi},
AxesLabel->{"x","y"},
AspectRatio->Automatic,
PlotStyle->{{Thickness[0.01],Red}}];
```

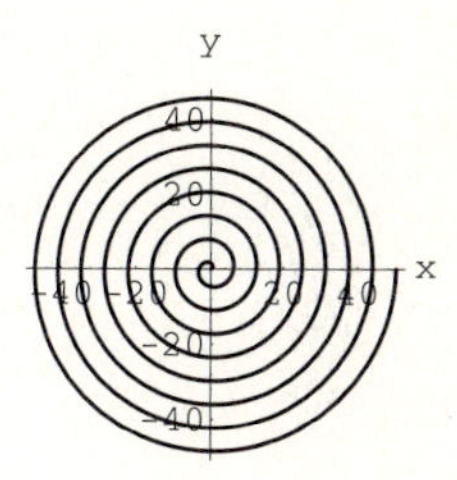

Bull's eye.

■ B.2) Parametric plots of curves in three dimensions

When you plot in two dimensions, you use two coordinate axes:

In[6]:=
```
h = 5; spacer = h/10;
twodims = Graphics[{
Line[{{-h,0},{h,0}}],
Text["x",{h + spacer,0}],
Line[{{0,-h},{0,h}}],
Text["y",{0, h + spacer}]}];
Show[twodims,PlotRange->All];
```

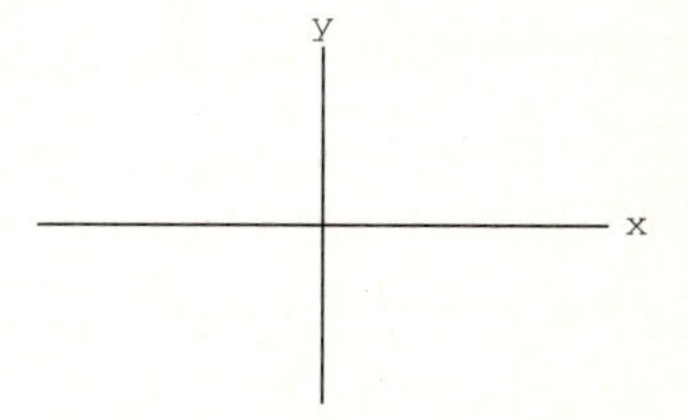

To plot a point, you specify an $\{x, y\}$ coordinate:

In[7]:=
```
{x,y} = {2,3};
Show[twodims,
Graphics[{Red,PointSize[0.03],Point[{x,y}]}],
PlotRange->All];
```

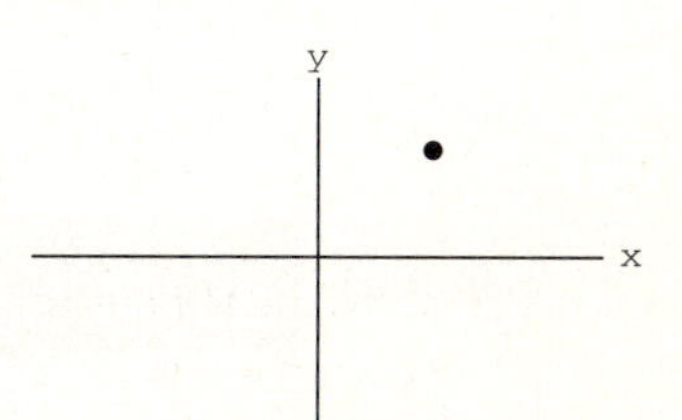

To plot in three dimensions, you need three coordinate axes:

In[8]:=
```
h = 5; spacer = h/10;
threedims = Graphics3D[{
{Blue,Line[{{-h,0,0},{h,0,0}}]},
Text["x",{h + spacer,0,0}],
{Blue,Line[{{0,-h,0},{0,h,0}}]},
Text["y",{0, h + spacer,0}],
{Blue,Line[{{0,0,-h},{0,0,h}}]},
Text["z",{0,0,h + spacer}]}];
Show[threedims,PlotRange->All,Boxed->False];
```

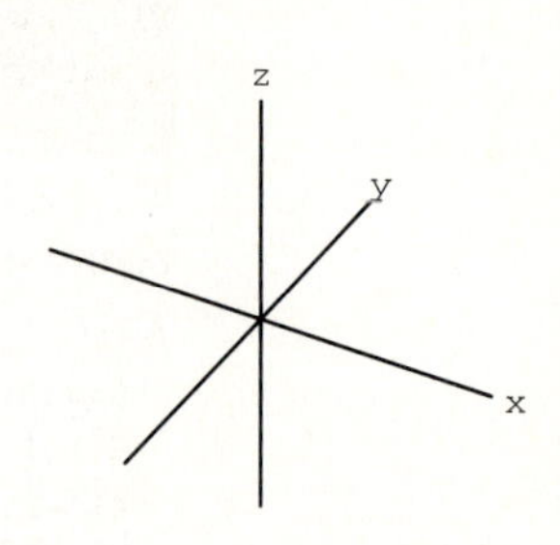

This is *Mathematica*'s default view. Usually Calculus&*Mathematica* uses the viewpoint CMView $= \{2.7, 1.6, 1.2\}$. Here is how the three axes look from this viewpoint:

In[9]:=
```
CMView = {2.7, 1.6, 1.2};
Show[threedims,PlotRange->All,ViewPoint->CMView,
Boxed->False];
```

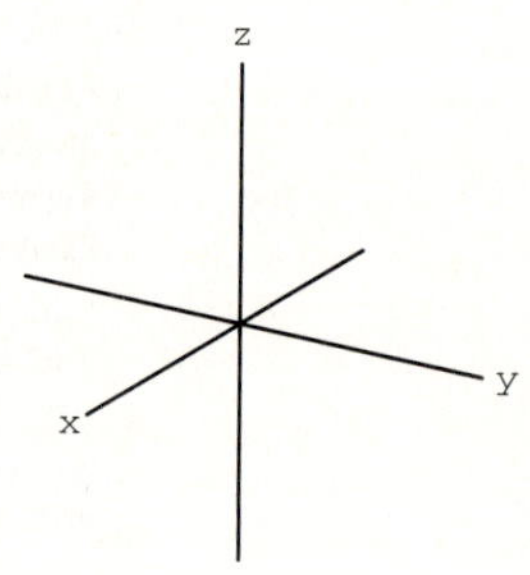

Add a framing box:

In[10]:=
```
Show[threedims,PlotRange->All,
ViewPoint->CMView];
```

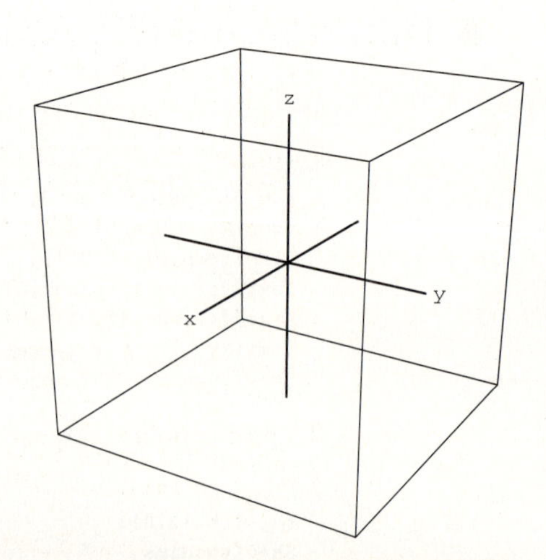

B.2.a) What is the advantage of using the CMView?

Answer: This way the x-axis points to the left, the y-axis points to the right, and the z-axis points up. It's the orientation we usually see in other math and science texts.

B.2.b) To plot a point in three dimensions, you specify an $\{x, y, z\}$ coordinate:

In[11]:=
```
{x,y,z} = {-1,-3,3};
CMView = {2.7, 1.6, 1.2};
Show[threedims,
Graphics3D[{Red,PointSize[0.03],Point[{x,y,z}]}],
PlotRange->All,ViewPoint->CMView,Boxed->False];
```

A three-dimensional curve related to a circle is a screw or helix. The x and y coordinates plot out to a circle, but the z coordinate lifts the curve higher and higher:

In[12]:=
```
Clear[x,y,z,t]
radius = 5;
x[t_] = radius Cos[t];
y[t_] = radius Sin[t];
z[t_] = t/2;
screw =
ParametricPlot3D[{x[t],y[t],z[t]},
{t,0,8 Pi},DisplayFunction->Identity];
Show[threedims,screw,PlotRange->All,
ViewPoint->CMView, Boxed->False,
DisplayFunction->$DisplayFunction,Axes->None];
```

Plot a three-dimensional curve related to a spiral.

Answer: The tornado. To get a tornado, you use the spiral in the x and y slots and use the z slot to lift the curve:

In[13]:=
```
Clear[x,y,z,radius,t]
radius[t_] = t/4;
x[t_] = radius[t] Cos[t];
y[t_] = radius[t] Sin[t];
z[t_] = t/2;
tornado = ParametricPlot3D[{x[t],y[t],z[t]},
{t,0,8 Pi},DisplayFunction->Identity];
Show[threedims,tornado,PlotRange->All,
ViewPoint->CMView, Boxed->False,
DisplayFunction->$DisplayFunction,Axes->None];
```

You can look down from above along the z-axis:

In[14]:=

```
Show[threedims,tornado,
PlotRange->All,ViewPoint->{0,0,4},
Boxed->False,
DisplayFunction->$DisplayFunction,Axes->None];
```

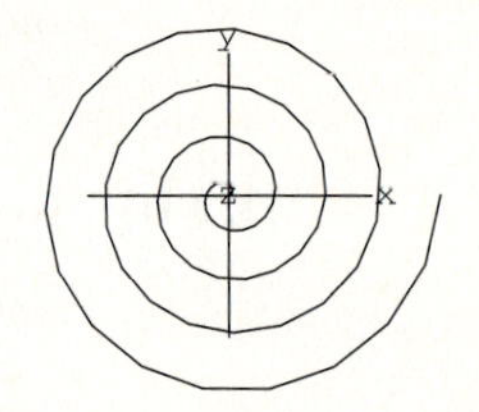

Or from a side along the x-axis:

In[15]:=

```
Show[threedims,tornado,
PlotRange->All,ViewPoint->{4,0,0},
Boxed->False,
DisplayFunction->$DisplayFunction,Axes->None];
```

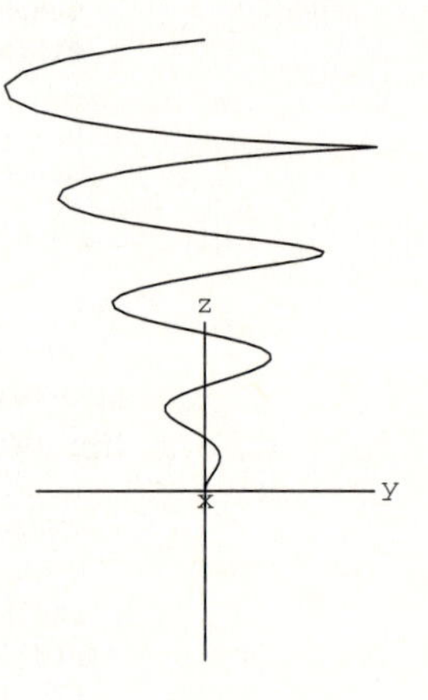

If you want to see actual numbers on the axes, eliminate the Axes->None option above:

In[16]:=

```
Show[threedims,tornado,
PlotRange->All,ViewPoint->CMView,
Boxed->False,
DisplayFunction->$DisplayFunction];
```

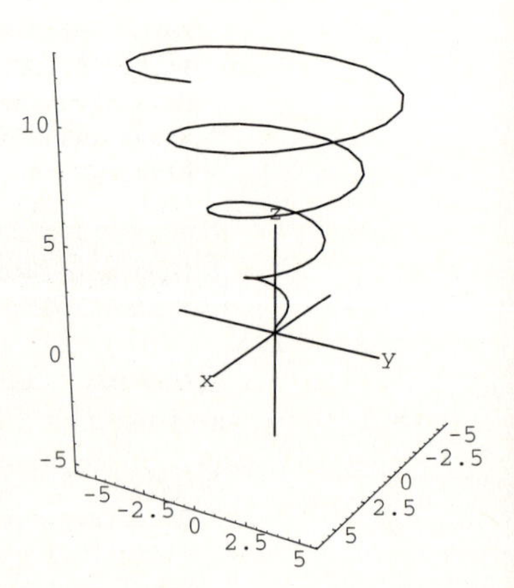

B.2.c) This problem appears only in the electronic version.

■ B.3) Parametric plots of surfaces in three dimensions

Here is a plot of the surface $z = x^2 + y^2$ in three dimensions:

In[17]:=

```
Clear[x,y,f]
f[x_,y_] = x^2 + y^2;
surface = Plot3D[f[x,y],{x,-2,2},{y,-2,2},
DisplayFunction->Identity];
threedims = ThreeAxes[3];
Show[threedims,surface,ViewPoint->CMView,
PlotRange->All,Boxed->False,
DisplayFunction->$DisplayFunction];
```

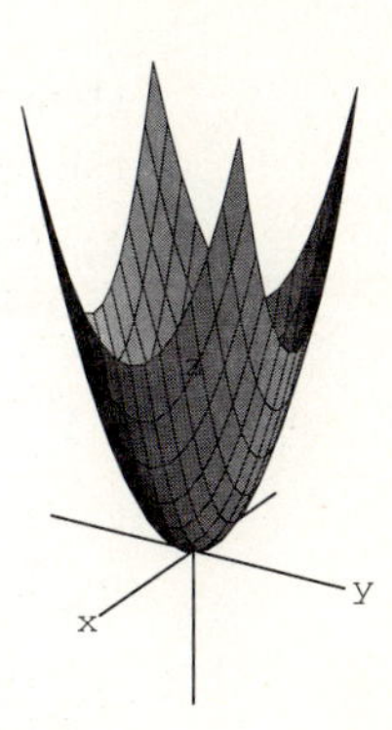

This plot consists of all points $\{x, y, f[x, y]\}$ as x and y run through the square $-2 \leq x \leq 2$ and $-2 \leq y \leq 2$. The surface plotted is over this square in the xy-plane.

This is not a parametric plot.

B.3.a) Use parametric plotting to plot the same surface over everything inside the circle $x^2 + y^2 = 4$ in the xy-plane.

Answer: Here is a plot of the circle $x^2 + y^2 = 4$ in the xy-plane:

In[18]:=

```
Clear[x,y,t]
x[t_] = 2 Cos[t];
y[t_] = 2 Sin[t];
circle = ParametricPlot3D[{x[t],y[t],0},
{t,0,2 Pi}, DisplayFunction->Identity];
threedims = ThreeAxes[3];
Show[threedims,circle,ViewPoint->CMView,
PlotRange->All,Boxed->False,
DisplayFunction->$DisplayFunction];
```

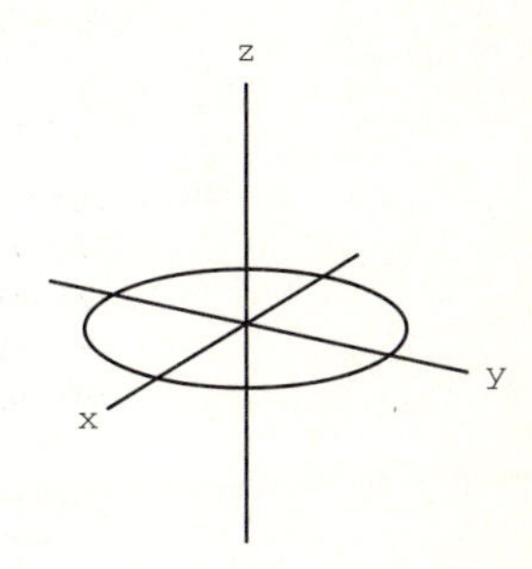

Here is a plot of everything on the xy-plane that is inside this circle:

In[19]:=

```
Clear[x,y,r,t]
x[r_,t_] = r Cos[t];
y[r_,t_] = r Sin[t];
disk =
ParametricPlot3D[{x[r,t],y[r,t],0},
{r,0,2},{t,0,2 Pi},
DisplayFunction->Identity];
Show[threedims,disk,ViewPoint->CMView,
PlotRange->All,Boxed->False,
DisplayFunction->$DisplayFunction];
```

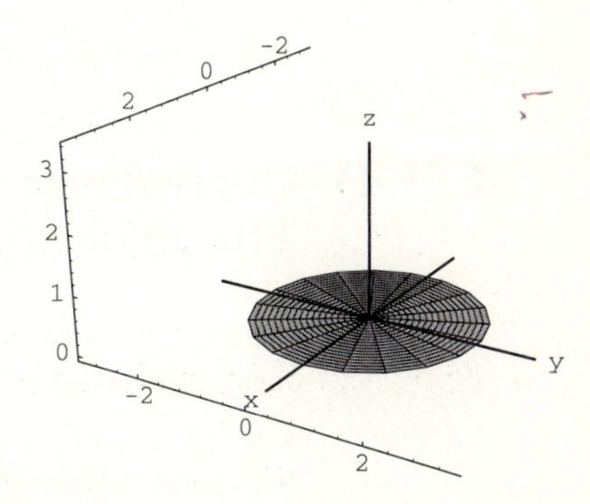

Here is a plot of the part of the surface $z = x^2 + y^2$ that is over everything inside the circle $x^2 + y^2 = 4$ in the xy-plane.

In[20]:=

```
Clear[x,y,z,r,t,f]
f[x_,y_] = x^2 + y^2;
x[r_,t_] = r Cos[t];
y[r_,t_] = r Sin[t];
z[r_,t_] = f[x[r,t],y[r,t]];
surface =
ParametricPlot3D[{x[r,t],y[r,t],z[r,t]},
{r,0,2},{t,0,2 Pi},DisplayFunction->Identity];
threedims = ThreeAxes[3];
Show[threedims,surface,ViewPoint->CMView,
PlotRange->All,Boxed->False,
DisplayFunction->$DisplayFunction];
```

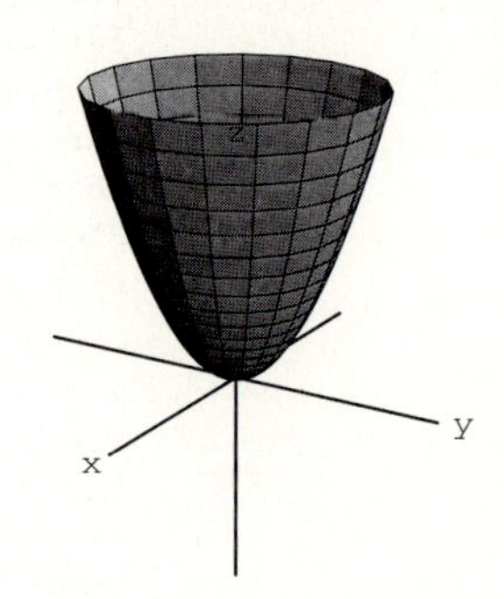

This is a parametric plot. And it's a very nice one.

B.3.b) Use parametric plotting to plot the surface

$$z = \sin[x^2 + y^2]$$

over everything inside the circle $x^2 + y^2 = 5$ in the xy-plane.

Answer: Sweat it not.

In[21]:=

```
Clear[x,y,z,r,t,f]
f[x_,y_] = Sin[x^2 + y^2];
x[r_,t_] = r Cos[t];
y[r_,t_] = r Sin[t];
z[r_,t_] = f[x[r,t],y[r,t]];
surface =
ParametricPlot3D[{x[r,t],y[r,t],z[r,t]},
{r,0,Sqrt[5]},{t,0,2 Pi},DisplayFunction->Identity];
threedims = ThreeAxes[2.5];
Show[threedims,surface,ViewPoint->CMView,
PlotRange->All,Boxed->False,
DisplayFunction->$DisplayFunction];
```

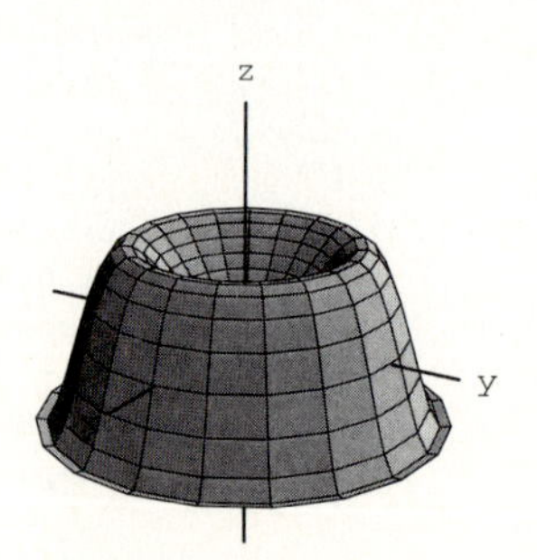

This is another parametric plot. Fill it with ice cream and enjoy.

■ B.4) Derivatives for curves given parametrically: The cycloid

B.4.a) Sometimes parametric formulas are the only viable way of setting up a precise description of a curve. Here is a case:

A circular wheel of radius r rolls along the x-axis. The path traced out by a point P marked on the circle is called a cycloid.

Give parametric formulas $\{x[t], y[t]\}$ for the cycloid.

Plot in the case that $r = 1$.

Answer: Set things up at the start like this:

In[22]:=

```
Clear[x,y,t];
origin = Graphics[{Red,PointSize[0.04],Point[{0,0}]}];
mark = Graphics[Text["mark",{0,0},{2,2}]];
wheel = Graphics[{Thickness[0.015],Blue,Circle[{0,1},1]}];
center = Graphics[{PointSize[0.03],Point[{0,1}]}];
Show[origin,wheel,center,mark,Axes->Automatic,
AxesOrigin->{0,0},AxesLabel->{"x","y"},
AspectRatio->Automatic,Ticks->None,
PlotRange->{{-1.5,1.5},{-0.5,2.5}}];
```

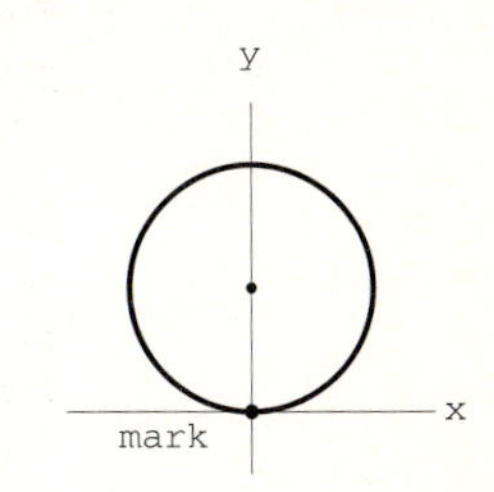

After the circle has rolled to the right a bit, the set-up looks like this:

In[23]:=

```
mark = Graphics[{Red,PointSize[0.04],
Point[{1.2 - Sin[1.2],1 - Cos[1.2]}]}];
wheel = Graphics[{Thickness[0.015],Blue,Circle[{1.2,1},1]}];
center = Graphics[{PointSize[0.03],Point[{1.2,1}]}];
radiusline = Graphics[Line[{{1.2 - Sin[1.2],
1 - Cos[1.2]},{1.2,1}}]];
rlabel = Graphics[Text["r",
{1.2 -Sin[1.2]/2,1.35 - Cos[1/2]/2}]];
Show[wheel,center,radiusline,rlabel,mark,
Axes->Automatic,AxesOrigin->{0,0},AxesLabel->{"x",""},
AspectRatio->Automatic,Ticks->None,
PlotRange->{{-.5,3},{-.5,2.5}}];
```

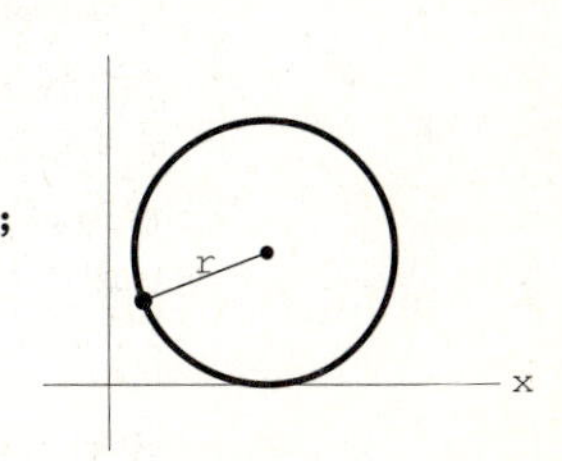

Now put in some labels: Do not worry about the graphic specifications. When the problem is over, then you'll realize how they were set.

In[24]:=

```
yline = Graphics[Line[{{0,1 - Cos[1.2]},
{1.2 - Sin[1.2], 1 - Cos[1.2]}}]];
ylabel = Graphics[Text["y",{-.1,1 - Cos[1.2]}]];
xline = Graphics[Line[{{1.2 - Sin[1.2],0},
{1.2 - Sin[1.2],1 - Cos[1.2]}}]];
xlabel = Graphics[Text["x",{1.2 - Sin[1.2],-.1}]];
mlabel = Graphics[Text["M",{1.2,0},{0,2}]];
tline = Graphics[Line[{{1.2,0},{1.2,1},
{1.2 - Sin[1.2],1 - Cos[1.2]}}]];
tarc = Graphics[Circle[{1.2,1},.2,
{3 Pi/2 -1.2,3 Pi/2}]];
tlabel = Graphics[Text["t",{1.05,.75}]];
rrlabel = Graphics[Text["r",{1.25,.5}]];
diagram = Show[wheel,mark, center,yline,ylabel,
xline, xlabel,mlabel,tline,tarc,tlabel,
rlabel, rrlabel, Axes->True,AxesOrigin->{0,0},
AxesLabel->{"x",""}, AspectRatio->Automatic,
Ticks->None, PlotRange->{{-0.5,3},{-0.2,2.1}}];
```

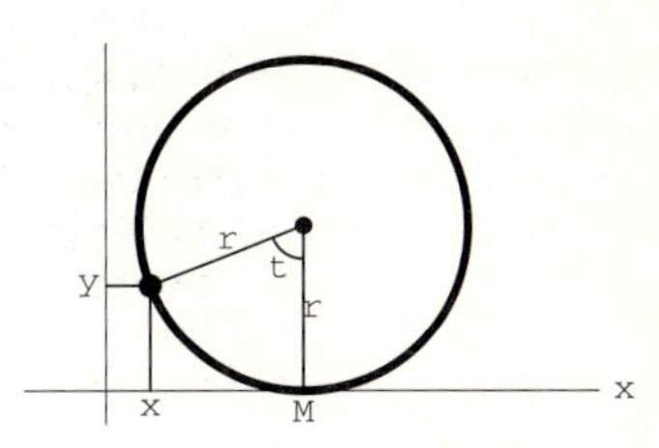

You can find the position $\{x[t], y[t]\}$ of the mark in terms of the angle measurement t:

Read off the parametric formula for y from the plot:

In[25]:=
```
Clear[r,x,y,t]
y[t_] = r - r Cos[t]
```

Out[25]=
```
r - r Cos[t]
```

Remember

$$\cos[\text{angle}] = \frac{\text{adjacent}}{\text{hypothenuse}}.$$

The center is sitting at $\{r\,t, r\}$ because the length of the arc on the circle from the point of contact M to the mark is $r\,t$. This tells you how to read off the parametric formula for x:

In[26]:=
```
x[t_] = r t - r Sin[t]
```

Out[26]=
```
r t - r Sin[t]
```

Here comes the plot of the curve traced out by the mark of the first roll in the case that $r = 1$:

In[27]:=
```
r = 1;
firstroll =
ParametricPlot[{x[t],y[t]},{t,0,2 Pi},
AspectRatio->Automatic,
PlotStyle->{{Red,Thickness[.008]}}];
```

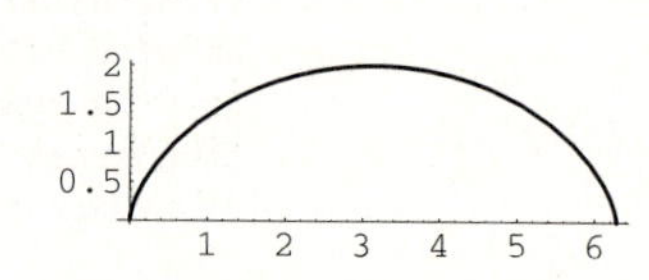

Two rolls:

In[28]:=
```
ParametricPlot[{x[t],y[t]},{t,0,4 Pi},
AspectRatio->Automatic,PlotStyle->
{{Red,Thickness[.008]}}];
```

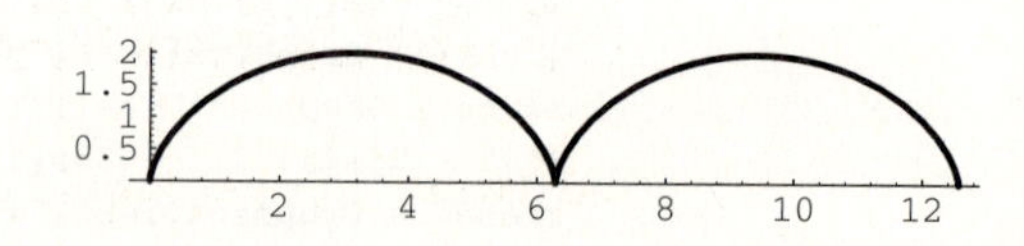

Lots of folks call that corner a cusp.

B.4.b) Sometimes a curve, like the cycloid above, comes via parametric equations

$$x = x[t] \qquad \text{and} \qquad y = y[t]$$

and it is clear that the curve is the graph of a function $y = f[x]$. This means that $y[t] = f[x[t]]$ for all t's.

Finding the explicit form of $f[x]$ may be impossible or too much trouble.

It might seem that in this situation you are out of luck in calculating the derivative $D[f[x], x] = f'[x]$, but there is a way to do it.

How?

Answer: If $y = f[x]$, then clearly $y[t] = f[x[t]]$. The chain rule says

$$y'[t] = f'[x[t]]\, x'[t].$$

Consequently

$$f'[x[t]] = \frac{y'[t]}{x'[t]}.$$

This tells you that to calculate $f'[x^*]$ for a given number x^*, you find a t^* with $x[t^*] = x^*$ and then you have

$$f'[x^*] = \frac{y'[t^*]}{x'[t^*]}.$$

B.4.c) Here is one roll of the cycloid for a wheel of radius $r = 2$:

In[29]:=
```
Clear[r,x,y,t]
x[t_] = r t - r Sin[t];
y[t_] = r - r Cos[t];
r = 2; firstroll =
ParametricPlot[{x[t],y[t]},{t,0,2 Pi},
AspectRatio->Automatic,AxesLabel->{"x","y"},
PlotStyle->{{Red,Thickness[.008]}}];
```

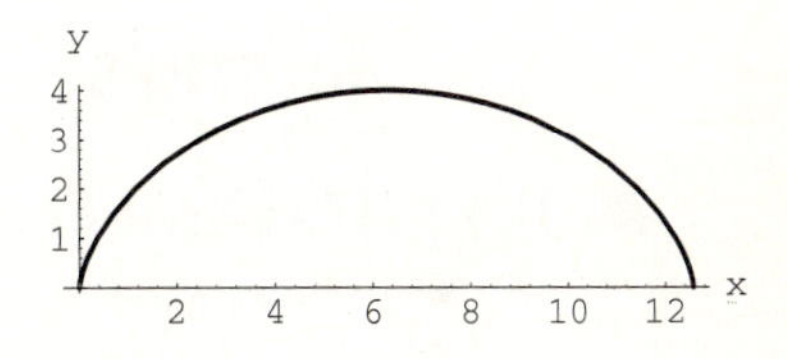

You can see that there is a function $f[x]$ so that this curve is the plot of $f[x]$; but you will not be able to find a formula for $f[x]$.

In spite of this, estimate the instantaneous growth rate $f'[3]$.

Answer: To get a hold of $f'[3]$, you've got to determine a t that gives $x[t] = 3$:

In[30]:=
```
a = 2; b = 3.5;
Plot[{x[t],3},{t,a,b},AxesOrigin->{a,x[a]},
AxesLabel->{"t","x[t]}"}];
```

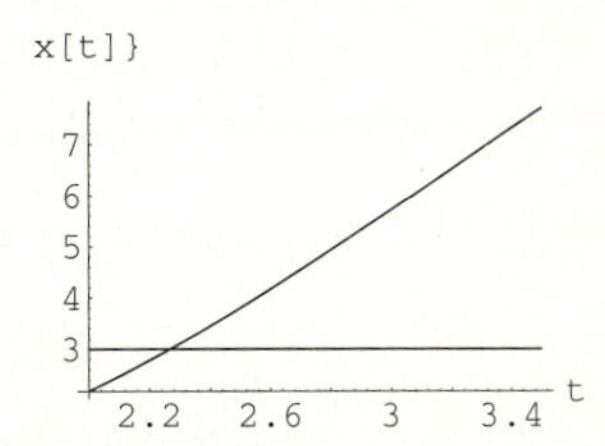

$t = 2.3$ is a reasonable first estimate; get a better estimate:

In[31]:=
```
FindRoot[x[t] == 3,{t,2.3}]
```

Out[31]=
```
{t -> 2.26717}
```

Try it out:

In[32]:=
```
tt = 2.26717;
x[tt]
```

Out[32]=
```
2.99999
```

Good enough. A good estimate of $f'[3]$ is:

In[33]:=
```
y'[tt]/x'[tt]
```

Out[33]=
```
0.467378
```

And you're out of here.

Tutorials

■ T.1) Parametric plotting for projectile motion

A projectile is fired from a cannon at ground level. The cannon is inclined at an angle a (here $0 < a < \pi/2$) with the horizontal, with a given muzzle speed determined by the explosive charge. If you neglect air resistance, it's not a big deal to come up with parametric formulas for the position $\{x[t], y[t]\}$ of the projectile t seconds after firing.

Here t measures time in seconds from the instant the cannon is fired.

$x[t]$ measures the horizontal distance in feet of the projectile down range from the cannon at time t.

$y[t]$ measures the height of the projectile in feet at time t.

Gravity acts on the y[t] component of the position but does not act on the $x[t]$ component. Thus

$$x''[t] = 0 \qquad \text{and} \qquad y''[t] = -32.$$

The muzzle speed, s, splits into horizontal and vertical components as follows:

$$x'[0] = s\cos[a] \qquad \text{and} \qquad y'[0] = s\sin[a].$$

Saying that the cannon is fired at ground level is to say

$$\{x[0], y[0]\} == 0.$$

See what $\{x[t], y[t]\}$ look like:

In[1]:=
```
Clear[x,y,t,a,s,Derivative]
DSolve[{x''[t] == 0,
x[0] == 0,
x'[0] == s Cos[a]},x[t],t]
```

Out[1]=
```
{{x[t] -> s t Cos[a]}}
```

In[2]:=
```
DSolve[{y''[t] == -32,
y[0] == 0,
y'[0] ==  s Sin[a]},y[t],t]
```

Out[2]=
```
                 2
{{y[t] -> -16 t  + s t Sin[a]}}
```

For a given muzzle speed s and cannon angle a:

In[3]:=
```
x[t_,s_,a_] = s t Cos[a]
y[t_,s_,a_] =  -16 t^2 + s t Sin[a]
```

Out[3]=
```
s t Cos[a]
```

Out[3]=
```
      2
-16 t  + s t Sin[a]
```

Now you can go to work.

T.1.a) Plot the trajectory of the projectile, given that the cannon is inclined at an angle $\pi/6$ with the horizontal and with muzzle speed 200 ft/sec.

Answer: See when the projectile is back on the ground:

In[4]:=
```
Solve[y[t,200,Pi/6] == 0,t]
```

Out[4]=
```
        25
{{t -> --}, {t -> 0}}
        4
```

Here comes the plot:

In[5]:=
```
Clear[trajectoryplotter]
trajectoryplotter[t_] =
{x[t,200,Pi/6],y[t,200,Pi/6]}
ParametricPlot[trajectoryplotter[t],{t,0,25/4},
PlotStyle->{{Red,Thickness[0.015]}}];
```

Out[5]=

```
                       2
{100 Sqrt[3] t, 100 t - 16 t }
```

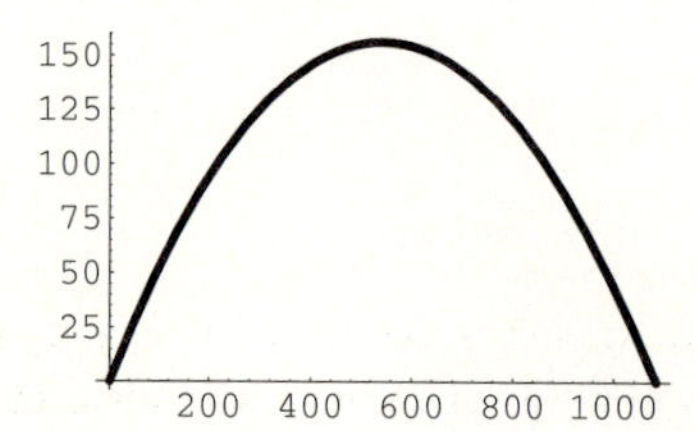

There it is, a beautiful parabolic arch.

T.1.b) Measure the horizontal range of the cannon as a function of muzzle speed and angle of inclination.

For a given muzzle speed, what angle of inclination maximizes the horizontal range of the cannon?

Answer: Go back to:

In[6]:=

```
Clear[x,y,t,s,a,t]
x[t_,s_,a_] = s t Cos[a]
y[t_,s_,a_] = -16 t^2 + s t Sin[a]
```

Out[6]=

```
s t Cos[a]
```

Out[6]=

```
     2
-16 t  + s t Sin[a]
```

Now find when the projectile hits the ground:

In[7]:=

```
Solve[y[t,s,a] == 0,t]
```

Out[7]=

```
        s Sin[a]
{{t -> --------}, {t -> 0}}
           16
```

For a given muzzle speed s and angle a, the horizontal range of the cannon is:

In[8]:=

```
x[s Sin[a]/16, s,a]
```

Out[8]=

```
 2
s  Cos[a] Sin[a]
----------------
       16
```

For a given muzzle speed, this is as big as it can be when the angle a is set so that $\cos[a]\sin[a]$ is as big as it can be.

Now examine the derivative of $\cos[a]\sin[a]$ with respect to a:

In[9]:=

```
Expand[D[Cos[a] Sin[a],a],Trig->True]
```

Out[9]=

```
Cos[2 a]
```

So $D[\cos[a]\sin[a], a] = 0$ for $\cos[2a] = 0$; which gives $2a = \pi/2$; this is the same as $a = \pi/4$. For a given muzzle speed, you get the greatest horizontal range by setting the firing angle a equal to $\pi/4$.

Not much of a surprise.

T.1.c) How do you use the parametric formulas:

In[10]:=

```
Clear[x,y,t,s,a,t]
x[t_,s_,a_] = s t Cos[a]
y[t_,s_,a_] = -16 t^2 + s t Sin[a]
```

Out[10]=

```
s t Cos[a]
```

Out[10]=

```
       2
-16 t   + s t Sin[a]
```

to explain why, no matter what the muzzle speed and the angle are, the projectile moves on a parabola?

Answer: Look at the parametric formulas:

In[11]:=

```
Clear[x,y,t,s,a,t]
x[t_,s_,a_] = s t Cos[a];
y[t_,s_,a_] = -16 t^2 + s t Sin[a];
```

If you put $x = x[t, s, a]$ and $y = y[t, s, a]$ and then you eliminate t by solving for t in terms of y and substituting the result into the formula for x, you get:

In[12]:=

```
ExpandAll[Eliminate[{x == x[t,s,a],
y == y[t,s,a]},t]]
```

Out[12]=

```
    2    2                          2          2
16 x  - s  x Cos[a] Sin[a] == -(s  y Cos[a] )
```

This tells you that when you fix a muzzle speed s and a firing angle a, then the projectile moves on the curve

$$y = -\frac{16\,x\sec[a]^2}{s^2} + x\tan[a].$$

For a given muzzle speed and angle, this curve is a parabola because no powers of x other than x^2, x^1, and x^0 are present.

■ T.2) Parametric plotting for designing a cam

This problem appears only in the electronic version.

■ T.3) Parametric plotting of the predator-prey model

Remember the predator-prey model from the previous lesson. In case you've forgotten about it, here's the scoop again:

This mathematical model was orginally studied by Lotka and Volterra. For a critical analysis of it, get hold of J. D. Murray, *Mathematical Biology*, Springer-Verlag, New York, 1990 and read.

Two species coexist in a closed environment. One species, the predator, feeds on the other, the prey. There is always plenty of food for the prey, but the predators eat nothing but the prey.

Put

$$\text{pred}[t] = \text{population of predators at time } t.$$
$$\text{prey}[t] = \text{population of prey at time } t.$$

It's reasonable to assume that there are positive constants a and b such that:

$$\text{prey}'[t] = a\,\text{prey}[t] - b\,\text{prey}[t]\,\text{pred}[t]$$

because the abundance of food for the prey allows the birth rate of the prey to be proportional to their current number, and the death rate of prey is proportional to both the current number of prey and the current number of predators.

It also makes some sense to assume that there are positive constants c and d such that:

$$\text{pred}'[t] = -c\,\text{pred}[t] + d\,\text{pred}[t]\,\text{prey}[t]$$

because it's reasonable to assume that the death rate of the predators is likely to be proportional to the current population of predators and that the birth rate of the predators is proportional to both the current number of the predators and the size of the food supply (the prey).

Here is what happens for a sample choice of a, b, c, d with the prey starting off with a population of four units and the predators starting out with a population of one unit. The units could be thousands or millions so that a population of 0.35 can make sense.

In[13]:=

```
endtime = 50; a = 0.7; b = 0.3; c = 0.3; d = 0.1;
Clear[pred,prey,t,Derivative]
approxsolutions = NDSolve[{prey'[t] ==  a prey[t] - b prey[t] pred[t],
pred'[t] == -c pred[t] + d pred[t] prey[t],
prey[0] == 4, pred[0] == 1},{prey[t],pred[t]},{t,0,endtime}];
Clear[fakepred,fakeprey]
fakepred[t_] = pred[t]/.approxsolutions[[1]];
fakeprey[t_] = prey[t]/.approxsolutions[[1]];
```

In[14]:=

```
preyplot =
Plot[fakeprey[t],{t,0,endtime},
PlotStyle->{{Blue,Thickness[0.015]}},
AxesLabel->{"t","prey"}];
```

And the predator population as a function of time t:

In[15]:=

```
predatorplot =
Plot[fakepred[t],{t,0,endtime},
PlotStyle->{{Red,Thickness[0.01]}},
AxesLabel->{"t","predators"}];
```

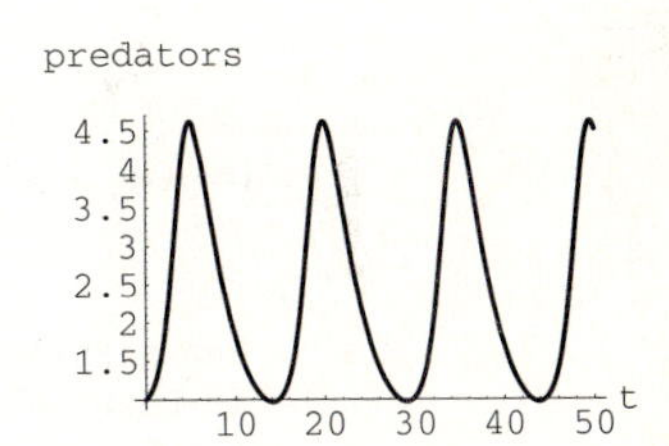

Here they are together:

In[16]:=

```
both = Show[predatorplot,preyplot
AxesLabel->{"t",""}];
```

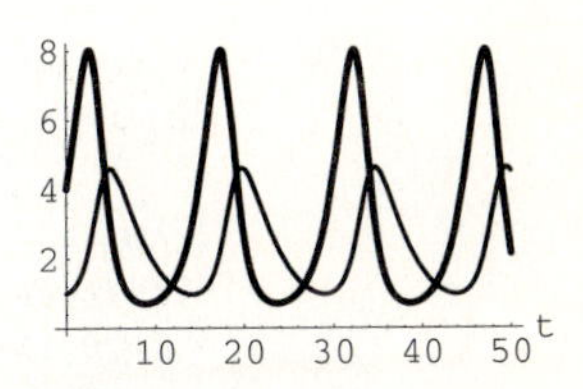

The plot of the prey population is thicker than the plot of the predator population.

> Note the relationship between the predator population's crests and dips and the prey population's crests and dips. Just after the predators start growing, the prey is just about eaten up.

T.3.a) Use parametric plotting to display the predator population as a function of the prey population and describe what you see.

Answer: Here it is:

In[17]:=
```
fullplot =
ParametricPlot[{fakeprey[t],fakepred[t]},
{t,0,endtime},PlotStyle->Thickness[0.01],
PlotRange->All,
AxesLabel->{"prey","predator"}];
```

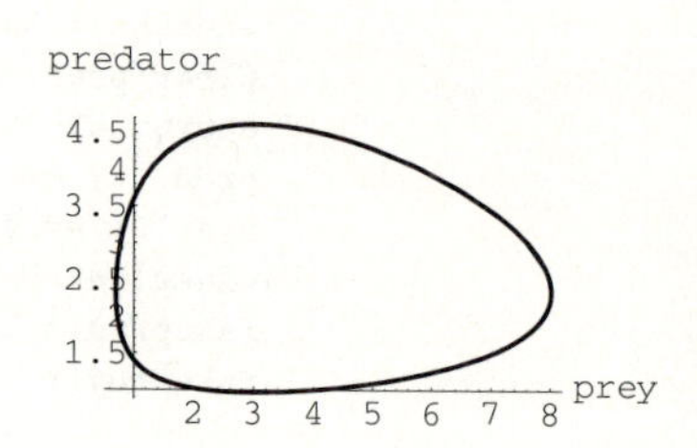

A distorted ellipse that some folks call a closed curve. This is a little hard to interpret. To get a grasp on it, see what happens as t goes from 0 to 5:

In[18]:=
```
five = ParametricPlot[{fakeprey[t],fakepred[t]},
{t,0,5},PlotStyle->Thickness[0.01],
AxesLabel->{"prey","predator"},
PlotRange->{{0,9},{0,5}}];
```

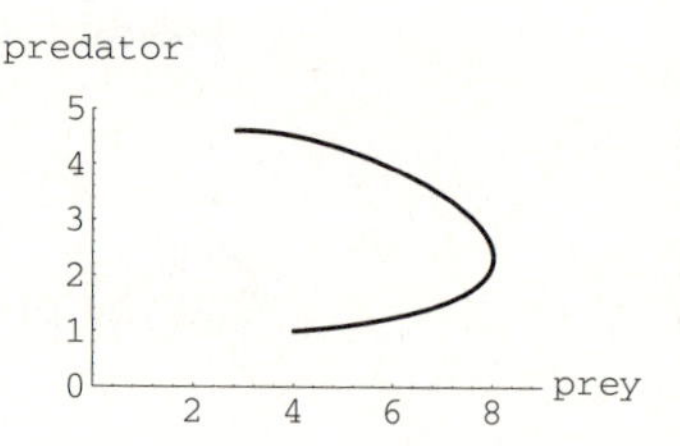

Now see what happens as t goes from 0 to 10:

In[19]:=
```
ten = ParametricPlot[{fakeprey[t],fakepred[t]},
{t,0,10},PlotStyle->Thickness[0.01],
AxesLabel->{"prey","predator"},
PlotRange->{{0,9},{0,5}}];
```

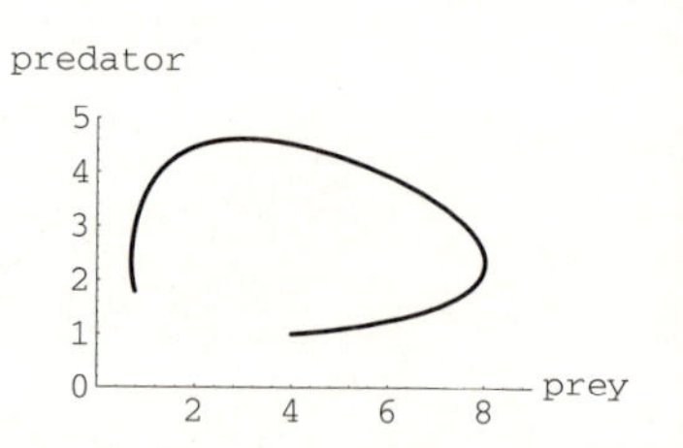

Now see what happens as t goes from 0 to 15:

In[20]:=
```
fifteen = ParametricPlot[{fakeprey[t],fakepred[t]},
{t,0,15},PlotStyle->Thickness[0.01],
AxesLabel->{"prey","predator"},
PlotRange->{{0,9},{0,5}}];
```

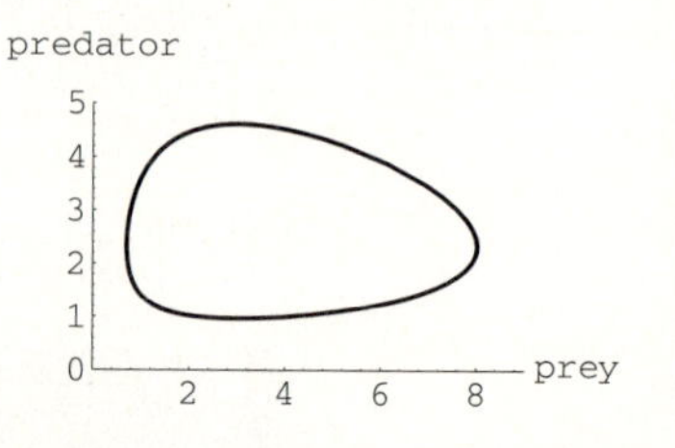

To get the full effect, grab all three plots, and animate, and run forward. Holy smoke! This is the same as:

In[21]:=
```
Show[fullplot,PlotRange->{{0,9},{0,5}}];
```

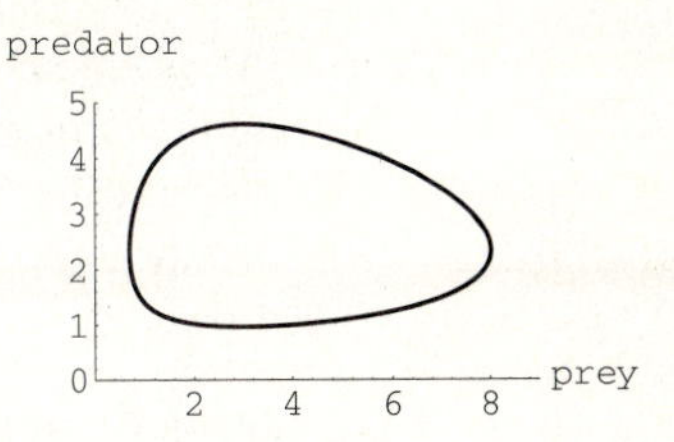

Evidently as time advances the parametric points {prey[t], pred[t]} advance around the closed curve in a counterclockwise way. As time advances on and on, the parametric points {prey[t], pred[t]} cycle around this curve over and over again.

Take another look but this time setting the axes' origin to $\{a/b, c/d\}$:

In[22]:=
```
Show[fullplot,PlotRange->{{0,9},{0,5}},
AxesOrigin->{c/d,a/b}];
```

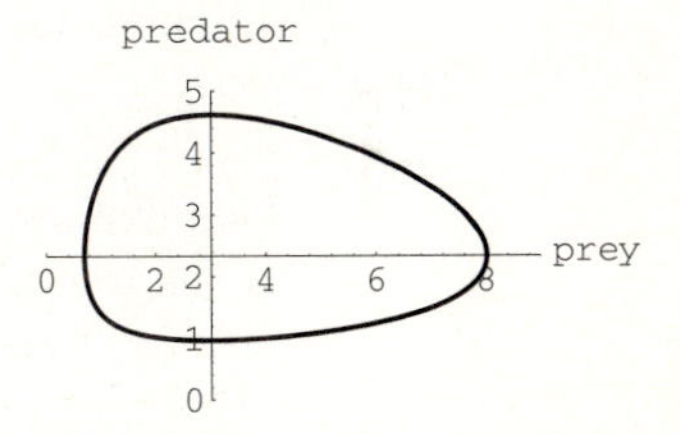

Again the closed curve tells you that predator and prey populations are periodic; they go through repeated cycles. And the time cycle for each is the same. The four sectors above defined by the axes indicate four phases depicting trends that reverse themselves as the curve crosses the axes at prey = c/d and predator = a/b.

T.3.b.i) Wait a minute! Take another look at the last plot.

Those axes pierce the curve at the maximum and minimum values of prey and predator populations.

> Is this just an accident?

Answer: Stupid question. In mathematics, there are no accidents.

T.3.b.ii)

> Explain why those axes pierce the plot at the maximum and minimum values of prey and predator populations.

Answer: Look at the original differential equations

$$\text{prey}'[t] = a\,\text{prey}[t] - b\,\text{prey}[t]\,\text{pred}[t],$$
$$\text{pred}'[t] = -c\,\text{pred}[t] + d\,\text{pred}[t]\,\text{prey}[t].$$

Take the first:

$$\text{prey}'[t] = a\,\text{prey}[t] - b\,\text{prey}[t]\,\text{pred}[t].$$

At the times t at which the prey population is at its maximum or at its minimum, you gotta have prey$'[t] = 0$. Consequently, at the times t at which the prey population is at its maximum or at its minimum,

$$0 = a\,\text{prey}[t] - b\,\text{prey}[t]\,\text{pred}[t] = \text{prey}[t]\,(a - b\,\text{pred}[t])\,.$$

This tells you that at the times t at which the prey population is at its maximum or at its minimum, pred$[t] = a/b$.

Similarly, use

$$\text{pred}'[t] = -c\,\text{pred}[t] + d\,\text{pred}[t]\,\text{prey}[t]$$

to see that for a maximum or minimum predator population, you have

$$0 = -c\,\text{pred}[t] + d\,\text{pred}[t]\,\text{prey}[t] = \text{pred}[t]\,(-c + d\,\text{prey}[t])\,,$$

so when the predator population is largest or smallest,

$$\text{prey}[t] = \frac{c}{d}.$$

This explains why when you use AxesOrigin->$\{c/d, a/b\}$, the axes pierce the plot at the maximum and minimum values of prey and predator populations.

In[23]:=
```
Show[fullplot,PlotRange->{{0,9},{0,5}},
AxesOrigin->{c/d,a/b}];
```

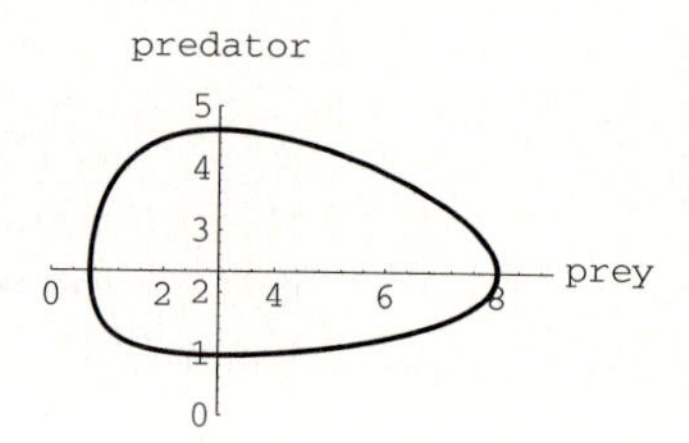

Math happens again.

■ T.4) Quick calculations

T.4.a) What is the instantaneous growth rate of the parametric curve

$$x = x[t] = e^t + t$$
$$y = y[t] = t\,(3 - t)\log[t]$$

at the point at which $x = 5$?

Answer: Take a look:

In[24]:=
```
Clear[x,y,t]
x[t_] = E^t + t;
y[t_] = t (3 - t) Log[t];
curveplot =
ParametricPlot[{x[t],y[t]},{t,0.01,5},
AxesLabel->{"x","y"}];
```

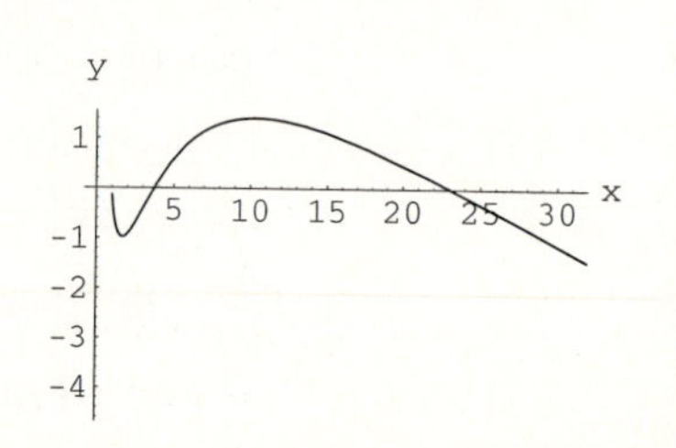

Now find the t for which $x[t] = 5$:

In[25]:=
```
Plot[{x[t],5},{t,0,4}];
```

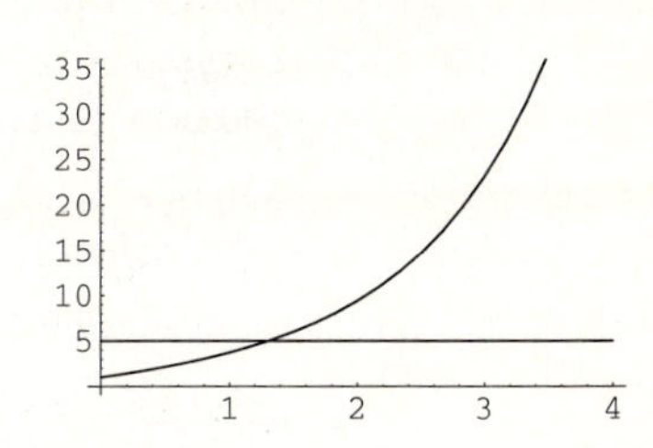

Start the search near $t = 1.2$:

In[26]:=
```
FindRoot[x[t] == 5,{t,1.2}]
```

Out[26]=
```
{t -> 1.30656}
```

The instantaneous growth rate at $x = 5$ is:

In[27]:=
```
(y'[t]/x'[t])/.t->1.30656
```

Out[27]=
```
0.382851
```

T.4.b) What is the highest point on the parametric curve

$$x = x[t] = e^t + t$$
$$y = y[t] = t\,(3 - t)\log[t]\,?$$

Answer: Evidently the curve stays negative for larger t's because

$$y[t] = t\,(3 - t)\log[t]$$

is negative for $t \geq 3$.

We search for a point t with $0 < t \leq 3$ at which $dy/dx = 0$. At $\{x[t], y[t]\}$, we know dy/dx is given by $y'[t]/x'[t]$. So we have to find the t's with $0 < t \leq 3$ at which $y'[t] = 0$.

In[28]:=
```
Plot[y'[t],{t,0.01,3}];
{FindRoot[y'[t] == 0,{t,.3}],
FindRoot[y'[t] == 0,{t,2.1}]}
```

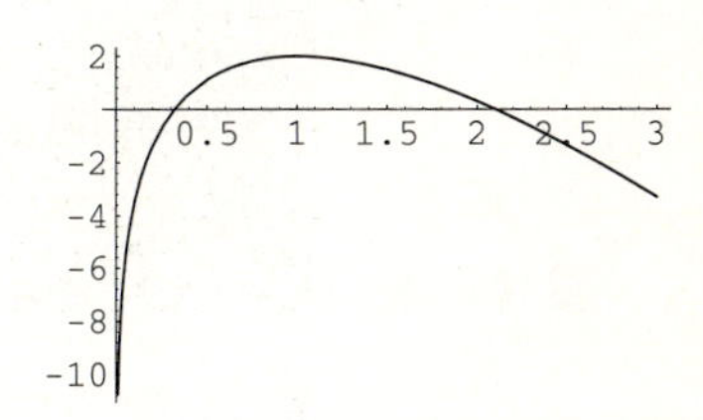

Out[28]=
```
{t -> 0.32105}
{t -> 2.10316}
```

Look at:

In[29]:=
```
{y[0.32105], y[2.10316]}
```

Out[29]=
```
{-0.977184, 1.40228}
```

The high point:

In[30]:=

```
high = {x[2.10316],y[2.10316]}
```

Out[30]=

```
{10.2952, 1.40228}
```

is the highest point on the curve. Check:

In[31]:=

```
pointplot = Graphics[RGBColor[1,0,0],PointSize[0.03],
Point[high]];
Show[curveplot,pointplot];
```

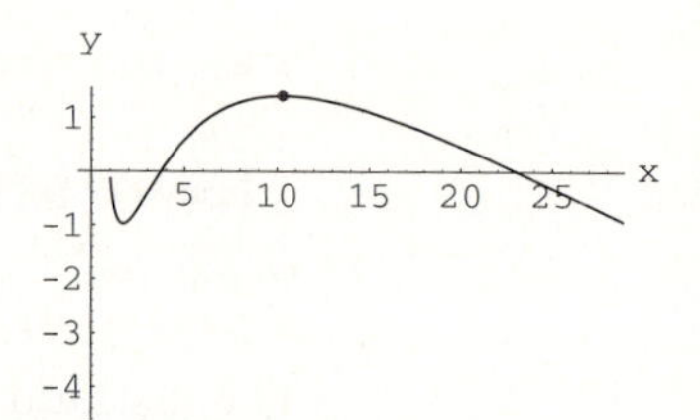

Nailed it.

T.4.c) If

$$x'[t] = \cos[t]\, y[t] \quad \text{with} \quad x[0] = 5 \quad \text{and}$$
$$y'[t] = 4\sin[t]\, y[t] \quad \text{with} \quad y[0] = 9$$

then what is dy/dx when $t = \pi$?

Answer: When $t = \pi$, then

$$\frac{dy}{dx} = \frac{y'[\pi]}{x'[\pi]} = \frac{4\,y[\pi]\sin[\pi]}{y[\pi]\cos[\pi]}$$
$$= \frac{4\sin[\pi]}{\cos[\pi]} = 4\left(\frac{0}{-1}\right) = 0.$$

Give It a Try

Experience with the starred ($\star$) problems will be especially beneficial for understanding later lessons.

■ G.1) Quick calculations

G.1.a) Plot the curve

$$x = x[t] = \frac{3\,t}{1+t^3}$$
$$y = y[t] = \frac{3\,t^2}{1+t^3}$$

for $1 \leq t \leq 9$.

> If you regard y to be a function $f[x]$ of x, then which x's give you $f'[x] = 0$? Which x gives rise to the largest value of $f[x]$?

G.1.b) When you go with

$$x = x[t] = r\,t - r\sin[t]$$
$$y = y[t] = r - r\cos[t],$$

as in the Basics, you get a cycloid. Calculating y in the form $y = f[x]$ is almost out of the question. Nevertheless, it's not a big deal to calculate $f'[x]$ when you are given a specific value of x. Here is a table in the form $\{x, f'[x]\}$ for selected x's.

In[1]:=

```
Clear[r,t,f]
Table[{N[r t - r Sin[t]],f'[N[r t - r Sin[t]]]},
{t,0,2 Pi,Pi/6}]
```

Out[1]=

```
{{0, f'[0]}, {0.0235988 r, f'[0.0235988 r]}, {0.181172 r, f'[0.181172 r]},
 {0.570796 r, f'[0.570796 r]}, {1.22837 r, f'[1.22837 r]}, {2.11799 r, f'[2.11799 r]},
 {3.14159 r, f'[3.14159 r]}, {4.16519 r, f'[4.16519 r]}, {5.05482 r, f'[5.05482 r]},
 {5.71239 r, f'[5.71239 r]}, {6.10201 r, f'[6.10201 r]}, {6.25959 r, f'[6.25959 r]},
 {6.28319 r, f'[6.28319 r]}}
```

Unfortunately $f'[x]$ has not been calculated simply because no clean formula for $f[x]$ is available.

> Copy, paste, and edit the last instruction so that each second slot exhibits the actual value of $f'[x]$ for the corresponding x in the first slot.

G.1.c) A curve is given in parametric form by

$$x = x[t] = t^2 + 5\,t - e^{-0.3t^2}$$
$$y = y[t] = 3\,t^2 + 54\,t + 8.$$

Here's a plot of part of this curve:

In[2]:=

```
Clear[x,y,t]
x[t_] = t^2 + 5 t - E^(-0.3 t^2);
y[t_] = 3 t^2 + 54 t + 8;
ParametricPlot[{x[t],y[t]},{t,-2,2},
PlotStyle->{{Thickness[0.01],Blue}},
AxesLabel->{"x","y"}];
```

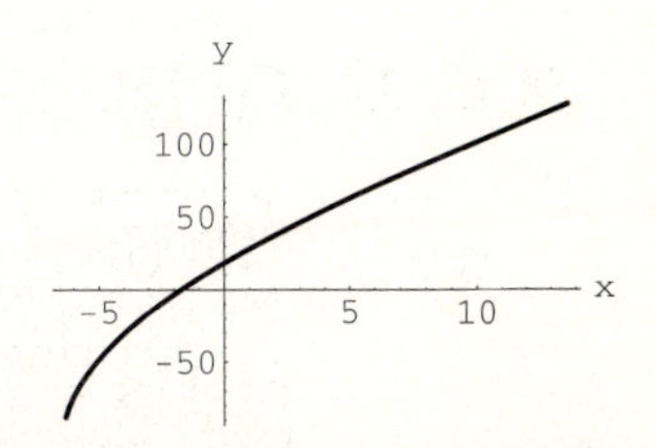

> Plot more and more of the curve until you get a pretty good idea of how this baby looks.

Use information from $x'[t]$ to give a good estimate of the smallest value of x on the curve.

Use information from $y'[t]$ to give a good estimate of the smallest value of y on the curve.

■ G.2) Parametric plotting of circles and ellipses in two dimensions★

A handy way to plot the circle $x^2 + y^2 = 1$ is to write

$$x[t] = \cos[t] \qquad \text{and} \qquad y[t] = \sin[t]$$

and then to plot the points

$$\{x[t], y[t]\} = \{\cos[t], \sin[t]\}$$

as t advances from 0 to 2π.

In[3]:=
```
Clear[x,y,t]
{x[t_],y[t_]} = {Cos[t], Sin[t]};
circle =
ParametricPlot[{x[t],y[t]},{t,0,2 Pi},
PlotStyle->{{Blue,Thickness[0.015]}},
AspectRatio->Automatic,AxesLabel->{"x","y"}];
```

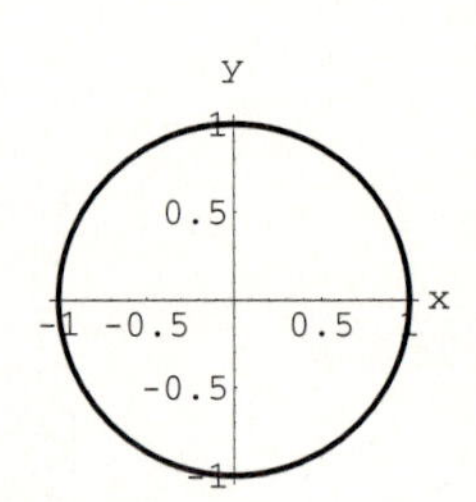

The option AspectRatio->Automatic tells *Mathematica* to try to give a true scale plot.

See what happens when you multiply $x[t] = \cos[t]$ by 3:

In[4]:=
```
Clear[x,y,t]
{x[t_],y[t_]} = {3 Cos[t], Sin[t]};
ParametricPlot[{x[t],y[t]} ,{t,0,2 Pi},
PlotStyle->{{Blue,Thickness[0.01]}},
AspectRatio->Automatic,AxesLabel->{"x","y"}];
```

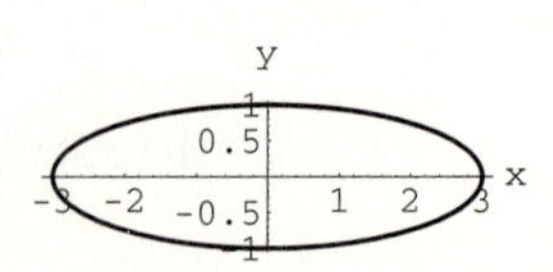

The plot looks quite a bit like the ellipse

$$\left(\frac{x}{3}\right)^2 + y^2 = 1.$$

Check this out:

In[5]:=
```
Clear[x,y,t]
Expand[((x/3)^2 + (y)^2)/. {x->3 Cos[t],y->Sin[t]}, Trig->True]
```

Out[5]=
```
1
```

The plot is the ellipse

$$\left(\frac{x}{3}\right)^2 + y^2 = 1.$$

The upshot:

$$\{x[t], y[t]\} = \{3\cos[t], \sin[t]\}$$

gives you a parametric formula for the ellipse

$$\left(\frac{x}{3}\right)^2 + y^2 = 1.$$

Now see what happens when you also multiply $y[t] = \sin[t]$ by 2:

In[6]:=
```
Clear[x,y,t]
{x[t_],y[t_]} = {3 Cos[t], 2 Sin[t]};
ParametricPlot[{x[t],y[t]} ,{t,0,2 Pi},
PlotStyle->{{Blue,Thickness[0.01]}},
AspectRatio->Automatic,AxesLabel->{"x","y"}];
```

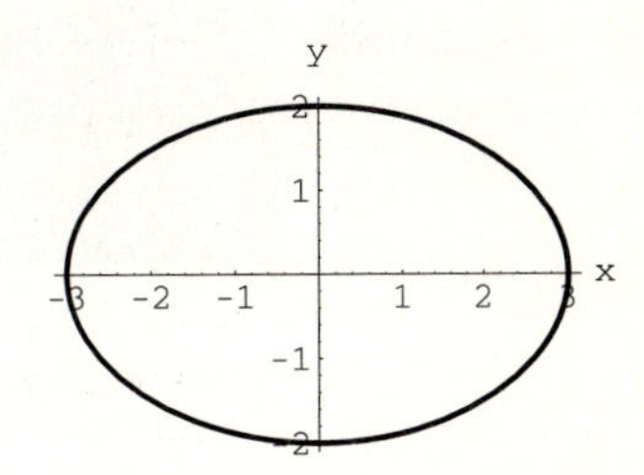

The plot looks quite a bit like the ellipse

$$\left(\frac{x}{3}\right)^2 + \left(\frac{y}{2}\right)^2 = 1.$$

Check this out:

In[7]:=
```
Clear[x,y,t]
Expand[((x/3)^2 + (y/2)^2)/.
{x->3 Cos[t],y->2 Sin[t]},
Trig->True]
```

Out[7]=
```
1
```

The plot is the ellipse

$$\left(\frac{x}{3}\right)^2 + \left(\frac{y}{2}\right)^2 = 1.$$

The upshot:

$$\{x[t], y[t]\} = \{3\cos[t], 2\sin[t]\}$$

gives you a parametric formula for the ellipse

$$\left(\frac{x}{3}\right)^2 + \left(\frac{y}{2}\right)^2 = 1.$$

G.2.a.i) Give parametric formulas for the ellipse

$$\left(\frac{x}{4}\right)^2 + \left(\frac{y}{3}\right)^2 = 1$$

and use your formulas to plot the curve.

G.2.a.ii) Give parametric formulas for the ellipse

$$\left(\frac{x}{2}\right)^2 + \left(\frac{y}{4}\right)^2 = 1$$

and use your formulas to plot the curve.

G.2.a.iii) Given positive constants a and b, give parametric formulas for the ellipse

$$\left(\frac{x}{a}\right)^2 + \left(\frac{y}{b}\right)^2 = 1.$$

G.2.a.iv) Look at this graphic:

In[8]:=
```
a = 5; b = 3;
Clear[acircleplotter,bcircleplotter,ellipseplotter,t]
acircleplotter[t_] = {a Cos[t],a Sin[t]};
ellipseplotter[t_] = {a Cos[t],b Sin[t]};
bcircleplotter[t_] = {b Cos[t],b Sin[t]};
CBS = ParametricPlot[
{acircleplotter[t],ellipseplotter[t],bcircleplotter[t]},
{t,0,2 Pi},
PlotStyle->{{Blue,Thickness[0.02]},
{Red,Thickness[0.02]},{Blue,Thickness[0.02]}},
AspectRatio->Automatic,AxesLabel->{"x","y"}];
```

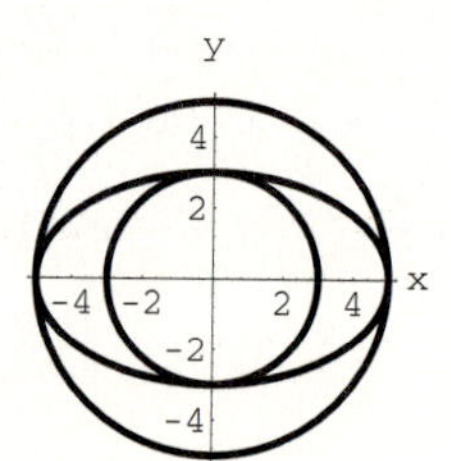

Here are the points plotted on each curve when you go with $t = \pi/6$:

In[9]:=
```
Clear[points,ray,t]
points[t_] = Graphics[{PointSize[0.05],
Point[{a Cos[t],a Sin[t]}],
Point[{a Cos[t],b Sin[t]}],
Point[{b Cos[t],b Sin[t]}]}];
ray[t_] = Graphics[
Line[{{0,0},{a Cos[t],a Sin[t]}}]];
Show[CBS,points[Pi/6],ray[Pi/6]];
```

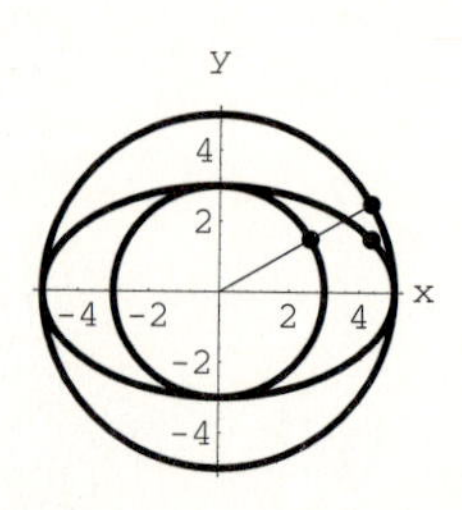

Here is how the two circles and the ellipse plot out as t runs from 0 to $\pi/2$:

In[10]:=

```
Table[Show[CBS,points[t],ray[t]],{t,0,Pi/2,Pi/10}];
```

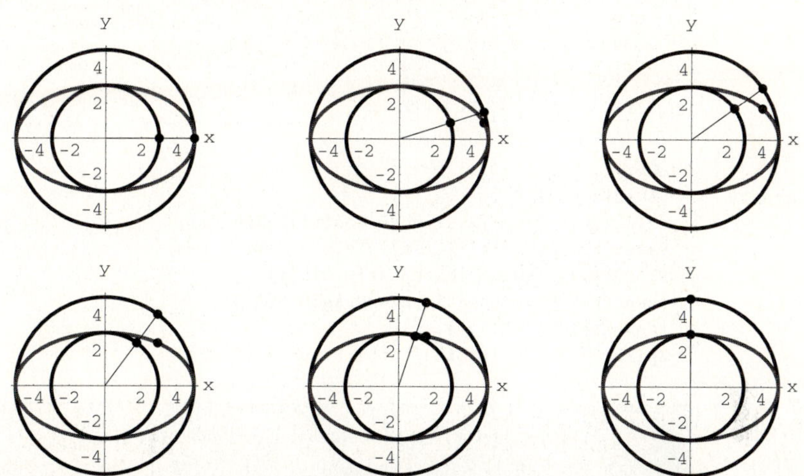

Your friend has no machine but has a ruler, a compass, a T-square, and a triangle. Tell your friend how to plot a lot of points on the ellipse

$$\left(\frac{x}{4}\right)^2 + \left(\frac{y}{2}\right)^2 = 1$$

by drawing concentric circles and then moving horizontally or vertically from well-chosen points on the circles.

G.2.b) A handy way to plot the circle of radius 1 centered at {2,3} whose equation is

$$(x-2)^2 + (y-3)^2 = 1$$

is to use:

In[11]:=

```
Clear[x,y,t]
{x[t_],y[t_]} = {2,3} + {Cos[t],Sin[t]};
ParametricPlot[{x[t],y[t]},{t,0,2 Pi},
PlotStyle->{{Blue,Thickness[0.015]}},
AspectRatio->Automatic,AxesOrigin->{2,3},
AxesLabel->{"x","y"},
Epilog->{Red,PointSize[0.06],Point[{2,3}]}];
```

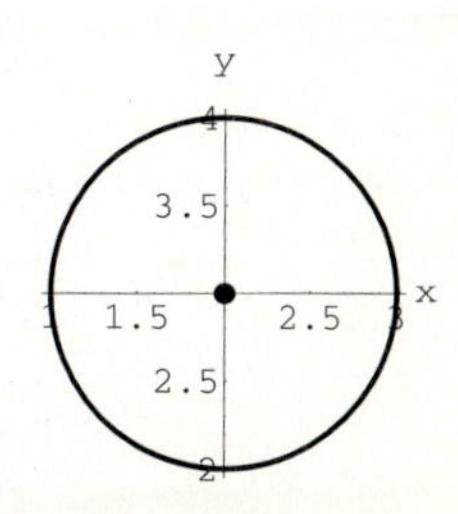

Set up parametric formulas for the circle of radius r centered at $\{h, k\}$

$$(x-h)^2 + (y-k)^2 = r^2.$$

Use your formulas to plot the circle

$$(x-1)^2 + (y-2)^2 = 5.$$

G.2.c) A handy way to plot the ellipse

$$\left(\frac{x-2}{5}\right)^2 + \left(\frac{y-4}{2}\right)^2 = 1$$

centered on $\{2, 4\}$ is to use:

In[12]:=
```
Clear[x,y,t]
{x[t_],y[t_]} = {2,4} + {5 Cos[t],3 Sin[t]};
ParametricPlot[{x[t],y[t]},{t,0,2 Pi},
PlotStyle->{{Blue,Thickness[0.015]}},
AspectRatio->Automatic,AxesOrigin->{2,4},
AxesLabel->{"x","y"},
Epilog->{Red,PointSize[0.05],Point[{2,4}]}];
```

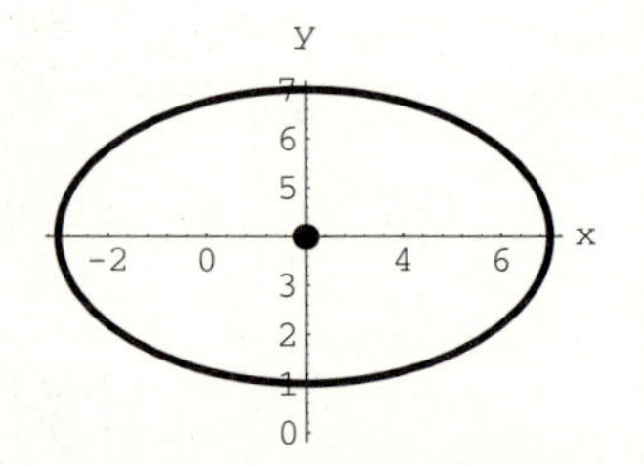

Use parametric formulas as above to plot the ellipse

$$\left(\frac{x+1}{2}\right)^2 + \left(\frac{y-4}{5}\right)^2 = 1.$$

At what point is this ellipse centered?

■ G.3) Elliptical orbits of planets and asteroids

G.3.a) You can plot the circle $x^2 + y^2 = r^2$ by using the parametric equations

$$x[t] = r\cos[t] \qquad \text{and} \qquad y[t] = r\sin[t]$$

and running t from 0 to $2\,\pi$.

You can plot the ellipse $(x/a)^2 + (y/b)^2 = r^2$ by using the parametric equations

$$x[t] = a\cos[t] \qquad \text{and} \qquad y[t] = b\sin[t]$$

and running t from 0 to $2\,\pi$.

Here you go with a true scale plot of the ellipse

$$\left(\frac{x}{9}\right)^2 + \left(\frac{y}{3}\right)^2 = 1:$$

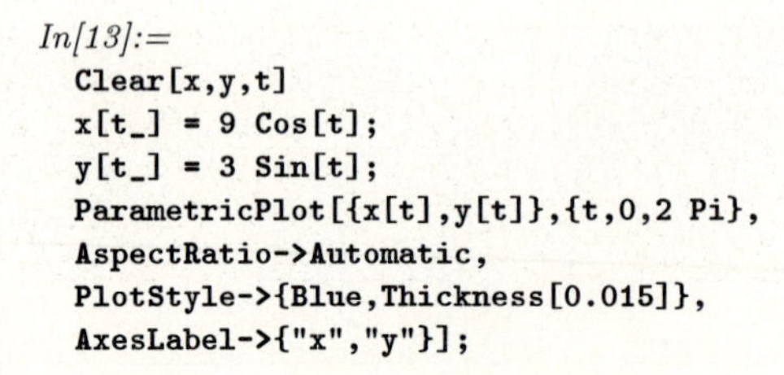

In[13]:=
```
Clear[x,y,t]
x[t_] = 9 Cos[t];
y[t_] = 3 Sin[t];
ParametricPlot[{x[t],y[t]},{t,0,2 Pi},
AspectRatio->Automatic,
PlotStyle->{Blue,Thickness[0.015]},
AxesLabel->{"x","y"}];
```

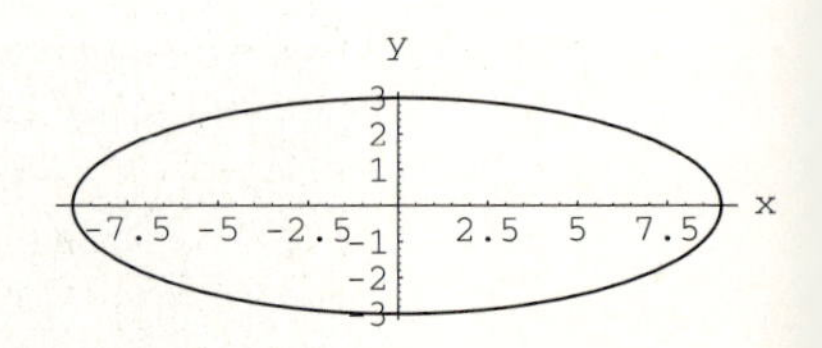

The option AspectRatio->Automatic tells *Mathematica* to try to give a true scale plot.

> Give a true scale plot of the ellipse
>
> $$\left(\frac{x}{4}\right)^2 + \left(\frac{y}{9}\right)^2 = 1.$$

G.3.b) Here is the ellipse $(x/4)^2 + (y/3)^2 = 1$ in true scale:

In[14]:=

```
Clear[x,y,t]
x[t_] = 4 Cos[t];
y[t_] = 3 Sin[t];
ParametricPlot[{x[t],y[t]},{t,0,2 Pi},
AspectRatio->Automatic,
PlotStyle->{Blue,Thickness[0.01]},
AxesLabel->{"x","y"}];
```

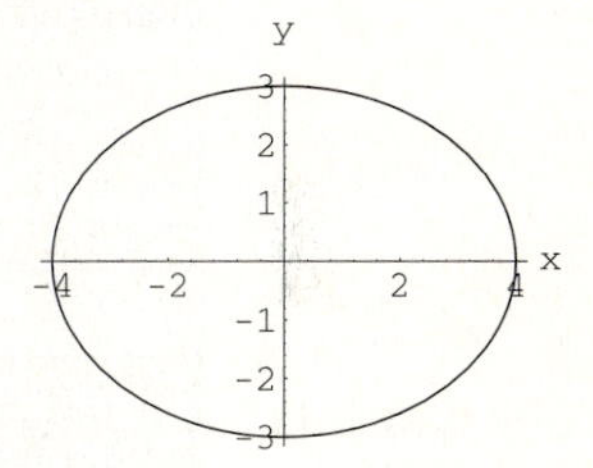

The long axis of this ellipse runs along the x-axis; its length is 8. The short axis of this ellipse runs along the y-axis; its length is 6.

Fancy folks call the long axis by the name "major axis" and they call the short axis by the name "minor axis."

To get what folks call the eccentricity of an ellipse, you take the ratio

$$\text{ratio} = \frac{\text{length of short axis}}{\text{length of long axis}}$$

and put

$$\text{eccentricity} = \sqrt{1 - \text{ratio}^2}.$$

Here is the eccentricity of the ellipse above:

In[15]:=

```
long = 8; short = 6; ratio = short/long;
eccentricity = N[Sqrt[1 - ratio^2]]
```

Out[15]=

```
0.661438
```

G.3.b.i)

> Plot the ellipse
>
> $$\left(\frac{x}{8}\right)^2 + \left(\frac{y}{3}\right)^2 = 1$$
>
> in true scale and calculate its eccentricity.

G.3.b.ii)

Plot the ellipse

$$\left(\frac{x}{2}\right)^2 + \left(\frac{y}{5}\right)^2 = 1$$

in true scale and calculate its eccentricity.

G.3.b.iii)

Given positive numbers a and b, how do you know that the eccentricity of the ellipse

$$\left(\frac{x}{a}\right)^2 + \left(\frac{y}{b}\right)^2 = 1$$

is a number no smaller than 0 but less than 1?

What do you call an ellipse whose eccentricity is 0?

G.3.b.iv)

Plot in true scale an ellipse whose long axis is 10 units long and whose eccentricity is 0.9.

G.3.b.v)

Plot in true scale an ellipse whose long axis is 10 units and whose eccentricity is 0.1.

G.3.c) If $a \geq b$, then the focuses of the ellipse

$$\left(\frac{x}{a}\right)^2 + \left(\frac{y}{b}\right)^2 = 1$$

are located at the points

$$\{-\text{eccentricity}\, a, 0\} \qquad \text{and} \qquad \{\text{eccentricity}\, a, 0\}.$$

But if $a < b$, then the focuses of the ellipse

$$\left(\frac{x}{a}\right)^2 + \left(\frac{y}{b}\right)^2 = 1$$

are located at the points

$$\{0, -\text{eccentricity}\, b\} \qquad \text{and} \qquad \{0, \text{eccentricity}\, b\}.$$

Here is a true scale plot of the ellipse

$$\left(\frac{x}{6}\right)^2 + \left(\frac{y}{3}\right)^2 = 1$$

shown along with its two focuses:

In[16]:=

```
Clear[x,y,a,b,t]; a = 6; b = 3;
x[t_] = a Cos[t]; y[t_] = b Sin[t];
eccentricity = Sqrt[1 -(3/6)^2];
ellipse = ParametricPlot[{x[t],y[t]},
{t,0,2 Pi},PlotStyle->{Blue,Thickness[0.01]},
AspectRatio->Automatic,AxesLabel->{"x","y"},
Epilog-> {{Red,PointSize[0.04],
Point[{-eccentricity a,0}]},
{Red,PointSize[0.04],
Point[{eccentricity a,0}]}}];
```

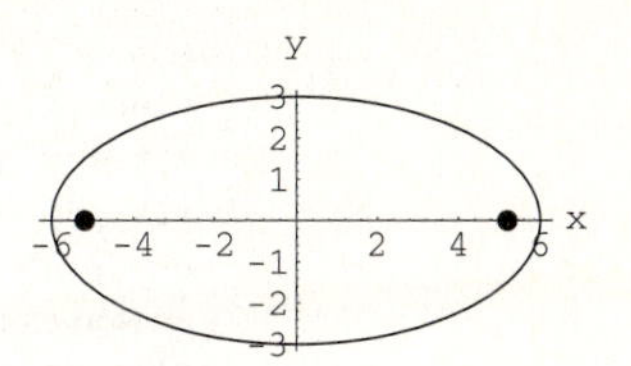

> Plot the ellipse
>
> $$\left(\frac{x}{2}\right)^2 + \left(\frac{y}{9}\right)^2 = 1$$
>
> and show it along with its two focuses.

G.3.d.i) Johannes Kepler (1571–1630) earned a permanent place in history by perfecting Copernicus's idea that the planets revolve about the sun. Greatly influenced by Copernicus and getting his start by working with other giants like Brahe and Galileo, he determined that planets, asteroids, and comets move in elliptical orbits around the sun with the sun at one of the focuses of the elliptical orbit.

Physicists call this fact "Kepler's first law." And this is big stuff.

Kepler's first law tells you that to give high-quality plots of the orbits of planets, asteroids, and comets, you are well advised to plot with the sun at a focus.

Here is the usual plot of

$$\left(\frac{x}{7}\right)^2 + \left(\frac{y}{3}\right)^2 = 1$$

with the center of the ellipse at the origin:

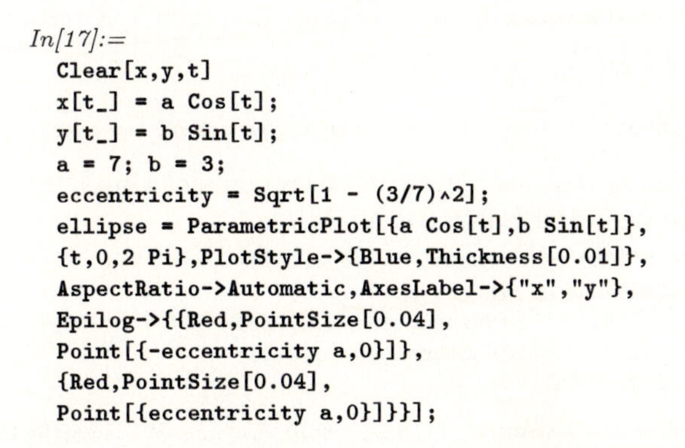

In[17]:=

```
Clear[x,y,t]
x[t_] = a Cos[t];
y[t_] = b Sin[t];
a = 7; b = 3;
eccentricity = Sqrt[1 - (3/7)^2];
ellipse = ParametricPlot[{a Cos[t],b Sin[t]},
{t,0,2 Pi},PlotStyle->{Blue,Thickness[0.01]},
AspectRatio->Automatic,AxesLabel->{"x","y"},
Epilog->{{Red,PointSize[0.04],
Point[{-eccentricity a,0}]},
{Red,PointSize[0.04],
Point[{eccentricity a,0}]}}];
```

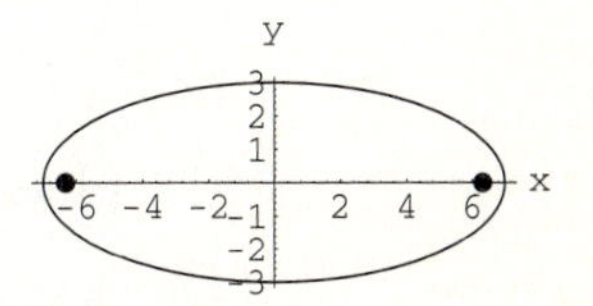

Here is how you plot the same curve shifted to the left so that the right focus is at the origin:

In[18]:=

```
Clear[x,y,t,a,eccentricity]
x[t_] = a Cos[t] - a eccentricity;
y[t_] = b Sin[t];
a = 7; b = 3;
eccentricity = Sqrt[1 - (3/7)^2];
ellipse = ParametricPlot[{x[t],y[t]},
{t,0,2 Pi},PlotStyle->{Blue,Thickness[0.01]},
AspectRatio->Automatic,AxesLabel->{"x","y"},
Epilog->{{Red,PointSize[0.04],
Point[{-2 eccentricity a ,0}]},
{Red,PointSize[0.04], Point[{0,0}]}}];
```

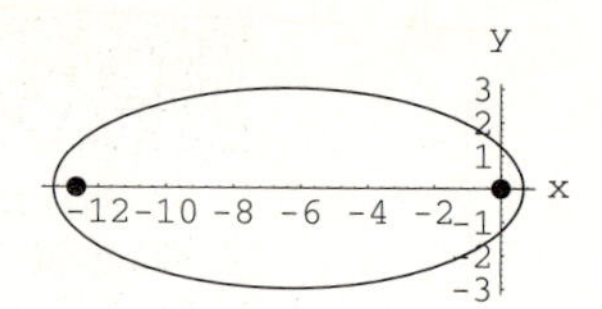

Get it?

> Plot the ellipse
>
> $$\left(\frac{x}{2}\right)^2 + \left(\frac{y}{6}\right)^2 = 1.$$
>
> Then plot the same curve adjusted so that the top focus is at the origin. Show the focuses in both plots.

G.3.d.ii)

> Plot an ellipse whose long axis is 20 units long and whose eccentricity is 0.65 with its center at the origin. Then plot the same curve adjusted so that the right focus is at the origin. Show the focuses in both plots.

G.3.d.iii) Kepler's first law says that planets, asteroids, and comets move in elliptical orbits around the sun with the sun at one of the focuses of the elliptical orbit.

One big mystery is that all these orbits lie roughly in the same plane.

This mystery is intriguing, but the fact that all these orbits lie in one plane makes it easy to give reasonable plots of the orbits of the planets, asteroids, and comets.

Here are the data for all the planets in the form

```
{planet name, length of long axis of elliptical orbit, eccentricity}
```

The lengths of the long axes are given in Astronomical Units. One Astronomical Unit is the distance from the Earth to the Sun.

```
{Mercury, 0.774,0.206}      {Venus, 1.446,0.007}
{Earth, 2.00,0.017}         {Mars, 3.04,0.093}
{Jupiter, 10.4,0.048}       {Saturn, 19.08,0.056}
{Uranus, 38.36,0.05}        {Neptune, 60.14,0.01}
{Pluto, 78.88,0.25}
```

Source: William K. Hartman, *Astronomy: The Cosmic Journey*, Wadsworth, 1991.

All but Mercury and Pluto have small eccentricities; so their orbits are nearly circular. No fun plotting these.

Plot the orbit of Mercury with the sun at one focus.

G.3.d.iv) To get a better kick out of this, give a plot that shows the orbits of Neptune and Pluto.

Is Pluto always the outmost planet?

G.3.d.v) To get an even better kick out of this, give a plot that shows the orbits of Earth and a typical member of the Apollo asteroid group.

In the form of the data above the Apollo data are {Apollo, 3.0, 0.56}.

■ G.4) Parametric plotting of circles, tubes, and horns in three dimensions★

Here are the three coordinate axes in three dimensions:

In[19]:=

```
h = 5; spacer = h/10;
threedims = Graphics3D[{
{Blue,Line[{{-h,0,0},{h,0,0}}]},
Text["x",{h + spacer,0,0}],
{Blue,Line[{{0,-h,0},{0,h,0}}]},
Text["y",{0, h + spacer,0}],
{Blue,Line[{{0,0,-h},{0,0,h}}]},
Text["z",{0,0,h + spacer}]}];
CMView = {2.7, 1.6, 1.2};
Show[threedims,PlotRange->All,
ViewPoint->CMView,Boxed->False];
```

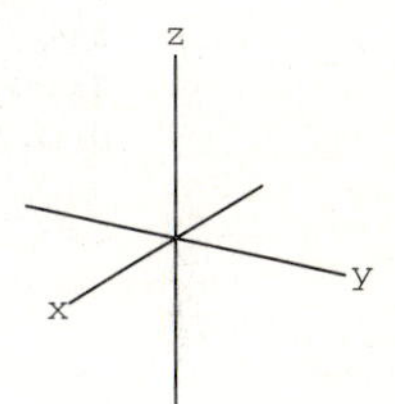

The xy-plane is the plane that you get by laying down a rigid flat sheet on the girders defined by the x-axis and the y-axis. Here is a piece of it:

In[20]:=

```
xyplane = Graphics3D[
Polygon[{{0,0,0},{h,0,0},{h,h,0},{0,h,0}}]];
Show[threedims,xyplane,PlotRange->All,
ViewPoint->CMView, Boxed->False];
```

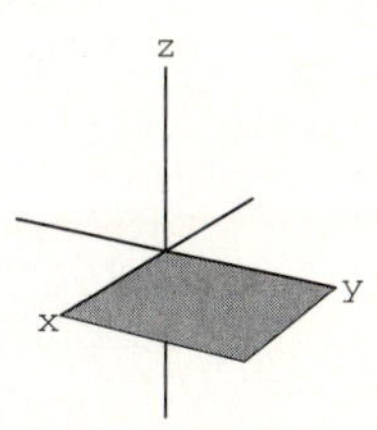

The xz-plane is the plane that you get by laying down a rigid flat sheet on the girders defined by the x-axis and the z-axis. Here is a piece of it:

In[21]:=

```
xzplane = Graphics3D[
Polygon[{{0,0,0},{0,0,h},{h,0,h},{h,0,0}}]];
Show[threedims,xzplane,PlotRange->All,
ViewPoint->CMView,Boxed->False];
```

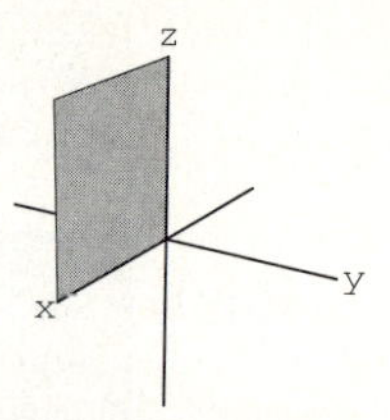

> Show a piece of the yz-plane together with the coordinate axes.
> Continue to use the option ViewPoint->CMView.

G.4.a) Here is the circle in the xy-plane of radius 1.5 centered at $\{2.5, 3, 0\}$:

In[22]:=

```
Clear[t]
circle = ParametricPlot3D[
Evaluate[{2.5,3,0} + 1.5 {Cos[t],Sin[t],0}],
{t,0,2 Pi},DisplayFunction->Identity];
h = 5;
xyplane =
Graphics3D[
Polygon[{{0,0,0},{h,0,0},{h,h,0},{0,h,0}}]];
Show[threedims,xyplane,circle,PlotRange->All,
ViewPoint->CMView,Boxed->False,
DisplayFunction->$DisplayFunction];
```

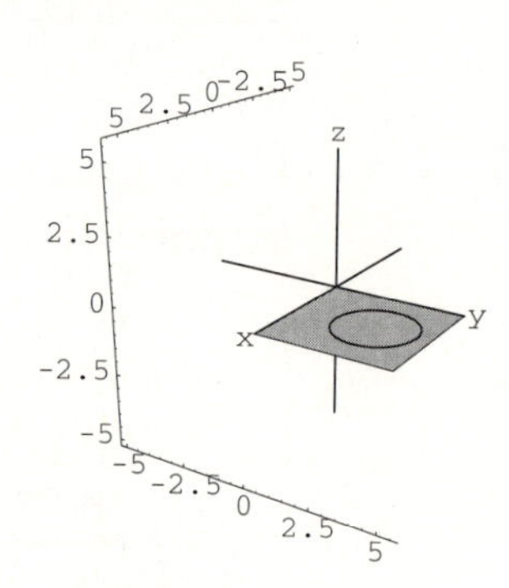

Here is the circle parallel to the xz-plane of radius 1.5 centered at $\{2.5, 3, 2\}$:

In[23]:=

```
Clear[t]
circle = ParametricPlot3D[
Evaluate[{2.5,3,2} + 1.5 {Cos[t],0,Sin[t]}],
{t,0,2 Pi},DisplayFunction->Identity];
h = 5;
xzplane = Graphics3D[
Polygon[{{0,0,0},{0,0,h},{h,0,h},{h,0,0}}]];
Show[threedims,xzplane,circle,PlotRange->All,
ViewPoint->CMView,Boxed->False,
DisplayFunction->$DisplayFunction];
```

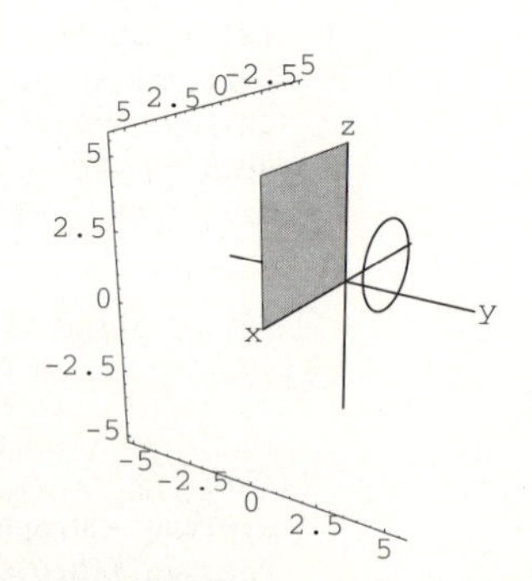

> Plot the circle parallel to the yz-plane of radius 1.5 centered at $\{2, 3, 2\}$.

G.4.b) Here are several circles of radius 3 parallel to the xz-plane centered on the y-axis:

In[24]:=

```
Clear[t]
circle1 = ParametricPlot3D[
Evaluate[{0,1,0} + 3 {Cos[t],0,Sin[t]}],
{t,0,2 Pi},DisplayFunction->Identity];
circle2 = ParametricPlot3D[
Evaluate[{0,2,0} + 3 {Cos[t],0,Sin[t]}],
{t,0,2 Pi},DisplayFunction->Identity];
circle3 = ParametricPlot3D[
Evaluate[{0,3,0} + 3 {Cos[t],0,Sin[t]}],
{t,0,2 Pi},DisplayFunction->Identity];
Show[threedims,circle1,circle2,circle3,
ViewPoint->CMView,Boxed->False,
PlotRange->All,
DisplayFunction->$DisplayFunction];
```

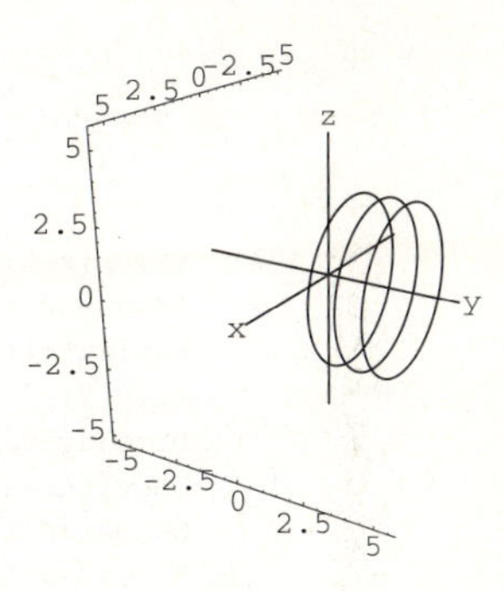

It's no fun to settle for a few crummy circles when you can get a whole tube composed of circles of radius 3 all parallel to the xz-plane and all centered on the y-axis:

In[25]:=

```
Clear[t,y]
tube = ParametricPlot3D[
Evaluate[{0,y,0} + 3 {Cos[t],0,Sin[t]}],
{t,0,2 Pi},{y,0,3},
DisplayFunction->Identity];
Show[threedims,tube,PlotRange->All,
ViewPoint->CMView,Boxed->False,
DisplayFunction->$DisplayFunction];
```

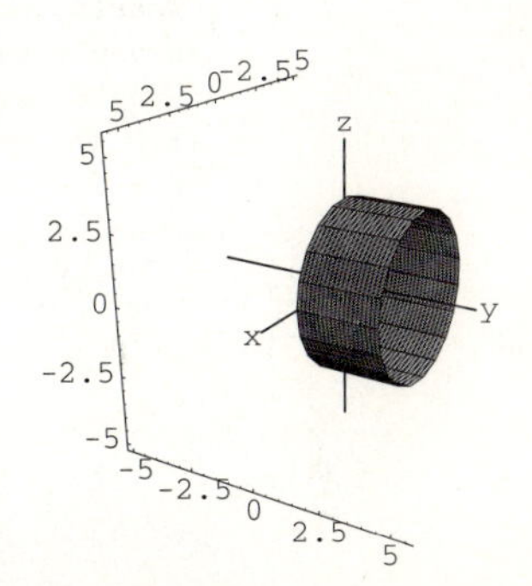

> Plot a tube composed of circles of radius 3.5 all parallel to the xy-plane and all centered on the x-axis.

G.4.c) Here is a modest curve in three dimensions:

In[26]:=

```
Clear[t,curveplotter]
curveplotter[t_] = {Cos[t],3 t Sin[t],3 t};
curve = ParametricPlot3D[Evaluate[curveplotter[t]],
{t,0,2}, DisplayFunction->Identity];
Show[threedims,curve,ViewPoint->CMView,
Boxed->False,
PlotRange->All,
DisplayFunction->$DisplayFunction];
```

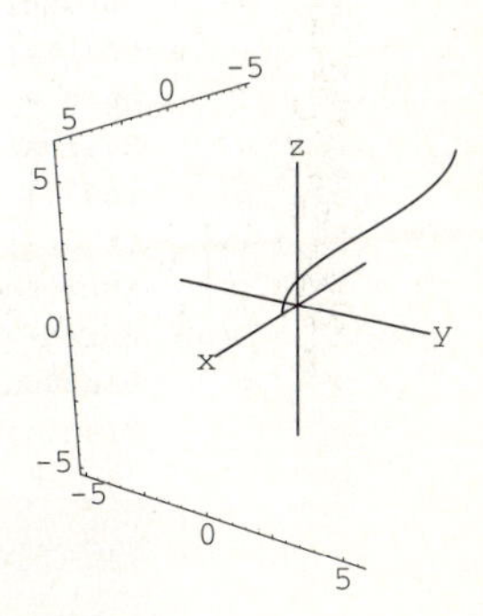

Here is what you get when you make a tube consisting of all circles parallel to the xy-plane of radius 2 centered at all points on this curve:

In[27]:=

```
Clear[s,t]
tube = ParametricPlot3D[
Evaluate[curveplotter[t] + 2 {Cos[s],Sin[s],0}],
{t,0,2},{s,0,2 Pi},
DisplayFunction->Identity];
Show[threedims,tube,ViewPoint->CMView,
Boxed->False,PlotRange->All,
DisplayFunction->$DisplayFunction];
```

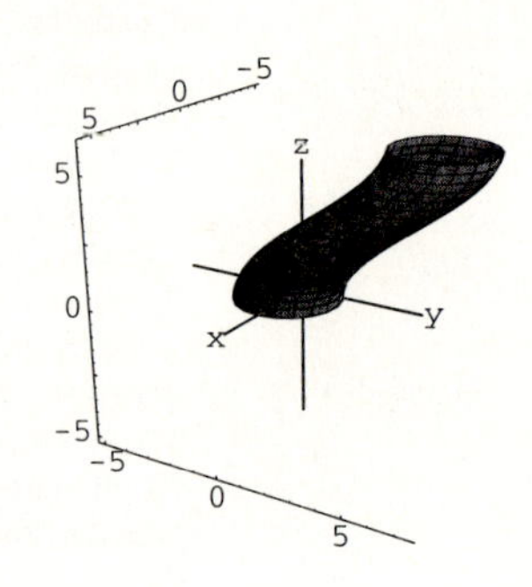

Here is another curve:

In[28]:=

```
Clear[t,curveplotter]
curveplotter[t_] = {t Sin[t],2 t,Cos[2 t]};
curve = ParametricPlot3D[Evaluate[curveplotter[t]],
{t,0,4}, DisplayFunction->Identity];
Show[threedims,curve,ViewPoint->CMView,
Boxed->False,PlotRange->All,
DisplayFunction->$DisplayFunction];
```

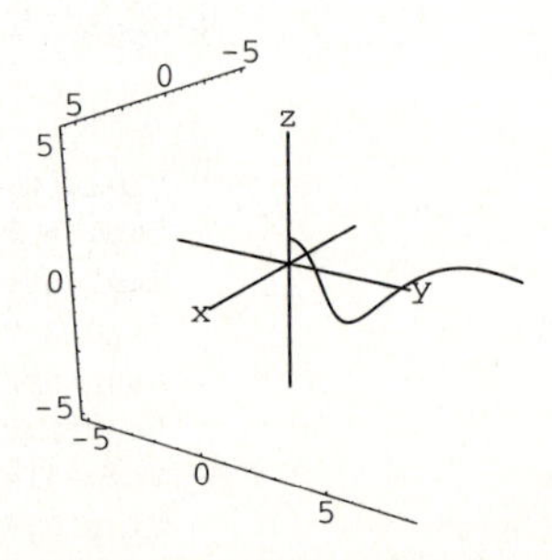

Plot the tube composed of all circles of radius 1.5 parallel to the xz-plane and centered on this curve.

G.4.d) Look at this plot:

In[29]:=

```
Clear[s,t,curveplotter,rad]
curveplotter[t_] = {t Sin[t],2 t,Cos[2 t]};
rad[t_] = (t^2)/6;
horn = ParametricPlot3D[
Evaluate[curveplotter[t] +
rad[t] {Cos[s],0,Sin[s]}],
{t,0,4},{s,0,2 Pi}, DisplayFunction->Identity];
Show[threedims,horn,ViewPoint->CMView,
Boxed->False,
PlotRange->All,
DisplayFunction->$DisplayFunction];
```

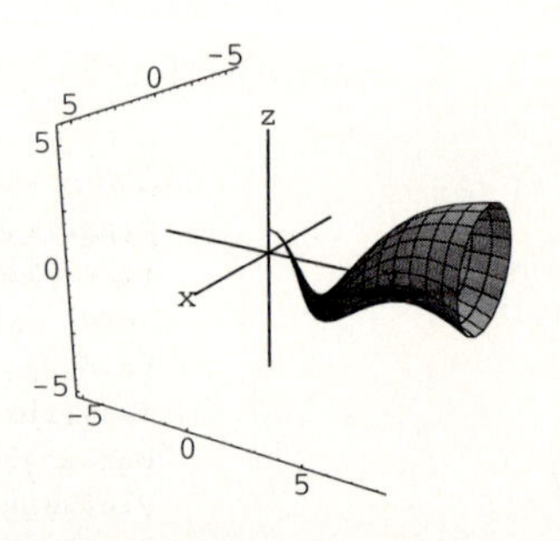

Describe the circles that this horn is composed of.

G.4.e) Do something that's either artistic or crazy.

■ G.5) Surfaces you can make by rotating curves

This problem appears only in the electronic version.

■ G.6) Projectile marksmanship

G.6.a) On flat terrain, a projectile with an initial velocity of 190 ft/sec is fired at a 20-foot-high vertical wall 1000 feet from the tip of the cannon inclined at an angle a with the horizontal.

G.6.a.i) If $a = \pi/6$, does the projectile hit the wall?

G.6.a.ii) For approximately what angles of inclination will the projectile hit the wall or soar over the wall?

G.6.a.iii) For approximately what angles of inclination will the projectile hit the wall?

G.6.b.i) You have a cannon with a muzzle speed of 190 ft/sec. At time $t = 0$, you fire the cannon with an angle of inclination $\pi/4 + z$. Then you wait d seconds and fire again with an angle of inclination $\pi/4 - z$.

Here is a plot which reveals what the second projectile is doing while the first is in flight in the case $z = \pi/8$ and $d = 1$.

In[30]:=

```
Clear[x,y,t,a,s,a]
x[t_,s_,a_] = s t Cos[a];
y[t_,s_,a_] = -16 t^2 + s t Sin[a];
z = Pi/8; d = 1;
ParametricPlot[
{{x[t,190,Pi/4 + z],y[t,190,Pi/4 + z]},
{x[t - d,190,Pi/4 - z],y[t - d,190,Pi/4 - z]}},
{t,0,190 Sin[Pi/4 +z]/16},
PlotRange->{-100,500},
PlotStyle->{{Thickness[0.015],Blue},
{Thickness[0.010],Red}}];
```

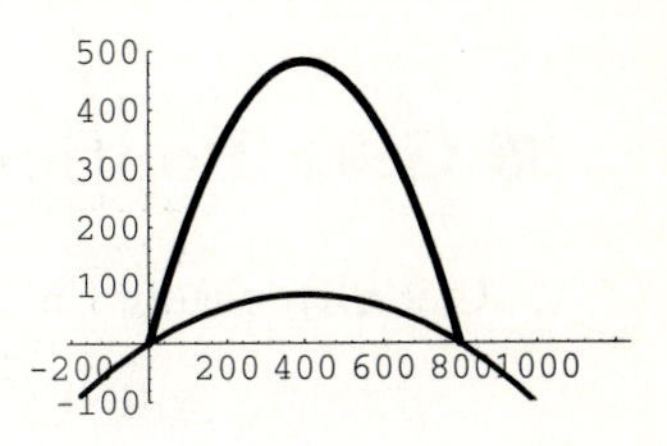

The trajectory of the second shot is thinner than the trajectory of the first shot. That its plot is below the ground on the right indicates that the second shot hit the ground before the first shot hit the ground.

That its plot is below the ground on the left indicates that the second shot was fired after the first shot was fired.

> Your mission, if you accept it, is to find a new delay time d so that both projectiles hit the ground at the same instant.

G.6.b.ii) You have a cannon with a muzzle speed of s ft/sec. At time $t = 0$, you fire a projectile with an angle of inclination $\pi/4 + z$. Then you wait $d[z]$ seconds and fire another projectile with an angle of inclination $\pi/4 - z$.

> This time your mission is:
>
> Determine a formula for the delay time
>
> $$d = d[s, z] \qquad (\text{for } 0 < z < \pi/4)$$
>
> as a function of z and s so that both projectiles hit the ground at the same spot and at the same instant.
>
> Show off your formula with a couple of sharp plots.

■ G.7) More cams

This problem appears only in the electronic version.

■ G.8) Parametric plotting of a predator-prey model in which the prey don't reproduce and the predators don't die

This problem appears only in the electronic version.

■ G.9) Politics and the environment

G.9.a.i) Study the Tutorial on the predator-prey model based on the differential equations

$$\text{prey}'[t] = a\,\text{prey}[t] - b\,\text{prey}[t]\,\text{pred}[t]$$
$$\text{pred}'[t] = -c\,\text{pred}[t] + d\,\text{pred}[t]\,\text{prey}[t].$$

Use the same a, b, c, and d as in the Tutorial but use starting data with

$$\text{prey}[0] = \frac{c}{d} + h,$$
$$\text{pred}[0] = \frac{a}{b} + h$$

for a relatively small value of h:

In[31]:=

```
h = 0.6; endtime = 20;
a = 0.7; b = 0.3; c = 0.3; d = 0.1;
Clear[pred,prey,t,Derivative]
approxsolutions =
NDSolve[{prey'[t] ==  a prey[t] - b prey[t] pred[t],
pred'[t] == -c pred[t] + d pred[t] prey[t],
prey[0] == c/d + h, pred[0] == a/b + h},
{prey[t],pred[t]},{t,0,endtime}];
Clear[fakepred,fakeprey]
fakepred[t_] = pred[t]/.approxsolutions[[1]];
fakeprey[t_] = prey[t]/.approxsolutions[[1]];
cycleplot = ParametricPlot[{fakeprey[t],fakepred[t]},
{t,0,endtime},PlotStyle->Thickness[0.01],
PlotRange->All, AxesLabel->{"prey","predator"},
AxesOrigin->{c/d,a/b}];
```

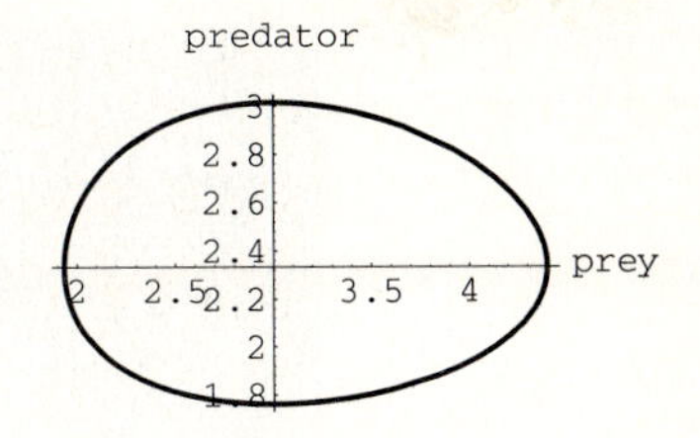

> Why was it a good idea to put the AxesOrigin at $\{c/d, a/b\}$?
>
> Reducing the size of h has what effect on the fluctuations of the two populations? What happens when you make $h = 0.01$?
>
> How about $h = 0$?
>
> Try to explain, in your own terms, why this turned out the way it did.

G.9.a.ii) Calculus&*Mathematica* is pleased to acknowledge that the idea for this problem came from the book *Elementary Differential Equations* by William Boyce and Richard DiPrima of Rensselaer Polytechnic Institute (Wiley, New York, 1977). These gentlemen wrote a really good book. You should be able to find it in your school's library.

Most folks call the populations

$$\text{prey} = \frac{c}{d} \qquad \text{and} \qquad \text{predators} = \frac{a}{b}$$

for the predator-prey model

$$\text{prey}'[t] = a\,\text{prey}[t] - b\,\text{prey}[t]\,\text{pred}[t]$$
$$\text{pred}'[t] = -c\,\text{pred}[t] + d\,\text{pred}[t]\,\text{prey}[t]$$

that you worked with above by the name "equilibrium populations."

Imagine that the prey are insects and that the predators are birds or fish. The government goes into action, spreading an insecticide like DDT all over the place. The insecticide kills both predator and prey in proportion to their current population, so the new model becomes

$$\text{prey}'[t] = a\,\text{prey}[t] - b\,\text{prey}[t]\,\text{pred}[t] - u\,\text{prey}[t]$$
$$\text{pred}'[t] = -c\,\text{pred}[t] + d\,\text{pred}[t]\,\text{prey}[t] - v\,\text{pred}[t]$$

where a, b, c, d, u, and v are all positive constants.

This is the same as

$$\text{prey}'[t] = (a - u)\,\text{prey}[t] - b\,\text{prey}[t]\,\text{pred}[t]$$
$$\text{pred}'[t] = -(c + v)\,\text{pred}[t] + d\,\text{pred}[t]\,\text{prey}[t]$$

What are the new equilibrium populations?

How do these new equilibrium populations tell you that the net effect of the government action was to raise the equilibrium population of the prey (insects) but to lower the equilibrium population of the predators?

Is that what the government had in mind?

G.9.b) This problem appears only in the electronic version.

■ G.10) Epidemics

This problem appears only in the electronic version.

■ G.11) Collision?

G.11.a) A Spartan missile and a Trojan missile are both flying at the same constant altitude. At time t, the Spartan missile is at the point:

In[32]:=

```
Clear[spartan,t]
spartan[t_] = {16.1 - 7 t + t^2, 13 t - 2 t^2};
```

At the same time t, the Trojan missile is at the point:

In[33]:=

```
Clear[trojan]
trojan[t_] = {26 - 13t + 2 t^2, 23 - 5t + t^2};
```

Here is a plot of the paths of the two missiles:

In[34]:=

```
spartanpath =
ParametricPlot[spartan[t],{t,0,6},
PlotStyle->{{Blue,Thickness[0.02]}},
DisplayFunction->Identity];
trojanpath =
ParametricPlot[trojan[t],{t,0,6},
PlotStyle->{{Red,Thickness[0.02]}},
DisplayFunction->Identity];
Show[spartanpath,trojanpath,
DisplayFunction->$DisplayFunction];
```

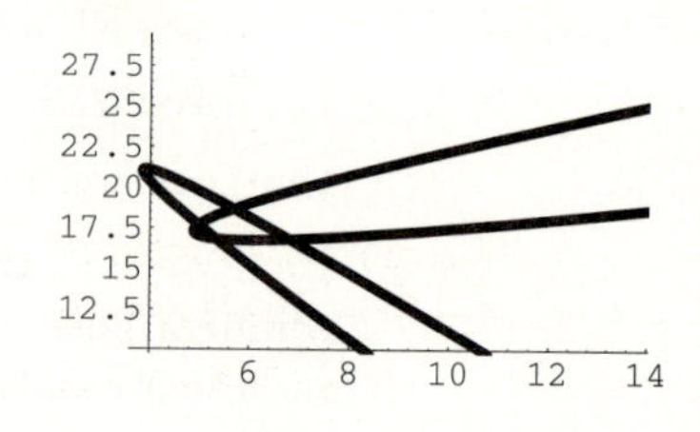

Their paths cross, but do they crash?

LITERACY SHEETS

Measuring Growth Rates

1.01 Growth Literacy Sheet

L.1) A function $f[x]$ starts out at $x = 0$ with a value of $f[0] = 1$ and goes up at a constant rate of three units on the y-axis for each unit on the x-axis. Give a formula for $f[x]$ and sketch its plot on the axes below:

The point plotted below is on the plot of $f[x]$ because $f[0] = 1$.

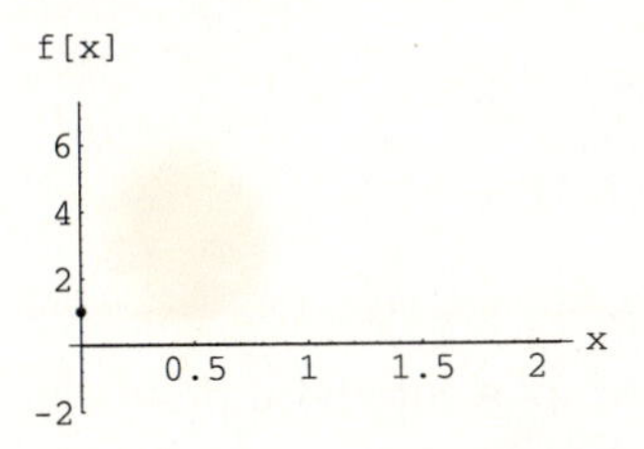

L.2) A function $f[x]$ starts out at $x = 2$ with a value of $f[2] = 6$ and goes up at a rate of -1.5 units on the y-axis for each unit on the x-axis. Give a formula for $f[x]$ and sketch its plot on the axes below:

The point plotted below is on the plot of $f[x]$ because $f[2] = 6$.

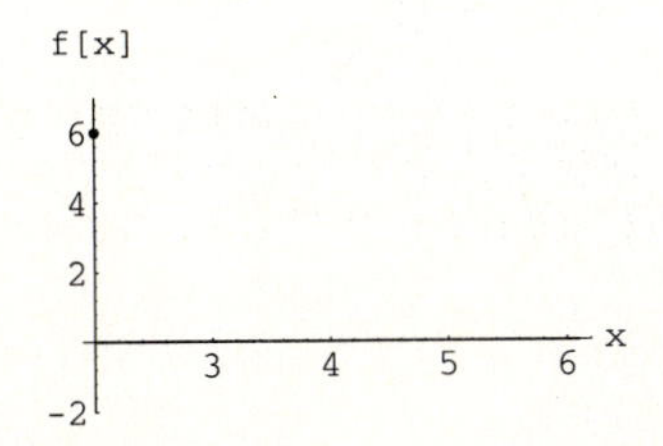

L.3) Given a constant number b, what happens to $f[x] = 3\,x + b$ every time x goes up by one unit? What is the growth rate of $f[x]$?

Given a constant number c, what happens to $g[x] = -2\,x + c$ every time x goes up by one unit? What is the growth rate of $g[x]$?

L.4) Linear growth is characterized by having a constant growth rate. Give a real-world example of a situation modeled by linear growth.

L.5) Most human bodies eliminate alcohol at the rate of 12 grams per hour. The typical 12-ounce American beer contains about 13.5 grams of alcohol. If you chug one 12-ounce beer at 6:00, another at 7:00, another at 8:00, another at 9:00, and another at 10:00 and you drink nothing else, then how many grams of alcohol will be in your body fluids at 11:00? Is this more or less than the grams of alcohol in your body fluids immediately after you chug the first beer? Should you expect to be a legal driver at 11:00?

If you chug one 12-ounce beer at 6:00, another at 6:30, another at 7:00, another at 7:30, another at 8:00, and another at 8:30 and you drink nothing else, then how many grams of alcohol will be in your body fluids at 10:00? Should you expect to be a legal driver at 10:00?

L.6) Here are two points $\{1, 2\}$ and $\{5, 4\}$ conveniently plotted on the axes below:

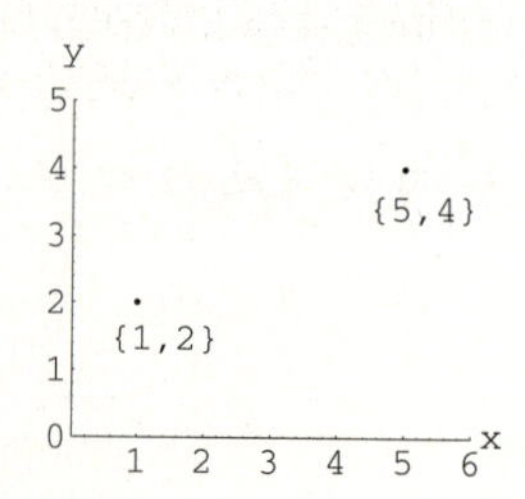

Find a formula for the line function whose plot hits both points and sketch its plot on the axes above. What happens to this function every time x goes up by one unit? What is the growth rate of this function?

L.7) Here are two points $\{2, 5\}$ and $\{4, 1\}$ conveniently plotted on the axes below:

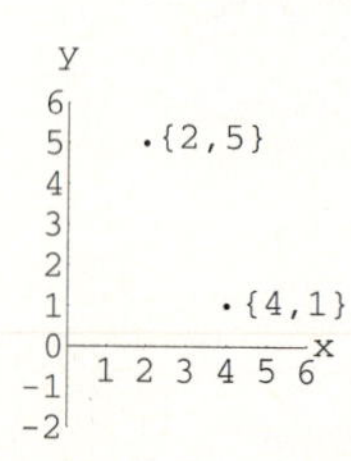

Find a formula for the line function whose plot hits both points and sketch its plot on the axes above. What happens to this function every time x goes up by one unit? What is the growth rate of this function?

L.8) Here are the plots of three line functions all with the same (negative) growth rate:

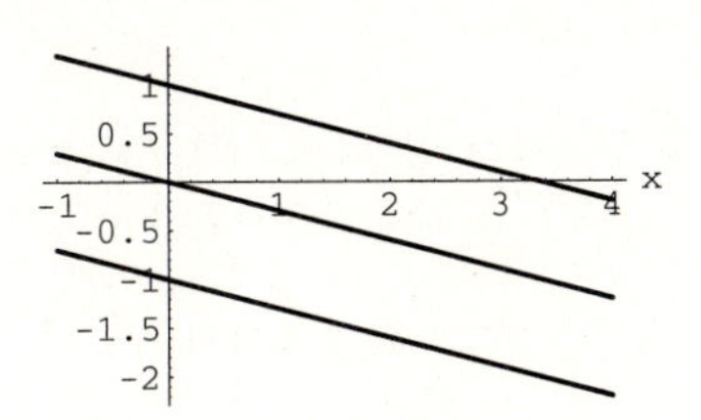

Look at the plot and fill in the blank: The plots of line functions with the same growth rate are ________ to each other.

L.9) Here are some points:

In[1]:=

```
data = {{-2.01,-2.59},{-1.74,-2.17},
{-1.5, -1.65},{-1.27,-1.31},{-1.04,-1.20},{-0.84,-1.08},
{-0.45,-0.74},{-0.35,-0.22},{0.,0.2},{0.27,0.37},{0.51, 0.45},
{0.73, 0.71},{0.98,1.19},{1.23,1.68},{1.58, 1.94},{1.69,2.02},{2.01,2.20}};
```

In[2]:=

```
dataplot = ListPlot[data,AxesLabel->{"x","y"},
PlotStyle->{PointSize[0.03]}];
```

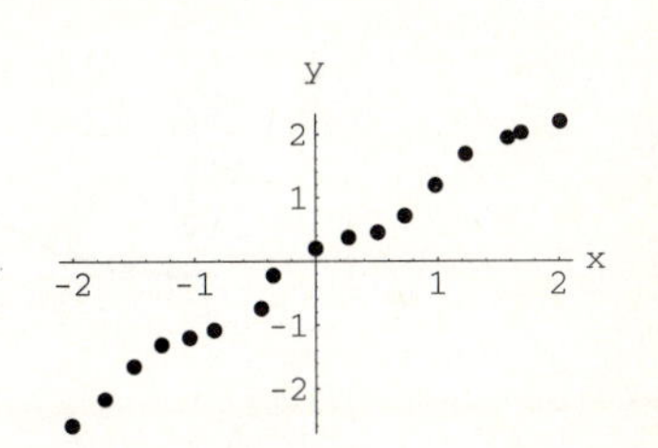

Explain why it is impossible to get a single line function whose plot runs through all of these points.

L.10) Here are the same points as in the last problem shown with plots of two line functions:

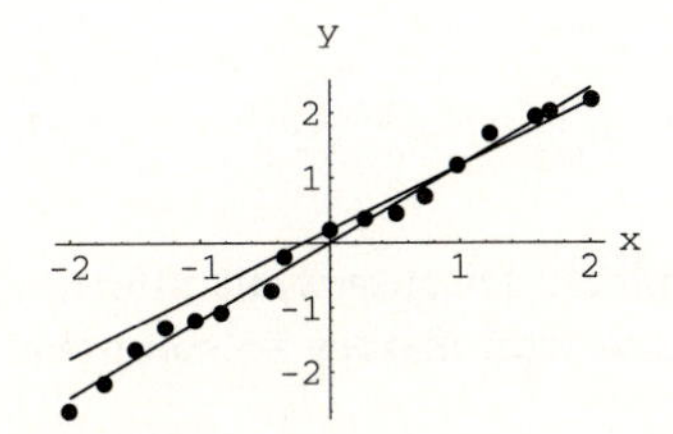

One of these line plots is the plot of *Mathematica*'s compromise line function whose plot flows with the data. The other is not. Which is which?

L.11) Here are the same points as in problem 9:

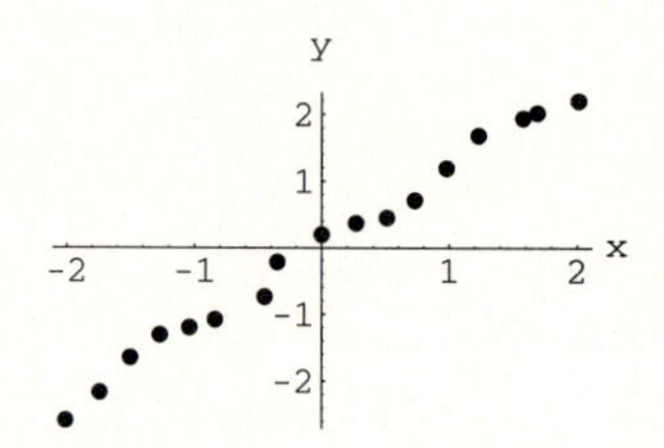

Sketch in the plot of a function, like *Mathematica*'s Interpolation function, that goes through all the plotted points and goes with the flow.

L.12) Here's a plot of all the results of all the censuses ever taken by the United States government in the form $\{x, y\}$ where x is a year and y is the U.S. population in millions for year x together with the plot of *Mathematica*'s Interpolation function, pop[x], through the data points.

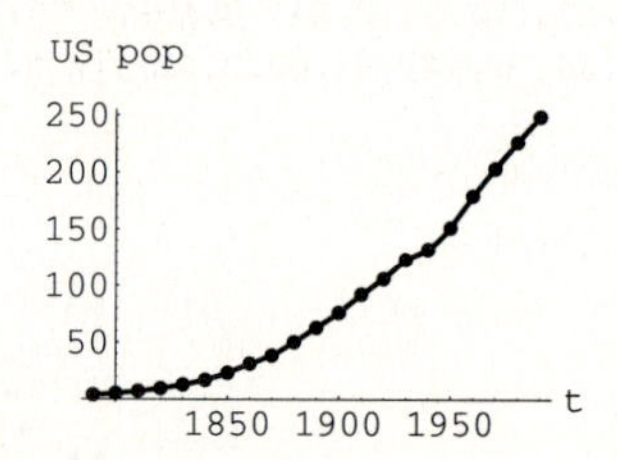

And here's a plot of pop[$x + 1$] − pop[x] (in millions) for $1790 \leq x \leq 1989$:

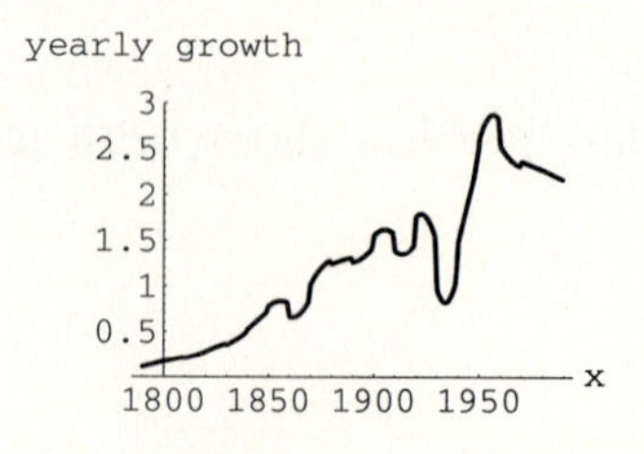

Discuss the relations between the two plots. Discuss what information the second plot exhibits. Use your knowledge of American history to comment on the reasons for the crests and dips.

L.13) Look at this plot of $f[x] = x^4 - 100{,}000\,x^2$:

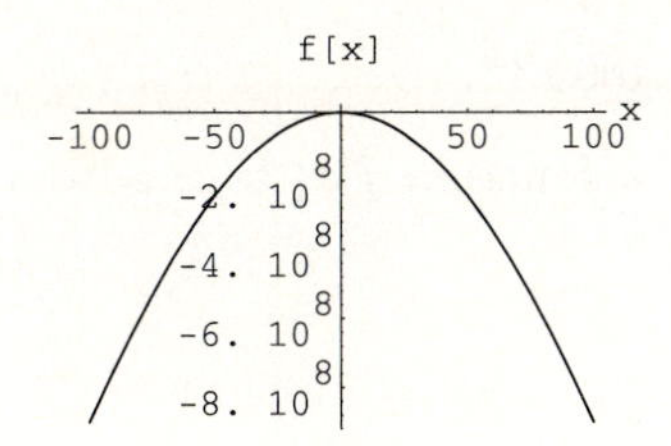

How do you know that this is not a good global scale plot of $f[x]$, and how do you know that the following plot is a good global scale plot of $f[x]$?

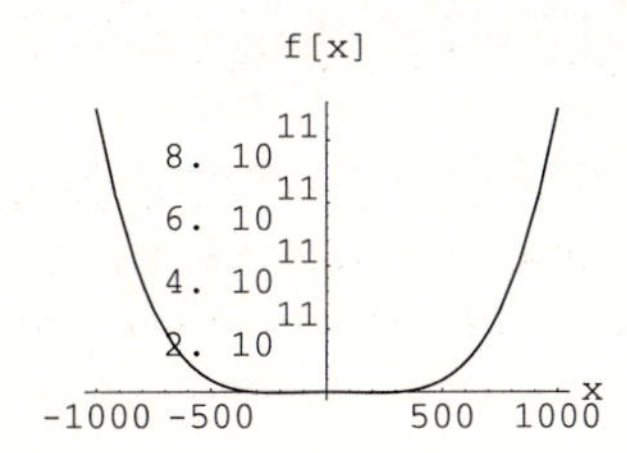

L.14) Put

$$f[x] = \frac{14\,x^4 + 50\,x^3 + 567\,x - 90}{2\,x^4 + 3\,x^2 + 1}.$$

What do you say are the limiting values

$$\lim_{x \to \infty} f[x] \qquad \text{and} \qquad \lim_{x \to -\infty} f[x]?$$

L.15) What do you say is the limiting value of

$$\lim_{x \to \infty} \frac{9\,x^9 + 12\,e^{0.3x}}{3\,x^{12} + 6\,e^{0.3x}}\,?$$

L.16) What do you say is the limiting value of

$$\lim_{x \to \infty} \frac{37\,x^8 - 123\,\sin[x] + 16\,x^2}{e^{0.1x}}\,?$$

Why?

L.17) What do you say is the limiting value of

$$\lim_{x \to \infty} e^{-0.1x}\left(1 + 15x^8\right)?$$

Why?

L.18) Rank the following functions in order of dominance as $x \to \infty$:

$$0.001\,x^{15}, \quad 0.0004\,e^{0.01x}, \quad 128\,x^2, \quad \sqrt{x},$$
$$23\,x, \quad 0.04\,x^3, \quad 0.00000013\,e^{2x}, \quad 100\,x^{0.6}.$$

L.19) The average percentage growth rate of a function $f[x]$ as x advances by one unit from x to $x+1$ is given by

$$\text{percen}[x] = 100\left(\frac{f[x+1]}{f[x]}\right) - 1.$$

Use this formula to measure the average percent growth rate of $f[x] = 3\,e^{0.7x}$ as x advances by one unit. Use this formula to measure the average percent growth rate of $f[x] = 0.7\,x + 3$ as x advances by one unit.

L.20) Agree that $f[x]$ is a power function and that $g[x] = 3\,e^{rx}$ where r is a positive number. Then which function $f[x]$ or $g[x]$ posts the larger average percent growth rate for large x's? How does this begin to account for the global scale dominance of the exponential functions over the power functions as $x \to \infty$?

1.02 Natural Logs and Exponentials Literacy Sheet

L.1) Write down e to four accurate decimals. Write down π to four accurate decimals. Has anyone ever succeeded in finding a pattern for all the decimals of either number?

L.2) Use the axes below to give hand sketches of the plots of

$$f[x] = e^x \qquad \text{and} \qquad g[x] = e^{-x}.$$

Label each curve.

Each of the plotted points is on at least one of the plots.

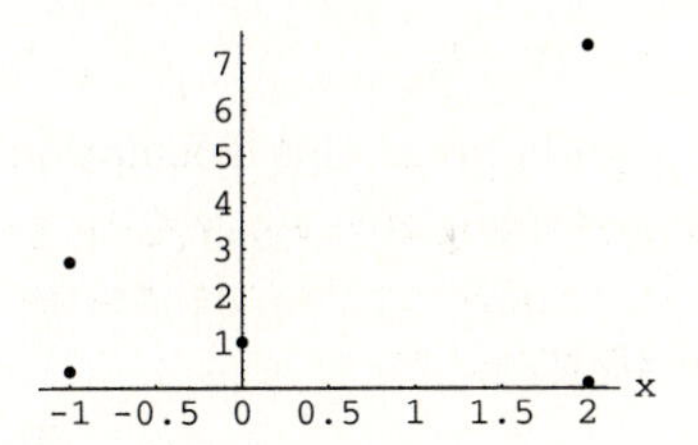

L.3) Are there numbers x that make e^x negative? Are there numbers x that make e^{-x} negative?

L.4) Use the axes below to give hand sketches of the plots of

$$f[x] = \log[e^x] \qquad \text{and} \qquad g[x] = \log[e^{-x}].$$

Label each curve.

Each of the plotted points is on at least one of the plots.

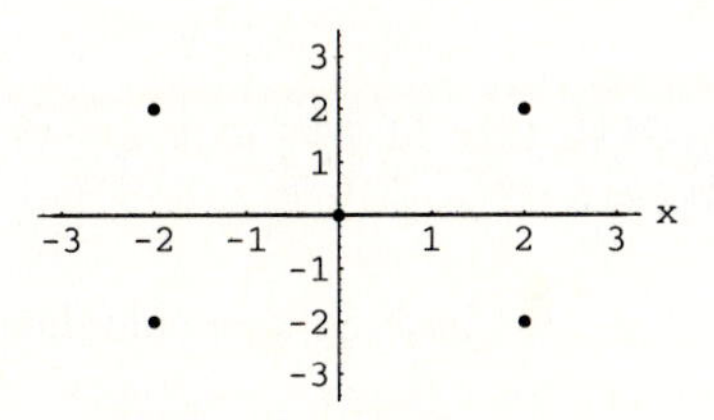

L.5) Why do lots of folks say that exponential growth is awesome? Give a real-world situation that is modeled by exponential growth.

L.6) What is exponential decay? Give a real-world situation that is modeled by exponential decay.

L.7) What does a constant inflation rate have to do with exponential growth? What is it about exponential growth that makes high inflation rates scary?

L.8) When an underwater light beam is turned on in sea water, the resulting intensity of the light decays exponentially along the direction of the beam. In fact, in clear sea water, the approximate percentage of the original intensity x meters from the source measured along the beam is given by intensity$[x] = 100\,e^{-1.4x}$. Here is a calculation of intensity[10]:

In[1]:=
```
Clear[intensity,x]
intensity[x_] = 100 E^(-1.4 x); intensity[10]
```
Out[1]=
```
0.0000831529
```

Use this calculation to help explain why almost no plants can grow in an ocean at a depth of more than 10 meters.

L.9) Go with the line function $f[x] = 2x+1$, and look at this calculation of $f[x+1]-f[x]$:

In[2]:=
```
Clear[f,x]
f[x_] = 2 x + 1; Simplify[f[x + 1] - f[x]]
```
Out[2]=
```
2
```

What does this tell you about the growth rate of $f[x]$ as x grows by one unit? Is this growth rate of $f[x]$ variable or constant?

Now go with the exponential function $f[x] = 3\,e^{0.5x}$, and look at this calculation of $f[x+1] - f[x]$:

In[3]:=
```
Clear[f,x]
f[x_] = 3 E^(0.5 x); Simplify[f[x + 1] - f[x]]
```
Out[3]=
```
           0.5 x
1.94616 E
```

What does this tell you about the growth rate of $f[x]$ as x grows by one unit? Is this growth rate of $f[x]$ variable or constant?

L.10) Go with the line function $f[x] = 2\,x + 1$, and look at this calculation of

$$100\left(\frac{f[x+1]}{f[x]} - 1\right):$$

In[4]:=
```
Clear[f,x]
f[x_] = 2 x + 1; Simplify[100(f[x + 1]/f[x] - 1)]
```
Out[4]=
```
  200
-------
1 + 2 x
```

What does this tell you about the percentage growth rate of $f[x]$ as x grows by one unit? Is this percentage growth rate of $f[x]$ variable or constant?

Now go with the exponential function $f[x] = 3\,e^{0.5x}$, and look at this calculation of

$$100\left(\frac{f[x+1]}{f[x]} - 1\right):$$

In[5]:=

```
Clear[f,x,r,a]
f[x_] = 3 E^(0.5 x); Simplify[100(f[x + 1]/f[x] - 1)]
```

Out[5]=

```
64.8721
```

What does this tell you about the percentage growth rate of $f[x]$ as x grows by one unit? Is this percentage growth rate of $f[x]$ variable or constant?

L.11) Someone tells you that the line functions $f[x] = a\,x + b$ are those that grow by a fixed amount every time x goes up by a fixed increment. Another person says that the exponential functions $f[x] = a\,e^{rx}$ are those that grow by a fixed percentage every time x goes up by a fixed increment. Is either person right?

L.12) The cost of a day in the hospital in the U.S. has caused lots of weeping, wailing, and gnashing of teeth. In fact, the Ford Motor Company reported that it spent the same amount on health plans as it spent on steel in 1992. The following table gives figures in the form $\{t, H[t]\}$ where t is measured in years since 1970 and $H[t]$ is the average dollar cost per patient day in a hospital in the U.S.

Source: *Statistical Abstracts*, 1986.

In[6]:=

```
bedcost ={{0,74},{2,95},{3,102},{4,114},{5,134},{6,153},
{7,174},{8,194},{9,217},{10,245},{11,284},{12,327},{13,369}};
```

In[7]:=

```
ListPlot[bedcost,PlotStyle->{PointSize[0.02]}];
```

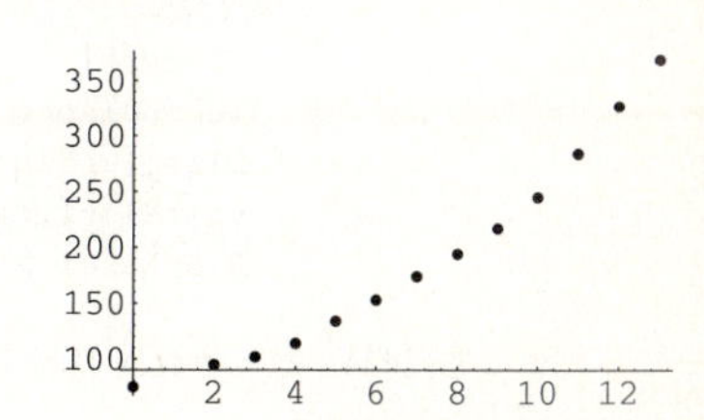

The following instruction exhibits the same data but in the form $\{t, \log[H[t]]\}$:

In[8]:=

```
logbed = Table[{bedcost[[j,1]],
N[Log[bedcost[[j,2]]]]},{j,1,Length[bedcost]}];
ListPlot[logbed,PlotStyle->{PointSize[0.02]},
PlotLabel->"Semi-log paper plot"];
```

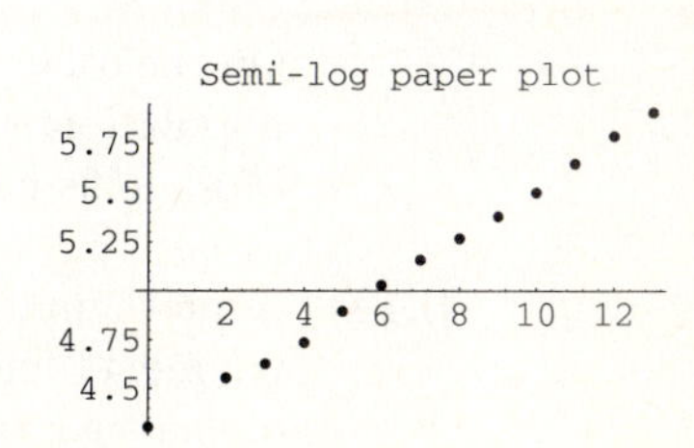

Why does this plot reveal vividly that the average dollar cost per patient day in a hospital in the U.S. is growing approximately exponentially? Why is this scary?

Discuss how to find a compromise exponential function $f[t] = a\, e^{rt}$ whose plot runs through or near the data.

L.13) Thomas Malthus, in his 1803 book *An Essay on the Principle of Population*, made quite a name for himself by arguing that population will grow exponentially and that this will lead to ultimate disaster. Here are some world population data $\{t, P[t]\}$ where t is the year, $t = 0$ corresponds to 1950, and $P[t]$ is the estimated world population in millions for the corresponding year t. Here is a plot:

In[9]:=

```
worldpop ={{0,2513},{10,3027},{20,3678},{30,4478},{35,4865},
{36,4942},{37,5026},{38,5128},{39,5234},{40,5321}};
```

In[10]:=

```
popdataplot = ListPlot[worldpop,
AxesLabel-> {"year","World population"},
PlotStyle->{PointSize[0.03]}];
```

World population

5000
4500
4000
3500
3000
10 20 30 40 year

Source: *Information, Please* almanac

This plot covers the years 1950–1990. Does this look like linear growth or exponential growth?

Here are the same data on semi-log paper:

In[11]:=

```
Clear[k]
logworldpop = Table[{worldpop[[k,1]],
N[Log[worldpop[[k,2]]]]},{k,1,Length[worldpop]}];
ListPlot[logworldpop,PlotStyle->{PointSize[0.03]},
PlotLabel->"Semi-log paper plot"];
```

Semi-log paper plot

8.5
8.4
8.3
8.2
8.1
7.9
10 20 30 40

Explain why this plot tells you that since 1950, the world population has been showing signs of exponential growth.

On the other hand, explain why it is possible to argue that the growth of the world's population is also showing signs of lagging behind exponential growth. The upshot: Most folks can feel free to interpret these plots as functions of their own biases.

L.14) A mathematically illiterate person charges $4,500 on MasterCard and makes no payments for a year. The advertised interest rate is 15.3% per year, so at the end of the year that person figures to owe:

In[12]:=
```
expected = 4500 + 0.153 4500
```
Out[12]=
```
5188.5
```

But at the end of the year, when the same person looks at the bill from the MasterCard people, that person is stunned because the bill is for more than \$5188.50. Help out this sorry excuse for a modern American.

L.15) From the Money section of the newspaper *USA Today*, June 18, 1992:

"Rates on credit card applications don't take into account monthly, and sometimes daily, compounding of finance charges. The average rate on credit cards is 18.5%, but most consumers pay effective rates of 20% or more because of compounding methods." What does this mean?

L.16) Discuss how to take an annual interest rate of 6% compounded monthly and figure out what the effective interest rate is.

L.17) Is this correct: $b = e^{\log[b]}$ as long as $b > 0$?

Is this correct: $b = \log[e^b]$ no matter whether $b > 0$, $b = 0$, or $b < 0$?

L.18) Lots of old-time math books make a big deal of studying exponential functions $f[x] = a\,b^{rx}$ for a positive base b other than $b = e$. Maybe this is to keep students confused. Given a positive number b, set s in terms of $\log[b]$ and r so that instead of calculating $f[x]$ by means of the formula $f[x] = a\,b^{rx}$, you can calculate $f[x]$ by means of the formula $f[x] = a\,e^{sx}$. How does this tell you that you don't really need to concern yourself with any base other than e?

L.19) Look at these *Mathematica* calculations and then say how old-fashioned common logarithms $\log[10, x]$ (base 10) are related to the natural logarithms $\log[x]$ (base e) used throughout modern science:

In[13]:=
```
Clear[x]
N[Log[10,x]]
```
Out[13]=
```
0.434294 Log[x]
```

In[14]:=
```
N[1/Log[10]]
```
Out[14]=
```
0.434294
```

Use your answer to tell someone how to convert a natural logarithm to a base 10 logarithm.

1.03 Instantaneous Growth Literacy Sheet

L.1) What happens to the plot of the $(\sin[x+h] - \sin[x])/h$ curve as h closes in on 0?

What happens to the plot of the $\left(e^{x+h} - e^x\right)/h$ curve as h closes in on 0?

What happens to the plot of the $(\log[x+h] - \log[x])/h$ curve as h closes in on 0?

L.2) For a function $f[x]$, the function $f'[x]$ is another function that measures what quantity?

L.3) Sketch on the axes below the graph of a function $f[x]$ such that $f'[x]$ is positive for all x's.

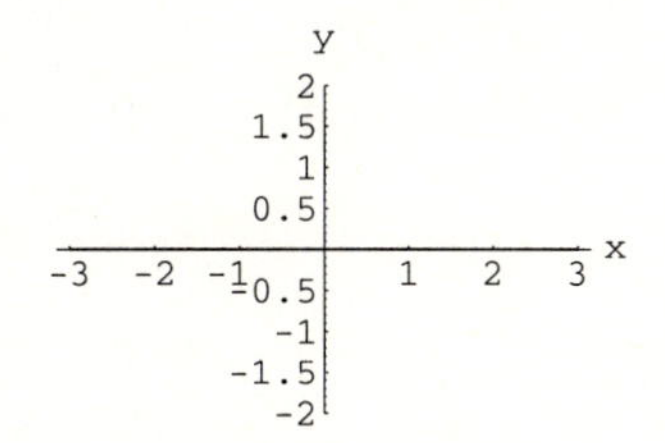

L.4) If $f'[x]$ is positive for all x with $0 \leq x \leq 1$, then which is the larger: $f[0]$ or $f[1]$?

L.5) Sketch on the axes below the graph of a function $f[x]$ such that $f'[x]$ is negative for all x's.

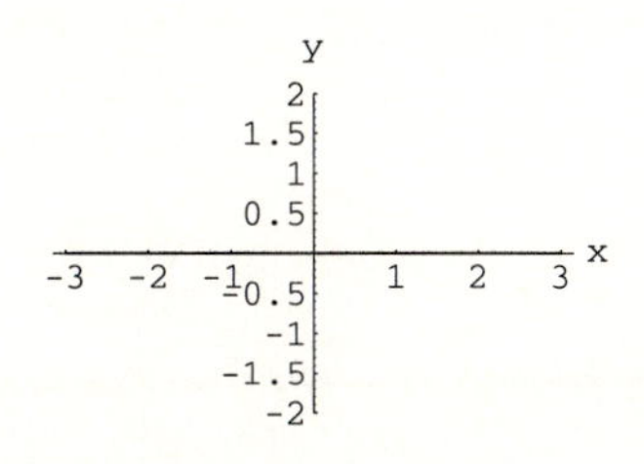

L.6) If $f'[x]$ is negative for all x with $0 \leq x \leq 1$, then which is the larger: $f[0]$ or $f[1]$?

L.7) Sketch on the axes below the graph of a function $f[x]$ such that $f'[x]$ is positive for all x with $0 \leq x \leq 1$, and $f'[x]$ is negative for $1 \leq x \leq 2$.

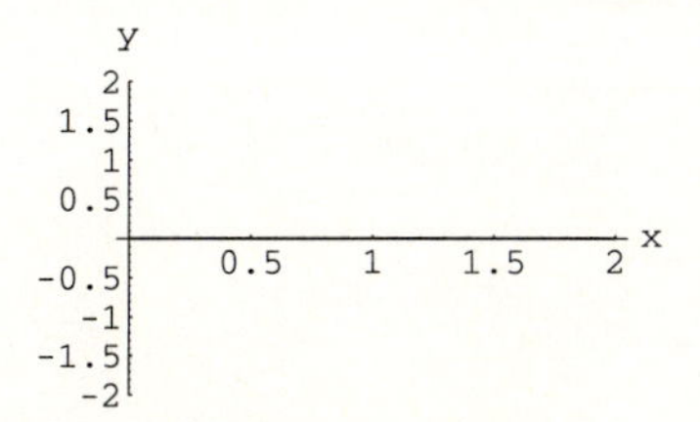

L.8) If $f[x]$ is a function such that $f'[x]$ is positive for all x with $0 \leq x < 1$, and $f'[x]$ is negative for $1 < x \leq 2$, then which x in $[0, 2]$ makes $f[x]$ the biggest?

L.9) Sketch on the axes below the graph of a function $f[x]$ such that $f[0.5] = 1$ and $f'[x] = 0$ for all x's.

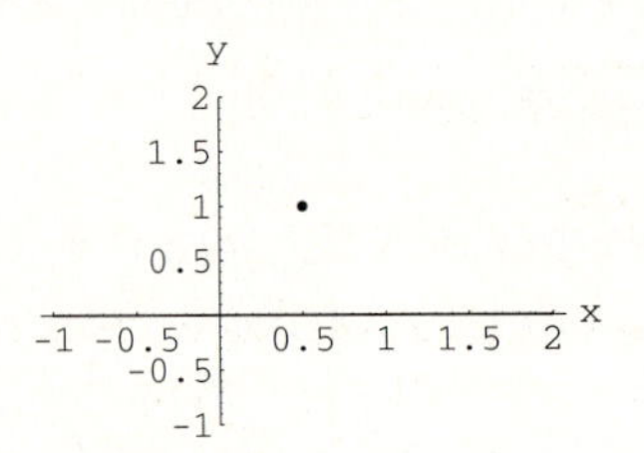

L.10) Sketch on the axes below the graph of a function $f[x]$ such that $f[1] = 0.5$ and $f'[x] = 1$ for all x's.

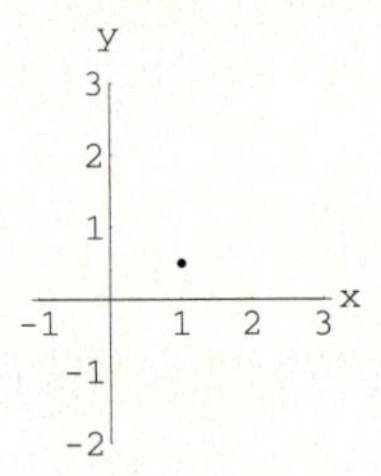

L.11) Sketch on the axes below the graph of a function $f[x]$ such that $f[1] = 0.5$ and $f'[x] = -1/2$ for all x's.

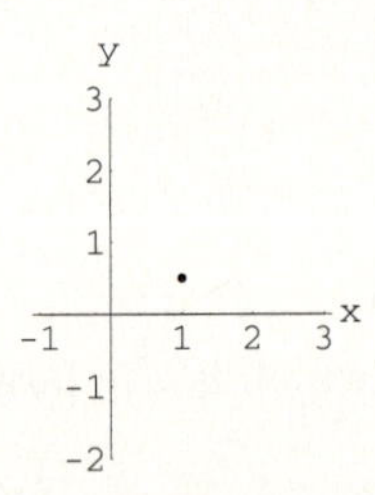

L.12) Fill in the blanks with the right formula:

If $f[x] = \sin[x]$, then $f'[x] =$ ____________ .

If $f[x] = \cos[x]$, then $f'[x] =$ ____________ .

If $f[x] = e^x$, then $f'[x] =$ ____________ .

If $f[x] = x^2$, then $f'[x] =$ ____________ .

If $f[x] = x$, then $f'[x] =$ ____________ .

If $f[x] = x^5$, then $f'[x] =$ ____________ .

L.13) Richard Feynman, the fellow who (among many other things) discovered the cause of the crash of the space shuttle Challenger, once said: "The base 10 was used only because we have 10 fingers and arithmetic of it is easy, but if we ask for a mathematically natural base (for logarithms), one that has nothing to do with the number of fingers on human beings, we might try to change our scale of logarithms in some convenient and natural manner."

The Feynman Lectures on Physics (in three volumes), by Richard P. Feynman, Robert B. Leighton, and Matthew Sands, Addison-Wesley, Reading, Massachusetts, 1963

What do you think Feynman was driving at?

L.14) You measure the instantaneous growth rate $f'[x]$ of $f[x]$ at each point of $[0, 1]$ and in so doing you observe that $f'[0] > 0$ but $f'[1] < 0$. Why can you be sure that there is at least one number s with $0 < s < 1$ such that instantaneous growth rate is zero ($f'[s] = 0$)?

L.15) Here are plots of four functions. Two of them are plots of the derivatives of the functions plotted in the other two. Match the plots of the functions with the plots of their derivatives.

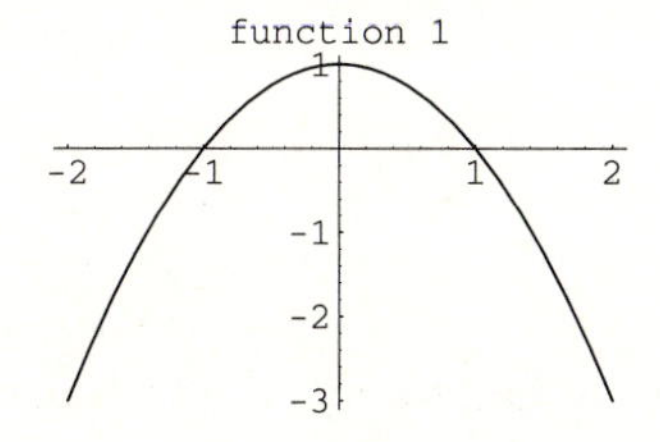

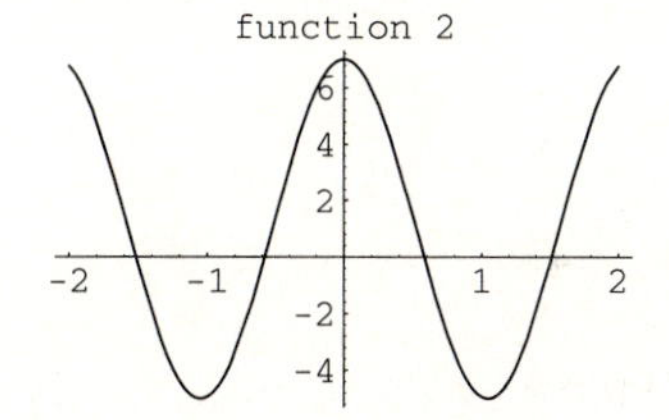

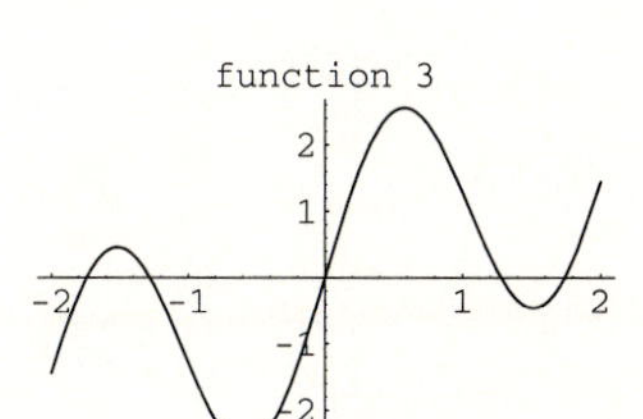

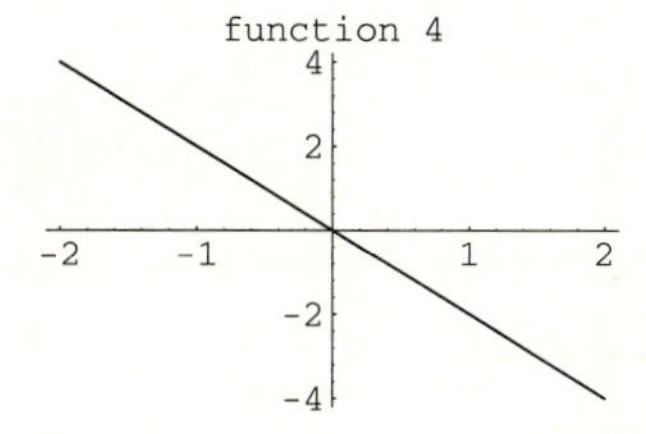

1.04 Rules of the Derivative Literacy Sheet

L.1) Differentiate the following functions with respect to x by hand:

a) e^{x}

b) e^{2x}

c) $x\,e^{-2x}$

d) $x\,\log[x]$

e) $\log\left[\frac{1}{x}\right]$ $(= -\log[x])$

f) $\cos[e^{x}]$

g) $\cos[e^{-3x}]$

h) $\sin[x^{2}]$

i) $\frac{\cos[x^{2}]}{x}$

j) $\sin[\left(\frac{2}{\sqrt{\pi}}\right)\,x^{2}]$

k) $\sin[\log[x]]$

l) $e^{-x}\,\sin[6\,x]$

m) $\sin[7\,x]^{2}$

n) $e^{\cos[x]}$

o) $e^{\cos[3x]}$

p) $\left(1 - x + x^{2}\right)^{24}$

q) $\log[1 - 3\,x]$

r) x^{e}

s) $\frac{1 + 8\,x - x^{3}}{1 + x}^{2}$

t) $\sqrt{x}$

u) $\frac{1}{1 - x}$

v) $\frac{1}{1 - x}^{5}$

w) $\frac{e^{ax}}{1 + e^{ax}}$

x) $x\,\log[x] - x$

y) $-\log[\cos[x]]$

L.2) If Jenny does trig identities A times faster than Sam and Sam does trig identities B times faster than Cal, then Jenny does trig identities AB times faster than Cal. How is this little story related to the chain rule?

L.3) If $f[x] = \sin[g[x]]$, then $f'[x] = \cos[g[x]]\, g'[x]$. How do you know this is correct?

L.4) Is the derivative of the sum given by the sum of the derivatives?

Is the derivative of the product given by the product of the derivatives?

L.5) Put

$$f[x] = \frac{4\,x^5 + 50\,\log[x]}{2\,x^5 + 3\,x^2 + 1}.$$

What is the limiting value

$$\lim_{x \to \infty} f[x]?$$

What is the limiting value

$$\lim_{x \to \infty} \frac{6\,x^{0.05} + 8\,\log[x]}{3\,x^{0.005} + 4\,\log[x]}?$$

What is the limiting value

$$\lim_{x \to \infty} \frac{720\,x^{81} - 123\,\cos[x] + 6\,\log[x]}{e^{0.04x}}?$$

What is the limiting value

$$\lim_{x \to \infty} \frac{\log[x]}{x^{0.003}}?$$

L.6) Rank the following functions in order of dominance as $x \to \infty$:

$$x^{152},\quad 0.0004\,e^{0.1x},\quad e^{0.02x}/x,\quad x\log[x],\quad 0.98\,x^2,\quad \sqrt{x},$$
$$100\log[x],\quad 17\,x,\quad 0.08\,x^3,\quad 0.000001\,e^{2x},\quad 10000000\,x^{0.004}.$$

L.7) Put $f[x] = e^{0.4x}$. Multiple choice:

a) Every time x advances by one unit, $f[x]$ goes up by 40%.

b) Every time x advances by one unit, $f[x]$ goes up by more than 40%.

c) Every time x advances by one unit, $f[x]$ goes up by less than 40%.

L.8) Take positive numbers a and r and put $f[x] = a\,e^{rx}$. Measure the instantaneous percentage growth rate of $f[x]$ at any point x in terms of r.

L.9) Put $f[x] = e^{-0.5x}$ and measure the instantaneous percentage growth rate of $f[x]$ at any point x.

L.10) Examine the derivative of $f[x] = x - \sin[x]$ to see whether the curve $y = f[x]$ ever goes down. Is it true that $x \geq \sin[x]$ for $x \geq 0$?

L.11) Examine the derivative of $f[x] = x^2 e^{(-x^2+1)}$ to help you give a reasonably good hand sketch of the curve $y = f[x]$ on the axes below. What is the maximum value $f[x]$ can have?

The two plotted points are on the curve.

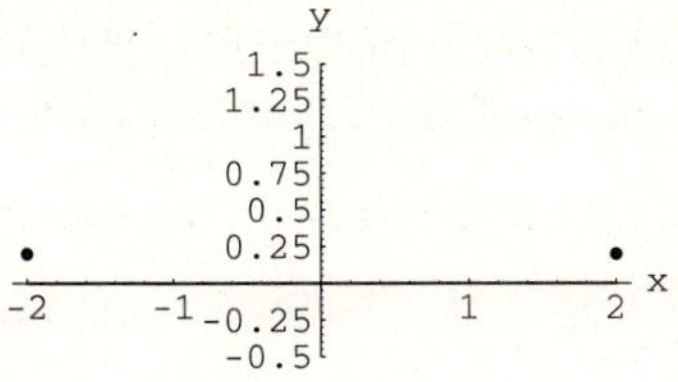

L.12) Start with the formulas $x = \log[e^x]$ and $D[\log[x], x] = 1/x$ and derive the formula $D[e^x, x] = e^x$.

L.13) Suppose $f[x]$ and $g[x]$ are two functions with $f'[x] > 0$ and $g'[x] > 0$ for all x's. This means both $f[x]$ and $g[x]$ go up as x advances from left to right. Does it also mean that the product $f[x]\,g[x]$ also goes up as x advances from left to right? To help you form your opinion, try $f[x] = x$ and $g[x] = x$. What happens if, in addition, $f[x] > 0$ and $g[x] > 0$ for all x's?

L.14) If x is measured in radians, then the derivative of $\sin[x]$ with respect to x is $\cos[x]$. Use the formula $\sin[x \text{ degrees}] = \sin[(2\,\pi/360)\,x \text{ radians}]$ to calculate the derivative of $\sin[x \text{ degrees}]$ with respect to x. Why does the resulting formula make calculus difficult if you insist on working with degrees instead of radians?

L.15) Differentiate both sides of the identity $\sin[2\,x]/2 = \sin[x]\cos[x]$ with respect to x and set the resulting expressions equal to each other. Look familiar?

Hold y constant and differentiate both sides of the identity

$$\sin[x + y] = \sin[x]\cos[y] - \sin[y]\cos[x]$$

with respect to x and set the resulting expressions equal to each other. Look familiar?

L.16) Explain the statement: Even though the computer has supplanted the logarithm for numerical calculations, the logarithm remains an important calculational tool.

L.17) Complete the sentence:

The derivative $f'[x]$ of $f[x]$ is the limiting case of ________ as ________.

L.18) You differentiate a certain function $f[x]$ and learn $f'[x] = (x - 1)\,e^{-x}$. You also know in advance that $f[0] = 0$. Give a rough sketch of the shape of the plot of the curve $y = f[x]$ for $0 \le x \le 2$.

L.19) Here are six plots. Three of them are plots of the derivatives of the functions plotted in the other three. Try to match the plots of the functions with the plots of their derivatives.

Function 1:

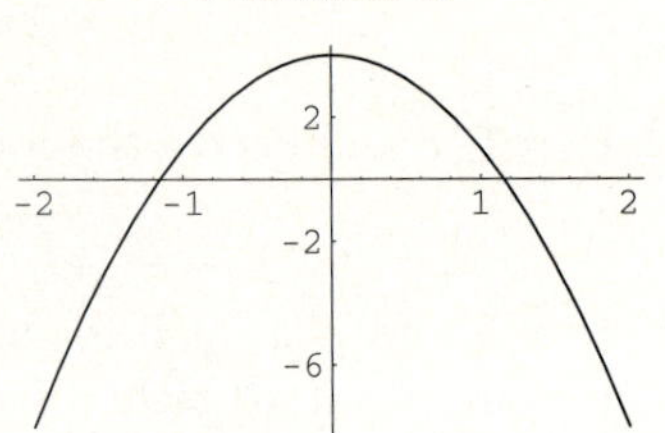

Function 2:

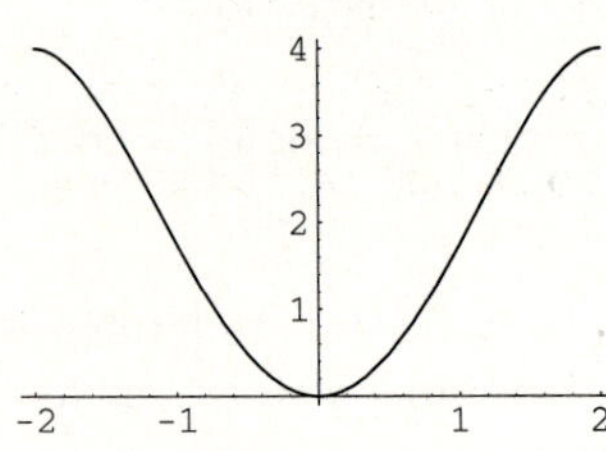

Function 3:

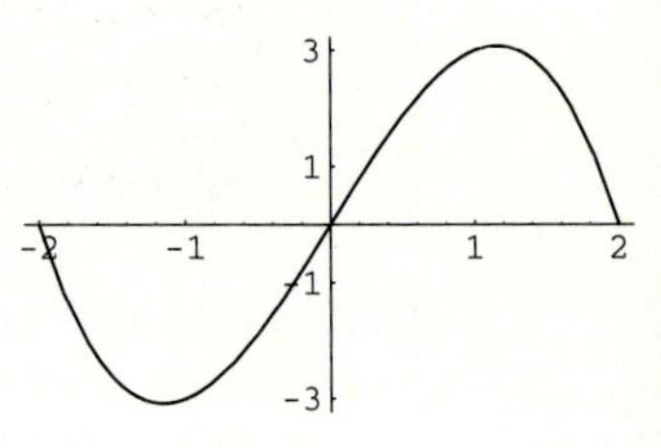

Function 4:

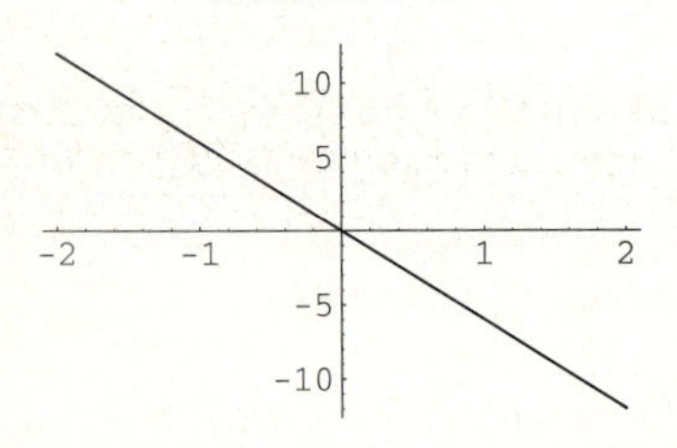

Function 5:

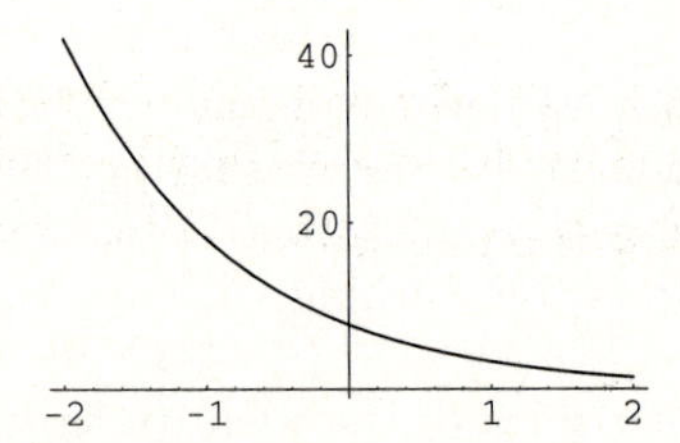

Function 6:

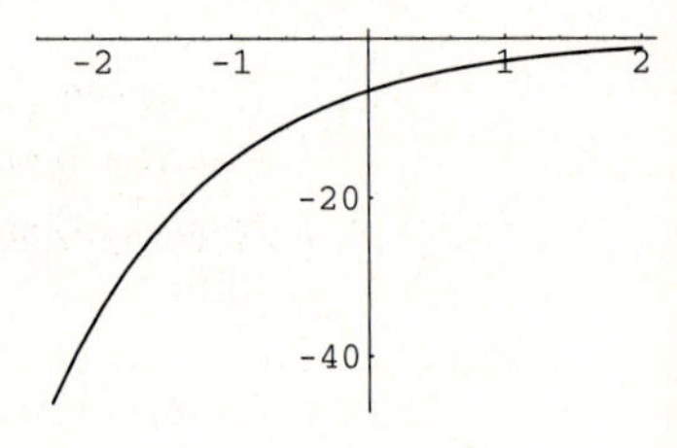

1.05 Using the Tools Literacy Sheet

L.1) Find the largest value of the product $x\,y$ given that x and y are related by $x+y=4$.

L.2) Examine the derivative of $f[x] = x\,e^{-x^2}$ to give a reasonably good hand sketch of the curve $y = f[x]$ on the axes below.

What are the maximum and minimum values $f[x]$ can have?

The two plotted points are on the curve.

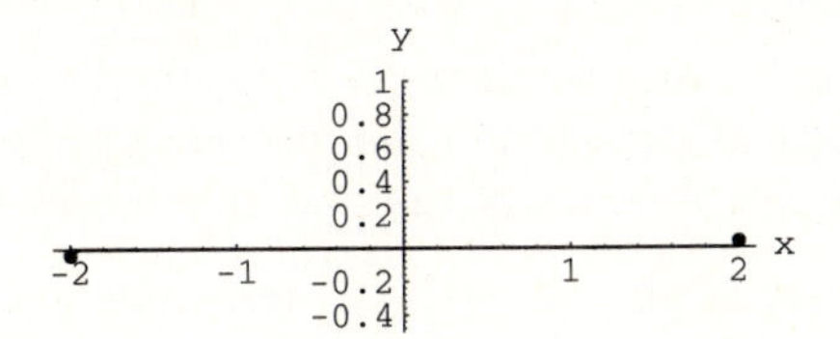

L.3) Examine the derivative of $f[x] = \left(x^2 - 2\log[x]\right)/2$ to give a reasonably good sketch of the curve $y = f[x]$ on the axes below. What is the minimum value $f[x]$ can have?

The plotted point is on the curve.

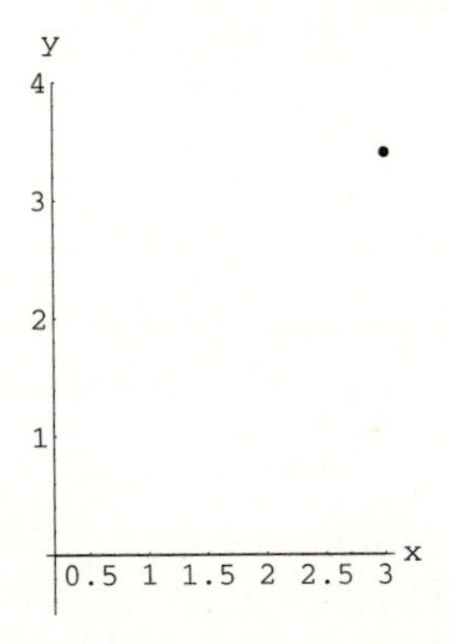

L.4) Use calculus to explain the statement: Of all rectangles with a fixed perimeter, the square measures out with the biggest area.

L.5) Given a positive number t, factor the derivative of $f[x] = x^t/e^x$ with respect to x to explain why the curve $y = f[x]$ first goes up as x advances from 0 and grows until x reaches a point x_t, after which the curve goes down. Give the exact value of the turning point x_t in terms of t. How does your result above reflect the fact that in the global scale as $x \to \infty$, the exponential growth of e^x dominates the power growth of x^t?

L.6) Good representative plots of functions exhibit all the dips and crests of the graph and give a strong flavor of the global scale behavior. If your plot of a function includes all points at which the derivative is 0, explain why you can be sure that your plot cannot miss any of the dips and crests of the graph of the function.

L.7) If $f'[x] = 0$ at exactly one point in $[0, 1]$ and $f'[x]$ is nonzero at all other points of $[0, 1]$, then at most how many points x can there be in $[0, 1]$ with $f[x] = 0$? Ask yourself: At most how many times can the curve go up and down?

L.8) R. M. Thrall and his University of Michigan colleagues (Report No. 40241-R-7, University of Michigan, 1967) gave the following crisp description of autocatalytic reaction of one substance into a new substance:

"Autocatalytic reaction progresses in such a way that the new substance catalyzes its own formation."

In the Thrall model, the reaction rate (with respect to time) is proportional to the amount x of the new substance at time t, and the reaction rate is also proportional to $(a - x)$ where a is the original amount of the first substance.

a) Why does this mean that $dx/dt = K x\,(a - x)$ for some positive constant K?

b) Differentiate with respect to x your expression for dx/dt and analyze it to determine the value of x for which the reaction rate is fastest.

For what x's is the reaction rate the slowest? Interpret.

L.9) a) What are the largest and smallest values of

$$2\,x^3 - 9\,x^2 + 12\,x - 1 \qquad \text{for } 0 \leq x \leq 2?$$

b) What are the largest and smallest values of

$$2\,x^3 - 9\,x^2 + 12\,x - 1 \qquad \text{for } 1 \leq x \leq 3?$$

L.10) Suppose A and B are positive constants and suppose $f[x] = A\,x + B/x$. Calculate a positive number x_0 such that $f[x_0] \leq f[x]$ for all other positive x's. What is the ratio $A\,x_0/\,(B/x_0)$?

L.11) New York, Denver, Madrid, Istanbul, and Beijing are all at latitude 40° North. Consequently, if you are going from New York to Beijing, the due east route will take you over Madrid and Istanbul, but the due west route will take you over Denver. The distances (in miles) by air are:

	Beijing	Denver	Istanbul	Madrid	New York
Beijing	—	6350	4380	5730	6830
Denver	6350	—	6150	5010	1630
Istanbul	4380	6150	—	1700	5010
Madrid	5730	5010	1700	—	3580
New York	6830	1630	5010	3580	—

Is Denver on the shortest route from New York to Beijing? Is either Madrid or Istanbul on that route? Can the due east route from New York to Beijing be the shortest route? Can the due west route from New York to Beijing be the shortest route?

1.06 Differential Equations of Calculus Literacy Sheet

L.1) You know that $y'[x] = a\,y[x]$ for all x's but DO NOT know the value of a. How many data points of the form $\{x, y[x]\}$ do you need to determine a formula for $y[x]$?

L.2) You know that $y'[x] = a\,y[x]$ for all x's and you DO know the value of a. How many data points of the form $\{x, y[x]\}$ do you need to determine a formula for $y[x]$?

L.3) You know that $y'[x] = a\,y[x]$ for all x's and you know that a is positive and that $y[x]$ is always positive. Does $y[x]$ go up or down as x advances from left to right?

L.4) Pencil in a rough sketch of the solution of $y'[x] = 0.2\,y[x]$ with $y[0] = 1$ on the axes below.

The plotted point is on the curve.

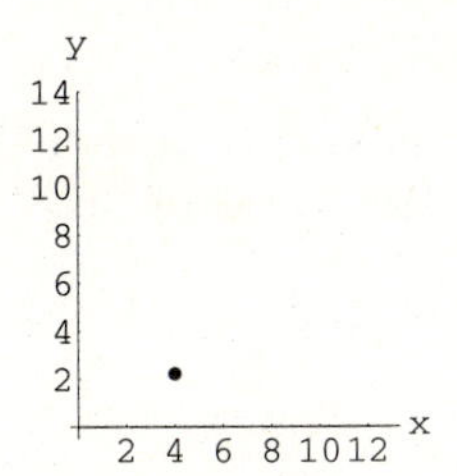

L.5) You know that $y'[x] = a\,y[x]$ for all x's and you know that a is negative but that $y[x]$ is always positive. Does $y[x]$ go up or down as x advances from left to right?

L.6) Pencil in a rough sketch of the solution of $y'[x] = -0.5\,y[x]$ with $y[0] = 6$ on the axes below.

The plotted point is on the curve.

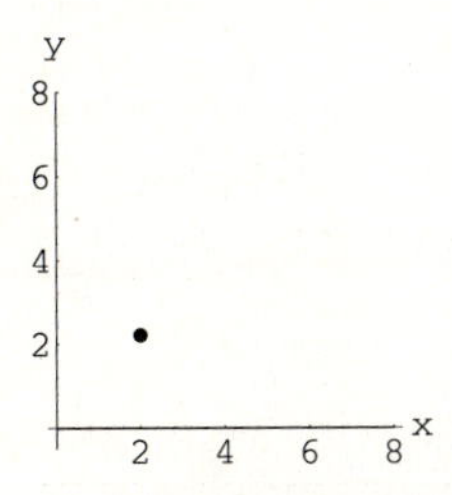

L.7) You know that $y'[x] = a\,y[x]$ for all x's and you know that $y[0] = 5$. Can $y[x]$ ever go negative?

L.8) Is there any x that makes e^x negative?

L.9) You know that $y'[x] = a\,y[x]$ for all x's and you know that a is positive. What is the value of

$$\lim_{x \to \infty} \frac{y[x]}{x^{10}}?$$

L.10) You know that $y'[x] = a\,y[x]$ for all x's and you know that a is negative. What is the value of

$$\lim_{x \to \infty} \frac{y[x]}{x^{10}}?$$

L.11) You know that $y'[x] = a\,y[x]$ for all x's and you know that $y[x]$ is positive for all x's. How does the $\log[y[x]]$ curve plot out?

L.12) You know that $y[t] = 83\,e^{0.017t}$. What differential equation does $y[t]$ satisfy? What is the instantaneous percentage growth rate of $y[t]$?

L.13) You know that your untouched bank account is accumulating interest compounded continuously. Why does this tell you that the instantaneous growth rate of the amount in the account is proportional to the amount in the account?

L.14) Given

$$y'[x] = 0.3\,y[x]\left(1 - \frac{y[x]}{100}\right) \qquad \text{with } y[0] = 1,$$

can $y[x]$ ever exceed 100?

L.15) Here are plots of the solutions of the differential equations

$$y'[x] = 0.6\,y[x] \qquad \text{with } y[0] = 2$$

and

$$y'[x] = 1.2\,y[x]\left(1 - \frac{y[x]}{100}\right) \qquad \text{with } y[0] = 2:$$

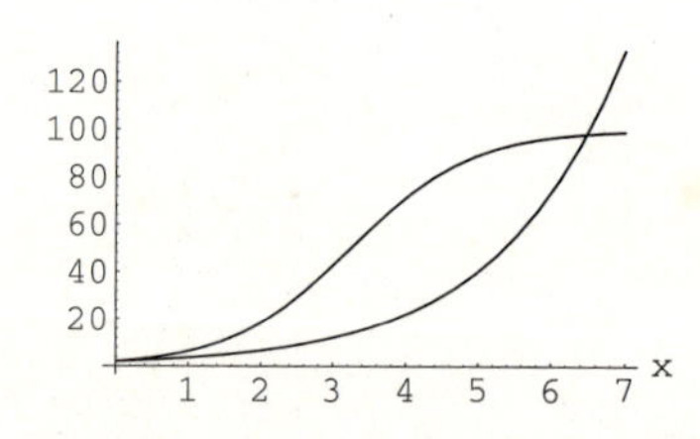

Notice that the plots start out together but share very little ink. Why do you think this happened?

Now look at plots of the solutions of the differential equations

$$y'[x] = 0.6\,y[x] \qquad \text{with } y[0] = 2$$

and

$$y'[x] = 0.6\,y[x]\left(1 - \frac{y\,[x]}{100}\right) \qquad \text{with } y[0] = 2:$$

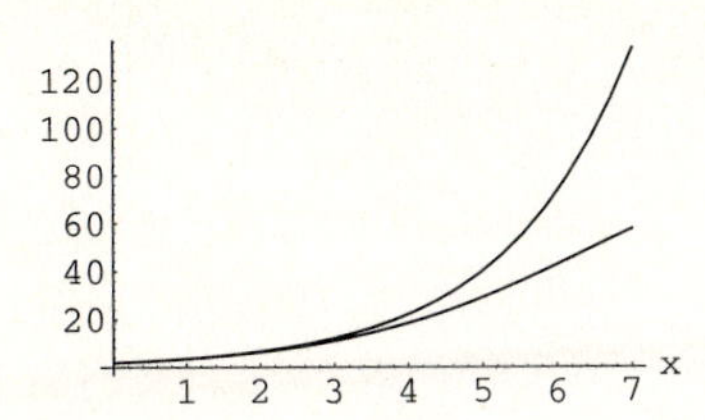

Explain why as x advances from 0, the plots of both of these solutions had no choice but to share a lot of ink initially.

Explain why for larger x's, the plots had no choice but to pull apart, with one plot eventually sailing way above the other.

L.16) Glucose in the blood is converted and excreted at a rate proportional to the present concentration of the glucose in the blood. If $C[t]$ is the concentration of glucose in the blood at time t, what differential equation does $C[t]$ satisfy? Give a formula for $C[t]$. What data do you need to set the constants in your formula for $C[t]$?

L.17) Here are six plots: They are plots of:

a) A solution of $y'[x] = r\,y[x]$ with $r > 0$.

b) A solution of $y'[x] = r\,y[x]$ with $r < 0$.

c) A solution of $y'[x] = r\,y[x]\,(1 - y[x]/b)$.

d) None of the above.

Which is which?

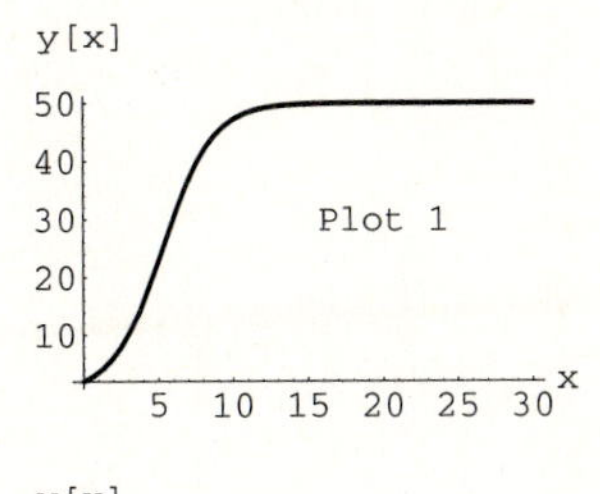

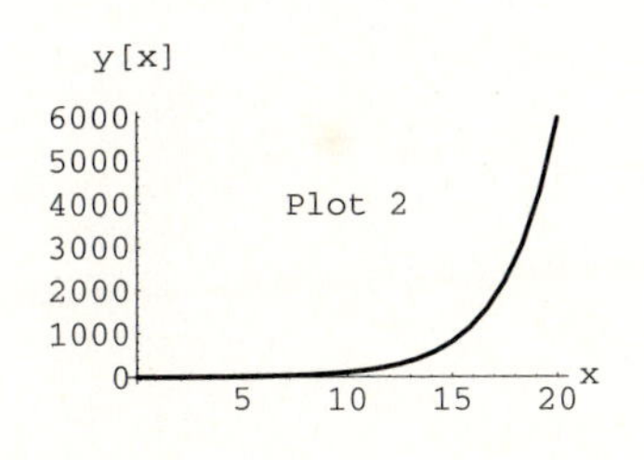

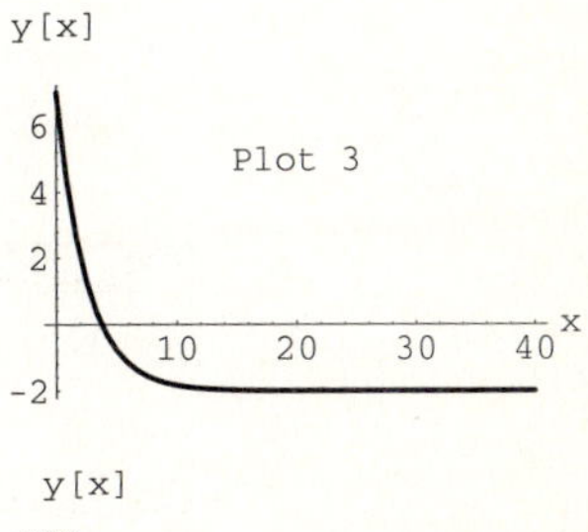

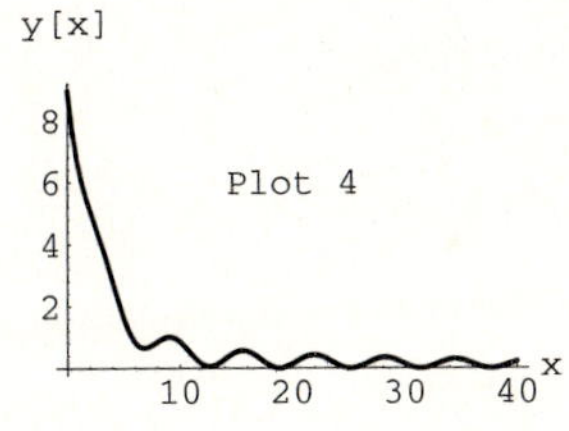

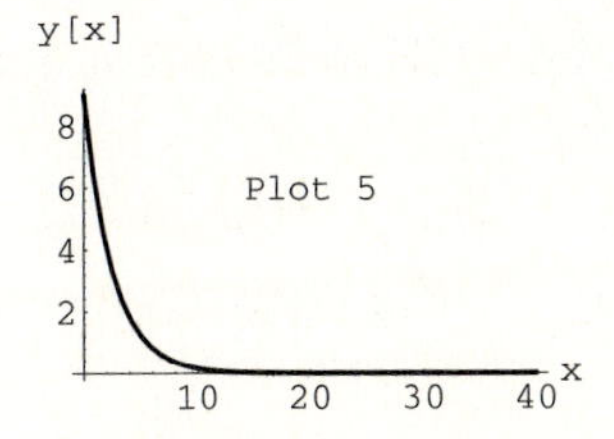

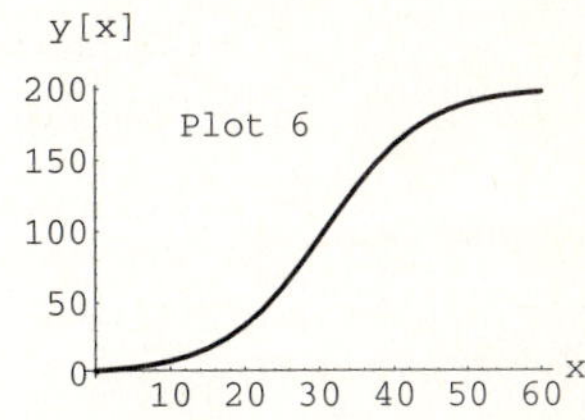

1.07 Race Track Principle Literacy Sheet

L.1) Here are plots of x and $\sin[x]$ for $0 \leq x \leq 2$:

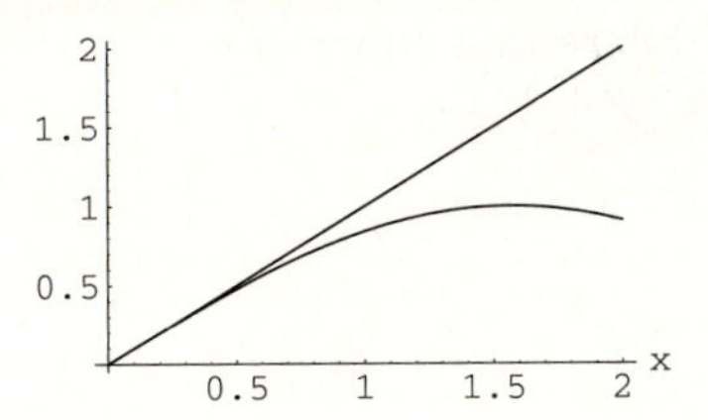

Use the Race Track Principle to explain why $x \geq \sin[x]$ for all $x \geq 0$. Remember that $-1 \leq \cos[x] \leq 1$ no matter what x is.

L.2) Here are plots of e^x and the line function $1 + x$ for $0 \leq x \leq 2$:

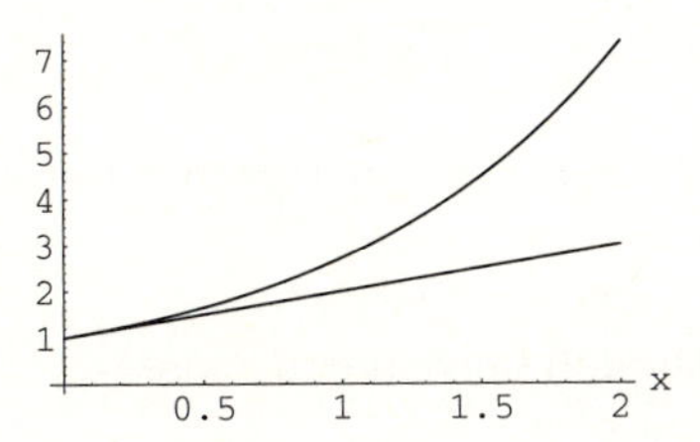

Use the Race Track Principle to explain why $e^x \geq 1 + x$ for all $x \geq 0$. Remember that $e^x \geq 1$ for $x \geq 0$.

L.3) Put $f[x] = \sin[x]^2$ and $g[x] = 1 - \cos[x]^2$ and look at $f[0]$ and $g[0]$:

In[1]:=
```
Clear[f,g,x]; f[x_] = Sin[x]^2;
g[x_] = 1 - Cos[x]^2;{f[0],g[0]}
```

Out[1]=
```
{0, 0}
```

Now look at $f'[x]$ and $g'[x]$:

In[2]:=
```
{f'[x],g'[x]}
```

Out[2]=
```
{2 Cos[x] Sin[x], 2 Cos[x] Sin[x]}
```

What version of the Race Track Principle confirms that

$$\sin[x]^2 = f[x] = g[x] = 1 - \cos[x]^2$$

for $x \geq 0$?

L.4) Here are plots of the solutions of the differential equations

$$y'[x] = 0.6\, y[x] \qquad \text{with } y[0] = 2$$

and

$$y'[x] = 1.2\, y[x] \left(1 - \frac{y[x]}{100}\right) \qquad \text{with } y[0] = 2:$$

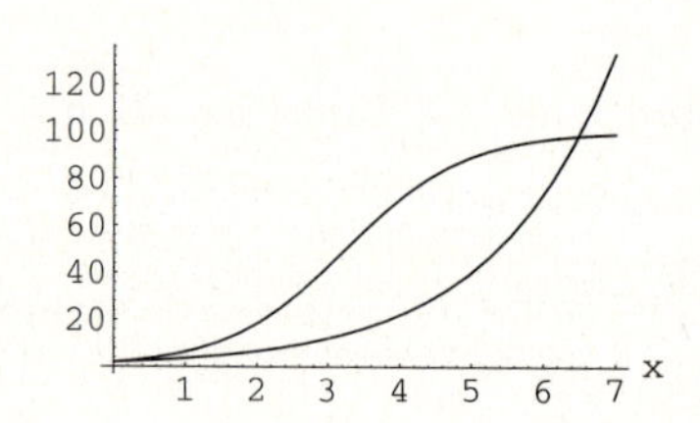

Identify the two plots. Notice that the plots start out together but share very little ink. Try to account for this.

L.5) Now look at plots of the solutions of the differential equations

$$y'[x] = 0.6\, y[x] \qquad \text{with } y[0] = 2$$

and

$$y'[x] = 0.6\, y[x] \left(1 - \frac{y[x]}{100}\right) \qquad \text{with } y[0] = 2:$$

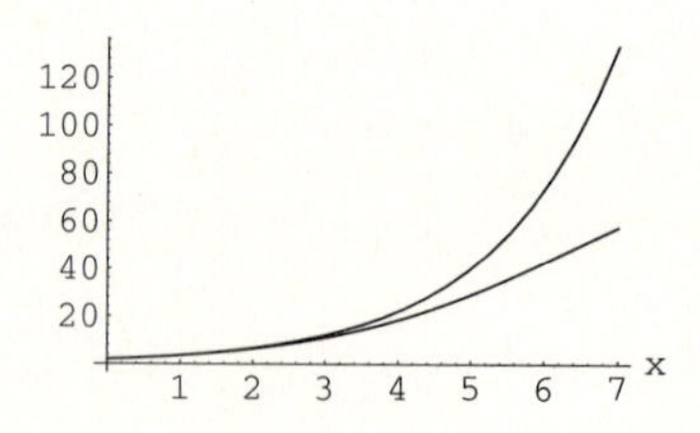

Identify the two plots. Explain why as x advances from 0, the plots of both of these solutions had no choice but to share a lot of ink initially. Explain why for larger x's, the plots had no choice but to pull apart, with one plot eventually sailing way above the other.

L.6) Here's an attempt to use Euler's faker (= Euler's method) to approximate

$$f[x] = \frac{x\sin[3x]}{2}$$

using only information about $f'[x]$ and the value of $f[0]$.

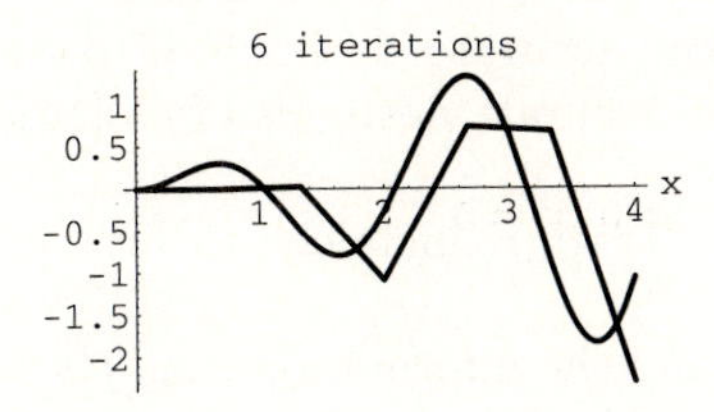

What you see above is what you get when you go with six iterations. Here's what you get when you go with 24 iterations:

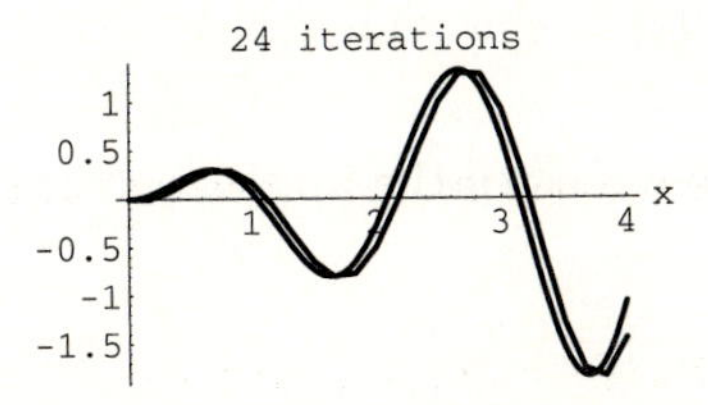

48 iterations:

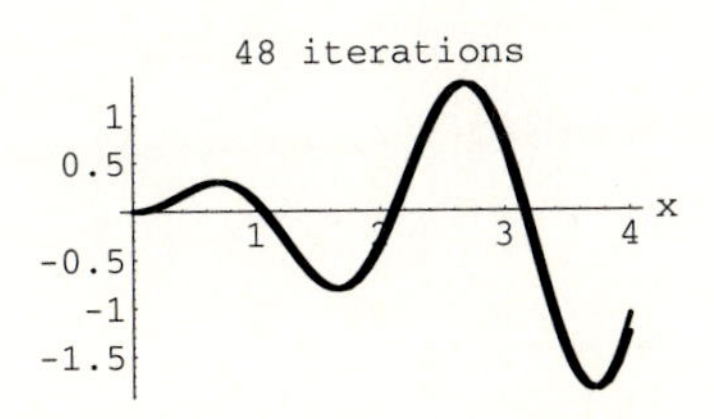

What is it about Euler's method that makes it so that you can always expect to get better results when you increase the number of iterations?

L.7) You are given $f[0] = 1$ and a formula for $f'[x]$, but neither you nor *Mathematica* can come up with a formula for $f[x]$. Explain the idea behind Euler's method for using $f'[x]$ to fake the plot of $f[x]$.

L.8) You are not given the formula for $f[x]$. In place of the formula, you are given that $f[0] = 1$ and you are given a formula for $f'[x]$. You calculate $f''[x]$ and find that $f''[x] > 0$ no matter what x is. When you use Euler's faker to give a fake plot of $f[x]$ for $x \geq 0$, you know in advance that the Euler fake plot must run under the true plot of $f[x]$. How do you know this in advance?

L.9) Use the fact that $-1 \leq \sin[x] \leq 1$; no matter what x is, to help you say how many accurate decimals of x guarantee eight accurate decimals of $\cos[x]$. How many accurate decimals of x guarantee eight accurate decimals of $\cos[10\,x]$?

L.10) Explain the ideas underlying the Race Track Principle.

L.11) You are given functions $f[x]$ and $g[x]$ and the information that $f[a] = g[a]$ and that $f'[x]$ and $g'[x]$ are very close for $a \leq x \leq b$. What can you say about $f[x]$ and $g[x]$ for $a \leq x \leq b$?

L.12) You are given functions $f[x]$ and $g[x]$ and the information that $f[a] = g[a]$ and that $f'[x] \leq g'[x]$ for $a \leq x \leq b$. What can you say about $f[x]$ and $g[x]$ for $a \leq x \leq b$?

L.13) Two functions $f[x]$ and $g[x]$ satisfy $f[a] = g[a]$ and $f'[x] \geq g'[x]$ for $x \leq a$. Why must it be true that $f[x] \leq g[x]$ for $x \leq a$?

Think of it this way: If the faster horse crosses the finish line in a tie, then which horse was ahead prior to the finish?

1.08 More Differential Equations Literacy Sheet

L.1) Look at the differential equation $y'[x] = y[x]\left(2 + \cos[x]^2 - y[x]\right)$ with $y[0] = 1$. How do you know before you do any plotting that as x advances from 0, the plot of the solution $y[x]$ initially must go up?

L.2) Look at the differential equation $y'[x] = y[x]\left(2 + \sin[x]^2 - y[x]\right)$ with $y[0] = 3$. How do you know before you do any plotting that as x advances from 0, the plot of the solution $y[x]$ initially must go down?

L.3) Look at the differential equation $y'[x] = y[x]\left(2 + \sin[x]^2 - y[x]\right)$ with $y[0] = 1$. How do you know before you do any plotting that as x advances from 0, the plot of the solution $y[x]$ goes up near x's for which $2 + \sin[x]^2 > y[x]$ and the plot of the solution $y[x]$ goes down near x's for which $2 + \sin[x]^2 < y[x]$? Why do you expect that the crests and dips of the plot of $y[x]$ are located at places where the $y[x]$ plot crosses the plot of $2 + \sin[x]^2$?

L.4) The guts of the predator-prey model involve the simultaneous differential equations

$$\text{prey}'[t] = a\,\text{prey}[t] - b\,\text{prey}[t]\,\text{pred}[t]$$
$$\text{pred}'[t] = -c\,\text{pred}[t] + d\,\text{pred}[t]\,\text{prey}[t]$$

where a, b, c, and d are given positive constants.

Why do you expect the crests and dips of the plot of $\text{prey}[t]$ to happen when $\text{pred}[t] = a/b$?

Why do you expect the crests and dips of the plot of $\text{pred}[t]$ to happen when $\text{prey}[t] = c/d$?

When $\text{prey}[t] < c/d$, do you expect the plot of $\text{pred}[t]$ to go up or down?

When $\text{pred}[t] > a/b$, do you expect the plot of $\text{prey}[t]$ to go up or down?

L.5) Pharmacokineticists have determined that if a given human being has a given concentration of a given drug at time $t = 0$, then there are always positive constants A and k such that the amount in grams $y[t]$ of the drug in the body fluids is governed by the differential equation

$$y'[t] = -\frac{k\,y[t]}{A + y[t]}$$

with $y[0] =$ concentration at time $t = 0$. In the case of alcohol, $y[t]$ is usually rather large in relation to A. Why do you think most pharmacokineticists are happy to replace the basic model

$$y'[t] = -\frac{k\,y[t]}{A + y[t]}$$

with $y[0] =$ concentration at time $t = 0$ by the simpler linear model

$$y'[t] = -k$$

with $y[0] =$ concentration at time $t = 0$ when they are studying alcohol concentration in human body fluids?

For drugs like cocaine, $y[t]$ is usually very small relative to A. For studying the decay of cocaine, why do you think that pharmacokineticists are happy to replace the basic model

$$y'[t] = -\frac{k\,y[t]}{A + y[t]} \qquad \text{with } y[0] = \text{concentration at time } t = 0$$

by the simpler exponential model

$$y'[t] = -\left(\frac{k}{A}\right) y[t] \qquad \text{with } y[0] = \text{concentration at time } t = 0?$$

L.6) The SIR infection model deals with disease in a closed population. This model is geared to diseases like certain strains of flu, which confer immunity to those who contract the disease and recover. Put

$$\begin{aligned} \text{Sus}[t] &= \text{the number of susceptible people at time } t \text{ after measurements begin;} \\ \text{Inf}[t] &= \text{the number of infected people at time } t \text{ after measurements begin;} \\ \text{Recov}[t] &= \text{the number of recovered people at time } t \text{ after measurements begin.} \end{aligned}$$

The model says that there are positive constants a and b such that

$$\text{Sus}'[t] = -a\,\text{Sus}[t]\,\text{Inf}[t] \qquad \text{and} \qquad \text{Recov}'[t] = b\,\text{Inf}[t].$$

This says that the number of susceptibles decreases at a rate jointly proportional to the number of susceptibles and the number of infecteds. And this says that the recovery rate is proportional to the number of infecteds. You also know that

$$\text{Sus}[t] + \text{Inf}[t] + \text{Recov}[t] = \text{total size of the population under study.}$$

For the purposes of this problem, assume that the total population stays constant. This tells you $\text{Sus}'[t] + \text{Inf}'[t] + \text{Recov}'[t] = 0$; so

$$\text{Inf}'[t] = -\text{Sus}'[t] - \text{Recov}'[t] = a\,\text{Sus}[t]\,\text{Inf}[t] - b\,\text{Inf}[t].$$

This gives you the full SIR epidemic model:

$$\begin{aligned} \text{Sus}'[t] &= -a\,\text{Sus}[t]\,\text{Inf}[t], \\ \text{Inf}'[t] &= a\,\text{Sus}[t]\,\text{Inf}[t] - b\,\text{Inf}[t] \\ \text{Recov}'[t] &= b\,\text{Inf}[t] \end{aligned}$$

with

$$\begin{aligned} \text{Sus}[0] &= \text{number of susceptibles,} \\ \text{Inf}[0] &= \text{number of infecteds,} \\ \text{Recov}[0] &= \text{number of recovered at the time the measurements begin.} \end{aligned}$$

Given that a and b are positive constants, how does this model reflect the facts that:

→ As t increases, $\text{Sus}[t]$ decreases?

→ The epidemic is at its worst (i.e., $\text{Inf}[t]$ is as big as possible) when $\text{Sus}[t] = b/a$?

→ The disease is spreading (i.e., $\text{Inf}'[t] > 0$) whenever $\text{Sus}[t] > b/a$?

→ The epidemic is into its end stage (i.e., $\text{Inf}'[t] < 0$) whenever $\text{Sus}[t] < b/a$?

1.09 Parametric Plotting Literacy Sheet

L.1) Put $x[t] = \cos[t]$ and $y[t] = \sin[t]$. Sketch on the axes below the curve traced out by $\{x[t], y[t]\}$ as t advances from 0 to $\pi/2$.

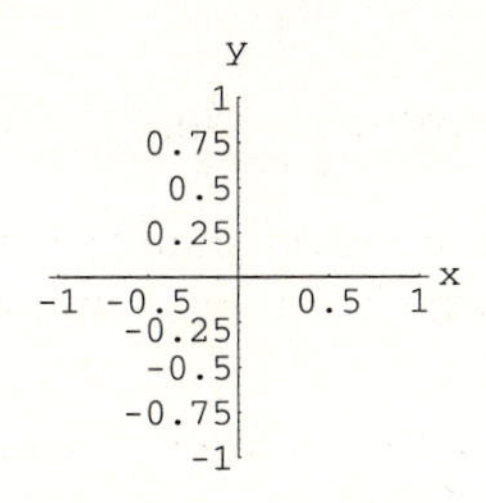

Next, sketch on the axes below the curve traced out by $\{x[t], y[t]\}$ as t advances from 0 to π.

Finally, sketch on the axes below the curve traced out by $\{x[t], y[t]\}$ as t advances from 0 to $2\,\pi$.

L.2) Put $x[t] = 2\cos[t]$ and $y[t] = \sin[t]$. Sketch with a pencil the curve traced out by $\{x[t], y[t]\}$ as t advances from 0 to $\pi/2$. Next, sketch the curve traced out by $\{x[t], y[t]\}$ as t advances from 0 to π. Finally, sketch the curve traced out by $\{x[t], y[t]\}$ as t advances from 0 to $2\,\pi$. Here are some axes for you to use:

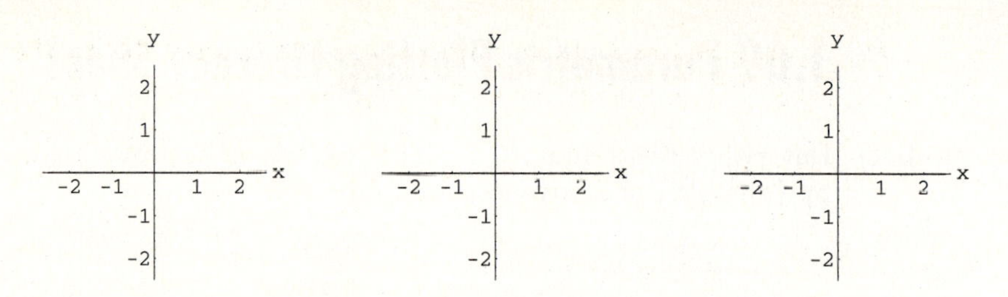

L.3) Put $x[t] = 2\cos[t]$ and $y[t] = 3\sin[t]$. Sketch on the axes below the curve traced out by $\{x[t], y[t]\}$ as t advances from 0 to $\pi/2$.

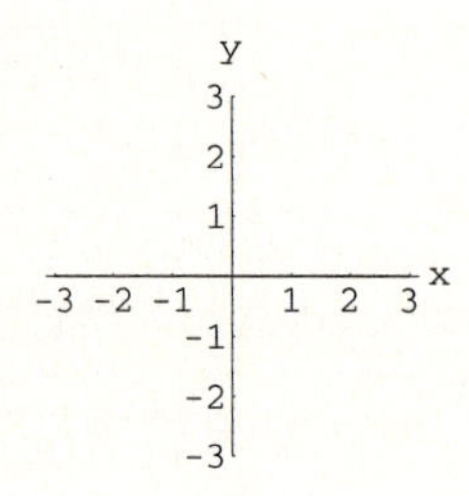

Next, sketch on the axes below the curve traced out by $\{x[t], y[t]\}$ as t advances from 0 to π.

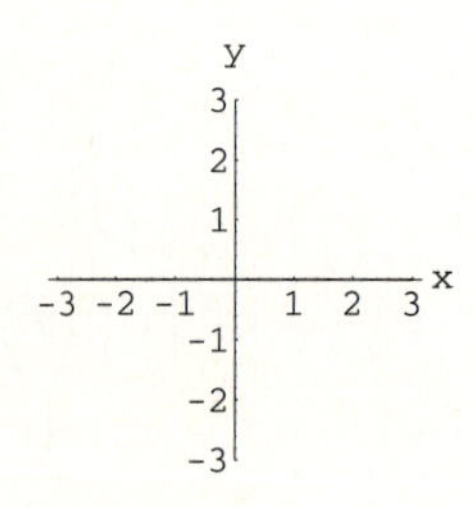

Finally, sketch on the axes below the curve traced out by $\{x[t], y[t]\}$ as t advances from 0 to $2\,\pi$.

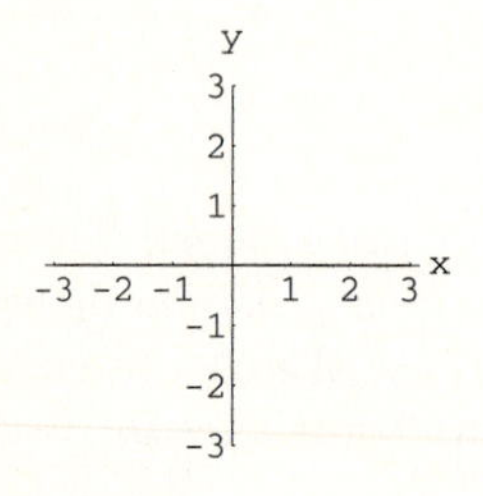

L.4) Put $x[t] = t$ and $y[t] = \sin[t]$. Sketch on the axes below the curve traced out by $\{x[t], y[t]\}$ as t advances from 0 to $2\,\pi$.

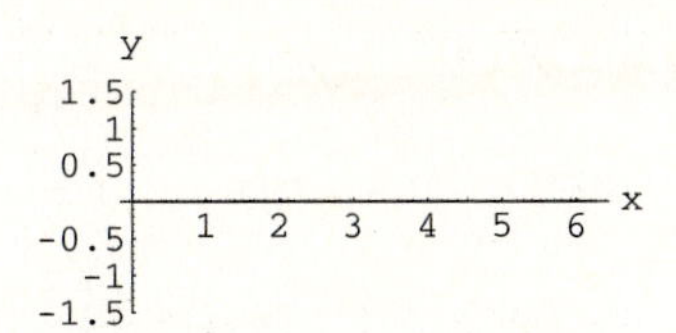

L.5) Put $x[t] = 1 + t$ and $y[t] = t/2$. Sketch on the axes below the curve traced out by $\{x[t], y[t]\}$ as t advances from 0 to 2.

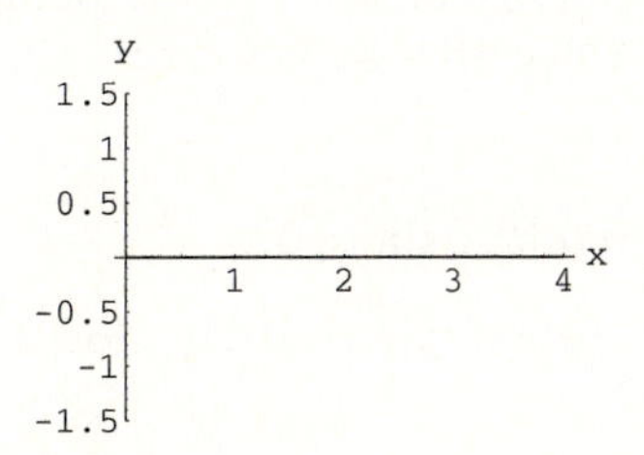

L.6) When you go with $x[t] = 1.2\,t + \sin[t]$ and $y[t] = 2\,(1 + \cos[t/2])$, then as t advances from 0, $\{x[t], y[t]\}$ sweeps out a curve:

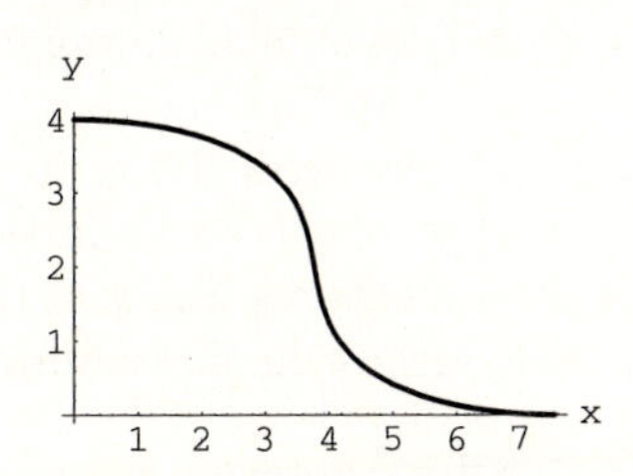

As you can see, this curve is the graph of a function $y = f[x]$. But it would be a whale of a lot of trouble (in fact, it would be impossible) to find a clean formula for $f[x]$. In spite of this, calculate $f'[x]$ by hand for:

a) $x = x[\pi/2] = (1.2)\,\pi/2 + 1$:

In[1]:=

```
N[(1.2) Pi/2 + 1]
```

Out[1]=

```
2.88496
```

b) $x = x[\pi] = (1.2)\,\pi$:

In[2]:=
```
N[(1.2) Pi]
```
Out[2]=
```
3.76991
```

For your information:

$$\begin{aligned}\sin[\pi/4] &= \cos[\pi/4] = 1/\sqrt{2},\\ \sin[\pi/2] &= 1,\\ \cos[\pi/2] &= 0,\\ \sin[\pi] &= 0,\\ \cos[\pi] &= -1.\end{aligned}$$

L.7) Come up with the specific coordinates $\{x, y\}$ of the highest point on the curve parameterized by $\{x[t], y[t]\} = \{te^t, 1 - t^2\}$ with $-\infty < t < \infty$.

L.8) Put

$$\{x[r,t], y[r,t], z[r,t]\} = \{1, 0, 1\} + \{r\cos[t], r\sin[t], 0\}.$$

Describe what you get when you plot $\{x[r,t], y[r,t], z[r,t]\}$ for $0 \le r \le 2$ and $0 \le t \le 2\pi$. Is this a curve or a surface?

Put

$$\{x[t], y[t], z[t]\} = \{1, 0, 1\} + \{2\cos[t], 2\sin[t], 0\}.$$

Describe what you get when you plot $\{x[t], y[t], z[t]\}$ for $0 \le t \le 2\pi$. Is this a curve or a surface? What relation does it have to what you said immediately above?

L.9) Curves can be thought of as one-dimensional creatures living in two or three dimensions. How many parameters do you need to plot a curve in three dimensions? Surfaces can be thought of as two-dimensional creatures living in three dimensions. How many parameters do you need to plot a surface in three dimensions?

L.10) You are given functions $x[t]$, $y[t]$, and $z[t]$. When you plot $\{x[t], y[t], z[t]\}$ in three dimensions for $0 \le t \le 4$, you get a curve.

What would you plot in three dimensions to get the tube consisting of the edges of all circles of radius 0.5 parallel to the xy-plane and centered on the curve?

What would you do to plot a sample of a horn consisting of edges of all circles parallel to the xy-plane and centered on the curve?

L.11) Comment on the statement: "Parametric plotting gives you a lot more freedom than nonparametric plotting."

Index

Entries are listed by lesson number and problem number.